[1.5] R_- the set of negative real numbers

R_+ the set of positive real numbers

$a < b$ a is less than b

$a > b$ a is greater than b

$a \leq b$ a is less than or equal to b

$a \geq b$ a is greater than or equal to b

$|a|$ the absolute value of a

[2.1] a^n the nth power of a, or a to the nth power

$P(x),\ D(y)$, etc. P of x, D of y, etc.

[3.1] a^0 1

a^{-n} $1/a^n$

[3.2] $a^{1/n}$ the real nth root of a for $a \in R$, or the positive one if there are two

$a^{m/n}$ $(a^{1/n})^m$

[3.3] $\sqrt[n]{a}$ the real nth root of a for $a \in R$, or the positive one if there are two

[3.5] $0.33\overline{3}$, etc.   repeating decimal

$a \approx b$ a is approximately equal to b

[4.3] $\pm$ plus or minus

[5.1] (a, b) the ordered pair of numbers whose first component is a and whose second component is b

$A \times B$ the Cartesian product of A and B

$R \times R$, or R^2 the Cartesian product of R and R

f, g, h, F, etc.   names of functions

$f(x)$ f of x, or value of f at x

[5.2] d the distance between two points

m the slope of a line

(*continued inside back cover*)

College Algebra   Third Edition

College Algebra   Third Edition

Edwin F. Beckenbach
University of California, Los Angeles

Irving Drooyan
Los Angeles Pierce College

William Wooton

Wadsworth Publishing Company, Inc. Belmont, California

ISBN-0-534-00183-1
L. C. Cat. Card No. 72-83470
Printed in the United States
of America

      5 6 7 8 9 10—
77 76 75

Cover painting: Ernest N. Posey

To the student

*A partially programmed study
guide based on the text is available
from your local bookstore under
the name* A Semi-Programmed
Study Guide for College Algebra, Third Edition, *by Bernard
Feldman.*

Preface

This third edition, like its predecessors, has been designed for the freshman college mathematics course that has come to mean many different things in different schools. Its organization allows the instructor to choose from a wide variety of topics that fall under the heading of Algebra, in order to achieve the desired level of mathematical sophistication for his course. Such flexibility seems imperative for a course which is seldom a terminal subject in mathematics but, rather, serves the transitional purpose of equipping the student for more advanced courses. The student is assumed to have had the equivalent of at least one year of high school algebra and one year of high school geometry.

In this edition, we have concentrated on making the material easier for the instructor to teach and, more important, easier for the student to understand. Problem sets have been revised in the light of classroom experience, and many sections have been rewritten to clarify central concepts.

> In Chapter 1 the formal statement-reason format for proofs has been changed to an informal paragraph style.
>
> Chapter 2 has been rewritten to put less emphasis on the structure of polynomials; synthetic division is now in a separate section, and a *new section* on *partial fractions* has been added.
>
> Chapter 3 now includes a separate section on sums and products of radical expressions.
>
> Chapter 4 has been revised to include separate sections on the quadratic formula and on word problems.
>
> Chapter 5, which has been completely rewritten, now includes a *new section* on *inverse relations*.
>
> Chapter 6 has a new treatment of the graphing of quadratic functions, based on the symmetry property of parabolas.
>
> Chapter 8 now includes *new sections* on *convex sets*, *polygonal regions*, and *linear programming*.
>
> Chapter 9 has been reorganized so that linear systems are solved by means of row-equivalent matrices in Section 9.3, right after the properties of matrices are introduced in the first two sections.

Chapter 10 on complex numbers and vectors has been completely rewritten.

Chapter 11 on the theory of equations has been revised to eliminate some of the formal proofs.

The section on power series has been eliminated in the rewriting and reorganizing of Chapter 12.

Revisions in Chapter 13 center on the sections dealing specifically with probability.

The Appendix now includes, under the heading *Mathematical Structure*, a compact summary of the basic concepts of the text and a brief treatment of mathematical systems as related to the number systems considered in the text.

A *new chapter-review set of problems* has been added to all chapters. Sectional problem sets, as before, include examples with solutions that enable the students to work through the exercises on their own.

The first chapter, which introduces the student to the complete ordered field of real numbers, is followed by three chapters—on polynomials, rational exponents, and open sentences in one variable—that may be optional for the student who shows a mastery of second-year high school algebra. Beginning with Chapter 5, discussions generally center around the function concept. Polynomial and rational functions (and their graphs) are covered in detail; variation is treated from the function standpoint; logarithms are developed from a consideration of exponential functions; determinants are presented as functions of matrices; sequences are treated as functions having sets of positive integers as domain; and probability is discussed from a set-function standpoint.

The book is designed for a semester course of three, four, or five units, or for two three-unit quarter courses. Chapters 7 through 13 are sufficiently independent of each other that any may be omitted for a short course.

A section at the end of the book provides answers for odd-numbered problems, along with graphs.

A semi-programmed study guide covering topics in the text is available for student use.

As in the second edition, a second color is used functionally to highlight key procedures in routine manipulations and to focus attention on key elements of figures. Marginal annotations direct the reader's attention to important ideas.

We sincerely thank David Wend of the mathematics department of Montana State University for his very careful reading of this new edition and his many good suggestions for improving it. Our special thanks go also to Edwin S. Beckenbach and Charles C. Carico for their assistance in the preparation of this edition and its ancillary materials and to our editor, Don Dellen, for his suggestions and encouragement.

Edwin F. Beckenbach
Irving Drooyan
William Wooton

Contents

Appendix

Odd-Numbered Answers   351

Indexes

1

Properties
of Real Numbers

1.1   *Definitions and Symbols*

A **set** is simply a collection of some kind. It may be a collection of people, colors, numbers, or anything else. In algebra, we are interested in sets of numbers of various sorts and in their relations to sets of points or lines in a plane or in space. Any one of the collection of things in a set is called a **member** or **element** of the set, and is said to be **contained in** or **included in** (or, sometimes, just **in**) the set. For example, the counting numbers 1, 2, 3, ... (where the dots indicate that the sequence continues indefinitely) are the elements of the set we call the set of **natural numbers**.

Set notation   Sets are usually designated by means of capital letters, A, B, C, etc. They are identified by means of **braces**, { }, with the members either listed or decribed. For example, the elements might be listed as in {1, 2, 3}, or described as in {first three natural numbers}. The expression "{1, 2, 3}" is read "the set whose elements are one, two, and three"; "{first three natural numbers}" is read "the set whose elements are the first three natural numbers."

Using the undefined notion of set membership, we can be more specific about some other terms we shall be using.

Definition 1.1   *Two sets A and B are **equal**, A = B, if and only if they have the same members—that is, if and only if every member of each is a member of the other.*

Thus, if A denotes {1, 2, 3}, B denotes {3, 2, 1}, C denotes {2, 3, 4}, and D denotes {natural numbers between 1 and 5}, then $A = B$ and $C = D$. The phrase "if and only if" used in this definition is simply the mathematician's way of making two statements at once. Definition 1.1 means: "Two sets are equal if they have the same members. Two sets are equal only if they have the same members." The second of these statements is logically equivalent to: "Two sets have the same members if they are equal."

1

Definition 1.2   *If the elements of a set A can be paired with the elements of a set B in such fashion that each element of A is paired with one and only one element of B, and conversely, then such a pairing is called a **one-to-one correspondence** between A and B.*

For example, if $A = \{a, b, c\}$ and $B = \{1, 2, 3\}$, then the sets A and B can be put into one-to-one correspondence in six different ways, two of which are shown here:

$$\{a, b, c\} \qquad \{a, b, c\}$$
$$\updownarrow \updownarrow \updownarrow \qquad \updownarrow \updownarrow \updownarrow$$
$$\{1, 2, 3\} \qquad \{2, 1, 3\}$$

Definition 1.3   *Two sets are **equivalent** if and only if a one-to-one correspondence exists between them.*

Equivalence symbol

The symbol $\sim$ is used to denote equivalence. Thus, $A \sim B$ is read "A is equivalent to B." Intuitively, equivalent sets are sets that contain the same number of members. Clearly, if two sets are equal, then they are equivalent, but the converse is not necessarily true—equality of sets requires that the members be identical, not merely that the sets be in one-to-one correspondence.

Definition 1.4   *If every member of a set A is a member of a set B, then A is a **subset** of B. If, in addition, B contains at least one member not in A, then A is a **proper subset** of B.*

Subset symbol

The symbol $\subset$ (read "is a subset of" or "is contained in") will be used to denote both the subset relationship and the proper-subset relationship. Thus

$$\{1, 2, 3\} \subset \{1, 2, 3, 4\} \quad \text{and} \quad \{1, 2, 3\} \subset \{1, 2, 3\}.$$

Notice that, by definition, every set is a subset of itself.

The set that contains no elements is called the **empty set**, or **null set**, and is denoted by the symbol $\emptyset$ (read "the empty set" or "the null set"); $\emptyset$ is a subset of every set, and it is a proper subset of every set except itself. If a set S is the null set or is equivalent to $\{1, 2, 3, \ldots, n\}$ for some fixed natural number n, then S is said to be **finite**. A set that is not finite is said to be **infinite**. For example, the set of *all* natural numbers, $\{1, 2, 3, \ldots\}$, is an infinite set.

Definition 1.5   *Two sets A and B are **disjoint** if and only if A and B contain no member in common.*

For example, if $A = \{1, 2, 3\}$ and $B = \{5, 6, 7\}$, then A and B are disjoint.

Set-membership notation

The symbol $\in$ (read "is a member of" or "is an element of") is used to denote membership in a set. Thus,

$$2 \in \{1, 2, 3\}.$$

Note that we write

$$\{2\} \subset \{1, 2, 3\} \quad \text{and} \quad 2 \in \{1, 2, 3\},$$

since $\{2\}$ is a *subset*, whereas 2 is an *element*, of $\{1, 2, 3\}$.

When discussing an individual but unspecified element of a set containing more than one member, we usually denote the element by a lowercase italic letter (for example, a, d, s, x), or sometimes by a letter from the Greek alphabet: α (alpha), β (beta), γ (gamma), and so on. Symbols used in this way are called **variables**.

Definition 1.6   *A **variable** is a symbol representing an unspecified element of a given set containing more than one element.*

The given set is called the **replacement set**, or **domain**, of the variable. If the domain is a set of numbers, then the variable represents a number. Thus

$$x \in A$$

means that the variable x represents an (unspecified) element of the set A. The members of the replacement set are called the **values** of the variable. A symbol with just one value is called a **constant**.

When discussing sets, it is often helpful to have in mind some general set from which the elements of all of the sets under consideration are drawn. For example, if we wish to talk about sets of college students, we may want to consider all college students in this country, or all students in general; or, taking a larger view, we may want to consider students as a special kind of human being—say, all those human beings who are consciously striving to increase their knowledge. Thus, we can draw sets of college students from any one of a number of different general sets. Such a general set is called the **universe of discourse**, or the **universal set**, and we shall usually denote it by the capital letter U. It follows that any set in a particular discussion is a subset of U for that discussion.

Negation symbol

The slant bar, /, drawn through certain symbols of relation, is used to indicate negation. Thus $\neq$ is read "is not equal to," $\not\subset$ is read "is not a subset of," and $\notin$ is read "is not an element of." For example,

$$\{1, 2\} \neq \{1, 2, 3\}, \quad \{1, 2, 3\} \not\subset \{1, 2\}, \quad \text{and} \quad 3 \notin \{1, 2\}.$$

Set-builder notation

Another symbolism useful in discussing sets is illustrated by

$$\{x \mid x \in A \text{ and } x \notin B\}$$

(read "the set of all x such that x is a member of A and is not a member of B"). This symbolism, called **set-builder notation**, is used extensively in this book. What it does is specify a variable (in this case, x) and, at the same time, state a condition on the variable (in this case, that x is contained in the set A and not in B).

Exercise 1.1

Designate each of the following sets by using braces and listing the members.

Example

{natural numbers between 8 and 12}

Solution

{9, 10, 11}

1. {natural numbers between 2 and 7}
2. {natural numbers between 13 and 20}
3. {natural numbers between 88 and 89}
4. {natural numbers between 100 and 102}
5. {days in the week}
6. {months in the year}

Replace the colored comma with either = or ≠ to make a true statement.

7. {natural numbers less than 3}, {1, 2}
8. {integers between −3 and 1}, {−2, −1}
9. {2}, {−2}   10. {4}, {7}
11. ∅, {0}   12. {5, 7, 9}, {7, 9, 5}

Replace the colored comma with either ∈ or ∉ to make a true statement.

13. 3, {2, 3, 4}   14. 15, {2, 4, 6, ...}
15. {2}, {2, 3, 4}   16. ∅, {2, 3, 4}

Replace the colored comma with either ⊂ or ⊄ to make a true statement.

17. 5, {4, 5, 6}   18. {5, 4, 6}, {4, 5, 6}
19. ∅, {4, 5, 6}   20. {3, 4}, {4, 5, 6}

21. Let $U = \{5, 6, 7\}$. List the subsets of U that contain

    a. three members   b. two members
    c. one member   d. no members

22. Let $U = \{1, 2, 3, 4\}$. List the subsets of U that contain

    a. four members   b. three members   c. two members
    d. one member   e. no members   f. 2 and one other member

23. Let $U = \{1, 2, 3, 4, 5, 6, 7, 8, 9\}$, $A = \{1, 2, 3, 4\}$, $B = \{4, 5, 6, 7\}$, and $C = \{6, 7\}$. Replace the comma with either ⊂ or ⊄ to make a true statement.

    a. A, U b. C, A c. A, B d. C, B

24. Let $U = \{\text{natural numbers}\}$, $A = \{\text{even natural numbers}\}$, $B = \{\text{odd natural numbers}\}$; furthermore, let $C = \{x \mid x \text{ is a natural number between 1 and 10}\}$, and $D = \{x \mid x \text{ is a natural number less than 9}\}$. Which of the following statements are true?

    a. $A \sim D$ b. $C = D$ c. $C \subset D$
    d. $A \sim B$ e. $A = B$ f. $D \subset C$
    g. $A \sim U$ h. A and B are disjoint.   i. $A \subset U$
    j. $C \sim D$ k. $C \subset A$ l. C and B are disjoint.

Designate each of the following sets by using set-builder notation.

Example   {even natural numbers}

Solution $\{x \mid x = 2n, n$ a natural number$\}$

25. {odd natural numbers}   26. {natural numbers}
27. {solutions of $2^x = 5$}   28. {solutions of $x^x = 5$}
29. {elements not in set A}   30. {elements in set B}

1.2   Operations on Sets

Ideas involving universal sets, subsets thereof, and certain operations on sets can be depicted using plane geometric figures called **Venn diagrams**. Figure 1.1 shows such a diagram representing a universe having as its elements all points of the rectangle and its interior. It also shows a number of subsets of the universe, denoted by circles and their interiors. In this figure, sets A, B, and C are disjoint, D is a subset of C, and E is neither a subset of C nor disjoint from C.

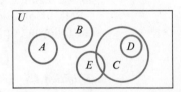

Figure 1.1

There are several mathematically important operations on the subsets of a given universe U. One such operation is defined as follows.

Definition 1.7   *The **union** of two subsets A and B of U is the set of all elements of U that belong either to A or to B or to both.*

Set-union symbol   The symbol $\cup$ is used to denote the union of sets. Thus $A \cup B$ (read "the union of A and B" or, sometimes, "A cup B") is the set of all elements that are in either A or B or both.

Example   If

$$A = \{1, 2, 3, 4, 5\} \text{ and } B = \{2, 3, 4, 5, 6\},$$

then

$$A \cup B = \{1, 2, 3, 4, 5, 6\}.$$

Notice that each element in $A \cup B$ is listed only once in this example, since repetition would be redundant. Figure 1.2 is a Venn diagram in which the shaded region depicts $A \cup B$.

A second set operation of interest is defined as follows.

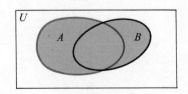

Figure 1.2

Definition 1.8   *The **intersection** of two subsets A and B of U is the set of all elements of U that belong to both A and B.*

Set-intersec-
tion symbol

The symbol $\cap$ is used to denote intersection. Thus $A \cap B$ (read "the intersection of A and B" or, sometimes, "A cap B") denotes the set of all elements of U that are in both A and B.

Example

If

$$A = \{1, 2, 3, 4, 5\} \text{ and } B = \{2, 3, 4, 5, 6\},$$

then

$$A \cap B = \{2, 3, 4, 5\}.$$

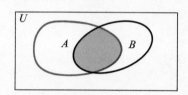

In the Venn diagram of Figure 1.3, the shaded region depicts $A \cap B$.

Figure 1.3

Both union and intersection are applied to two sets in relation to each other. The following operation on sets, however, applies to only one set in relation to U.

Definition 1.9   *The **complement** of a set A in U is the set of all elements of U that do not belong to A.*

The symbol A' (or, sometimes, $\bar{A}$, $\sim A$, or $\tilde{A}$) denotes the complement of A in U.

Example

If

$$U = \{1, 2, 3, 4, 5\} \text{ and } A = \{2, 4\},$$

then

$$A' = \{1, 3, 5\}.$$

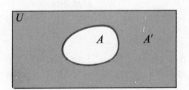

The shaded part of the Venn diagram in Figure 1.4 represents A'.

Figure 1.4

Exercise 1.2

Let $U = \{1, 2, 3, 4, 5, 6, 7, 8, 9, 10\}$, $A = \{2, 4, 6, 8, 10\}$, $B = \{1, 2, 3, 4, 5\}$,   *and* $C = \{1, 3, 5, 7, 9\}$. *List the members of each of the following sets.*

1. A'	**2.** B'	**3.** C'	**4.** $A \cap B$
5. $A \cup B$	**6.** $A \cup C$	**7.** $A \cap C$	**8.** $A' \cap B'$
9. $A' \cup C'$	**10.** $(A \cap B)'$	**11.** $A' \cup C$	**12.** $C' \cap B$

For each of the Problems 13–24, copy the Venn diagram shown here on a sheet of paper. Shade the part of the diagram corresponding to each of the following sets.

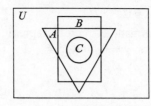

13. $A \cap B$	**14.** $C \cap A$	**15.** $C \cap B$
16. $A \cap C'$	**17.** $B \cap C'$	**18.** $B' \cup A'$
19. $B' \cap A'$	**20.** $B' \cap A$	**21.** $A' \cap B$
22. $(C \cap B)'$	**23.** $(C' \cap A)'$	**24.** $(A' \cap B)'$

Using Venn diagrams if necessary, complete each of the following equations.

25. $(A')' =$	**26.** $A \cap A' =$	**27.** $A \cap U =$	**28.** $A \cup A' =$
29. $A \cup U =$	**30.** $A \cap \emptyset =$	**31.** $A \cup \emptyset =$	**32.** $A \cap A =$
33. $\emptyset' \cap \emptyset =$	**34.** $A \cup A =$	**35.** $\emptyset \cup \emptyset =$	**36.** $U \cap U =$

37. Under what conditions would each of the following statements be true?

a. $A \cup B = \emptyset$	**b.** $A \cup \emptyset = \emptyset$	**c.** $A \cap U = U$
d. $A \cup B = A$	**e.** $A \cup \emptyset = U$	**f.** $A' \cap U = U$
g. $A \cap B = A$	**h.** $A' \cup \emptyset = \emptyset$	**i.** $A \cup B = A \cap B$

1.3   The Field Postulates

You should already be somewhat familiar with the seven sets of numbers described below.

1. The set N of **natural numbers**, whose elements are the counting numbers:

$$N = \{1, 2, 3, \ldots\}.$$

2. The set J of **integers**, whose elements are the counting numbers, their negatives, and zero:

$$J = \{\ldots, -2, -1, 0, 1, 2, \ldots\}.$$

3. The set Q of **rational numbers**, whose elements are all those numbers that can be represented as the quotient of two integers $\frac{a}{b}$ (or a/b, or $a \div b$), where b is not 0. Among the elements of Q are such numbers as $-3/4$, $18/27$, $3/1$, and $-6/1$. In symbols,

$$Q = \left\{ x \mid x = \frac{a}{b}, \quad a, b \in J, \quad b \neq 0 \right\}.$$

The members of the set of rational numbers are the numbers with terminating or repeating decimal representations.

4. The set H of **irrational numbers**, whose elements are the numbers with decimal representations that are nonterminating and nonrepeating. Among the elements of this set are such numbers as $\sqrt{2}, \pi$, and $-\sqrt{7}$. An irrational number cannot be represented in the form a/b, where a and b are integers. In symbols,

$$H = \{\text{irrational numbers}\}.$$

5. The set R of **real numbers**, which is the union of the set of all rational numbers and the set of all irrational numbers:

$$R = \{x \mid x \in (Q \cup H)\}.$$

6. The set I of **imaginary numbers**, whose members can be represented in the form $x + yi$, where x and y are real numbers, $y \neq 0$, and $i^2 = -1$:

$$I = \{x + yi \mid x, y \in R,\ y \neq 0,\ i^2 = -1\}.$$

If $x = 0$, then the imaginary number $x + yi$ is written yi, with $y \in R$, $y \neq 0$. Such a number is called a **pure imaginary number**.

7. The set C of **complex numbers**, whose members can be represented in the form $x + yi$, where x and y are real numbers and $i^2 = -1$:

$$C = \{x + yi \mid x, y \in R,\ i^2 = -1\}.$$

Relations between sets of numbers

The foregoing sets of numbers are related as indicated in Figure 1.5. We can

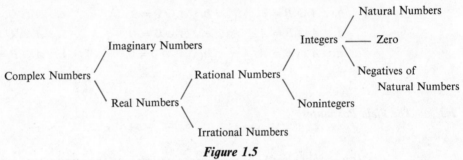

Figure 1.5

characterize each set by stating properties that we assume it to have. In mathematics, when we make formal assumptions about members of a set or about their properties, we call the assumptions **axioms**, or **postulates**. The words **property**, **law**, and **principle** are sometimes used in referring to assumptions, although these words may also be applied to certain consequences thereof.

The first assumptions to be considered here have to do with *equality*. An **equality**, or an "is equal to" assertion, is simply a mathematical statement that two symbols or words, or two groups of symbols or words, are names for the same thing.

Equality postulates

We postulate that, for any members a, b, and c of any set S, the equality $(=)$ relationship satisfies the following laws:

E-1	$a = a$.	*Reflexive law for equality.*
E-2	If $a = b$, then $b = a$.	*Symmetric law for equality.*
E-3	If $a = b$ and $b = c$, then $a = c$.	*Transitive law for equality.*
E-4	If $a = b$, then a may be replaced by b and b by a in any mathematical statement without altering the truth or falsity of the statement.†	*Substitution law for equality.*

† We have adopted a very powerful postulate in E-4, one that subsumes E-2 and E-3 as special cases. By wording E-4 as we have, we can eliminate a great deal of fussiness in later arguments, and, because E-2 and E-3 are fundamental properties of the equality relation, we have elected to retain them as postulates.

The equality axioms indicate how we use symbols and warn that we must neither change the meaning of a symbol in the middle of a discussion nor use the same symbol for two different things in the same context. Other axioms, somewhat different conceptually, are used to characterize mathematical systems. They are assumptions made about the behavior of elements of sets under binary operations.

Definition 1.10   *A **binary operation** in a set A is an operation that assigns to each pair a and b of elements of A, taken in a definite order (a first and b second), a corresponding element of A.*

Addition and multiplication postulates

Since the set R of real numbers is the set of greatest interest to us at the moment, we postulate the basic laws for the elements of R in relation to the binary operations of addition and multiplication. You are probably familiar with these laws from your earlier work. They are listed on page 10 for reference. Parentheses are used in stating some of the relations to indicate that symbols within parentheses are to be viewed as representing a single entity.

To give meaning to expressions $a + b + c$, $a \times b \times c$, $a + b + c + d$, and so on, let us make the following agreement.

Definition 1.11   *If $a, b, c, d, \ldots \in R$, then*

$$a + b + c = a + (b + c), \qquad a + b + c + d = a + (b + c + d), \ldots,$$

and

$$a \times b \times c = a \times (b \times c), \qquad a \times b \times c \times d = a \times (b \times c \times d), \ldots.$$

If, for a given set and a given pair of binary operations (not necessarily the set of real numbers or ordinary addition and multiplication) Postulates F-1 through F-11 are satisfied by the elements of the set, then the set is called a **field** under these operations. Thus, we speak of the **field R of real numbers**.

Inverse operations

In terms of the operations of addition and multiplication, let us now define two new operations on real numbers.

Definition 1.12   *The **difference** of elements $a \in R$ and $b \in R$, denoted by $a - b$, is given by*

$$a - b = a + (^-b).$$

The operation of finding a difference is called **subtraction**. This definition explains why we ordinarily write $-b$ for ^-b (the negative of b).

Definition 1.13   *The **quotient** of elements $a \in R$ and $b \in R$, $b \neq 0$, denoted by $\frac{a}{b}$, a/b, or $a \div b$, is given by*

$$\frac{a}{b} = a \times (b^{-1}).$$

The operation of finding a quotient is called **division**. In particular, for $a = 1$ we have, by the substitution law for equality,

$$\frac{1}{b} = 1 \times b^{-1},$$

from which we obtain

$$\frac{1}{b} = b^{-1}.$$

In all that follows, we shall ordinarily write $-a$ for the additive inverse of a and $\frac{1}{b}$ or $1/b$ for the multiplicative inverse of b. We shall also follow the customary practice of writing ab or $a \cdot b$ for $a \times b$.

Field postulates of the set *R* of real numbers

Let a, b, c, be arbitrary elements of R.

F-1 $a + b$ is a unique element of R.   *Closure law for addition.*

F-2 $(a + b) + c = a + (b + c)$.   *Associative law for addition.*

F-3   There exists an element $0 \in R$ (called   *Additive-identity law.*
the **identity element for addition**) with
the property

   $$a + 0 = a \quad \text{and} \quad 0 + a = a$$

for each $a \in R$.

F-4   For each $a \in R$, there exists an element   *Additive-inverse law.*
$^-a \in R$ (called the **additive inverse**, or
negative, of a) with the property

   $$a + (^-a) = 0 \quad \text{and} \quad (^-a) + a = 0.$$

F-5 $a + b = b + a$.   *Commutative law for addition.*

F-6 $a \times b$ is a unique element of R.   *Closure law for multiplication.*

F-7 $(a \times b) \times c = a \times (b \times c)$.   *Associative law for multiplication.*

F-8 $a \times (b + c) = a \times b + a \times c$ and   *Distributive law.*
$(b + c) \times a = b \times a + c \times a$.

F-9   There exists an element $1 \in R$ (called   *Multiplicative-identity law.*
the **identity element for multiplication**),
$1 \neq 0$, with the property

   $$a \times 1 = a \quad \text{and} \quad 1 \times a = a$$

for each $a \in R$.

F-10 $a \times b = b \times a$.   *Commutative law for multiplication*

F-11   For each element $a \in R$, $a \neq 0$, there   *Multiplicative-inverse law.*
exists an element $a^{-1} \in R$ (called the
multiplicative inverse, or **reciprocal**,
of a) with the property

   $$a \times (a^{-1}) = 1 \quad \text{and} \quad (a^{-1}) \times a = 1.$$

Exercise 1.3

Let $N = \{natural\ numbers\}$, $J = \{integers\}$, $Q = \{rational\ numbers\}$, $H = \{irrational\ numbers\}$, $R = \{real\ numbers\}$, $I = \{imaginary\ numbers\}$, and $C = \{complex\ numbers\}$. *State whether each statement is true or false.*

1. $-5 \in N$ 2. $0 \in Q$ 3. $\sqrt{3} \in R$
4. $\pi \in C$ 5. $-3 \in R$ 6. $-7 \in H$
7. $\{-1, 1\} \subset N$ 8. $\{-1, 1\} \subset J$ 9. $\{-1, 1\} \subset Q$
10. $\{-1, 1\} \subset H$ 11. $\{-1, 1\} \subset R$ 12. $\{-1, 1\} \subset C$

Represent each set, listing members if finite, and using ellipsis dots, ..., *if infinite.*

Examples   a. {natural numbers between 5 and 9}   b. {integers greater than 3}

Solutions   a. $\{6, 7, 8\}$ b. $\{4, 5, 6, \dots\}$

13. {first five natural numbers}   14. {integers between -4 and 5}
15. {natural numbers less than 7}   16. {integers greater than -5}
17. {integers between -10 and -5}   18. {nonnegative integers}

Use set-builder notation, $\{x \mid condition\ on\ x\}$, *to represent each set.*

Example   {natural numbers greater than 12}

Solution $\{x \mid x \in N$ and x is greater than $12\}$

19. {natural numbers}   20. {integers}
21. {real numbers}   22. {integers less than -6}
23. {real numbers between -4 and 3}   24. {nonnegative real numbers}

25. Let $A = \{4, -2, 2/5, \sqrt{-7}, 0, -3/4, \sqrt{2}, \sqrt{7}, \sqrt{-1}\}$. Designate each of the following sets by using braces and listing the members.
   a. {natural numbers in A}   b. {integers in A}
   c. {rational numbers in A}   d. {real numbers in A}

26. Let $B = \{-6, 3, \sqrt{5}, -3/4, \sqrt{-2}, 0, 5, -1, \sqrt{3}\}$. Designate each of the following sets by using braces and listing the members.
   a. {natural numbers in B}   b. {integers in B}
   c. {irrational numbers in B}   d. {real numbers in B}

Each variable in Problems 27–50 denotes a real number.

Each of the statements 27–32 is an application of one of the postulates E-1 *through* E-4. *Justify each statement by citing an appropriate postulate.*

Example   If $3 = a$, then $a = 3$.

Solution   Symmetric law for equality

27. If $a + 3 = b$ and $b = 7$, then $a + 3 = 7$.

28. If $x = 5$ and $y = x + 2$, then $y = 5 + 2$.

29. If $2 + y = 6$, then $6 = 2 + y$.

30. If $a = 2c$ and $c = 6$, then $a = 2 \cdot 6$.

31. If $y = 7 + x$ and $7 + x = z$, then $y = z$.

32. $x + 8 = x + 8$.

Each of the statements 33–50 is an application of one of the postulates F-1 *through* F-11. *Justify each statement by citing the appropriate postulate.*

Example $2(3 + 1) = 2 \cdot 3 + 2 \cdot 1$

Solution   Distributive law

33. ab is a real number   **34.** $7 + 0 = 7$

35. $(2 \cdot 3) \cdot 4 = 2 \cdot (3 \cdot 4)$ **36.** $(5 + 4) + 1 = (4 + 5) + 1$

37. $3 + (-3) = 0$ **38.** $5 + (-2) = (-2) + 5$

39. $a\left(\dfrac{1}{a}\right) = \left(\dfrac{1}{a}\right)a \quad (a \neq 0)$ **40.** $\dfrac{1}{c}(a + b) = \dfrac{1}{c} \cdot a + \dfrac{1}{c} \cdot b \quad (c \neq 0)$

41. $(a + b) + c = c + (a + b)$ **42.** $(a + b) + c = (b + a) + c$

43. $a + (b + c)d = a + bd + cd$ **44.** $a + (b + c)d = a + d(b + c)$

45. $a + (b + c)d = (b + c)d + a$ **46.** $(a + b) + [-(a + b)] = 0$

47. $a(b + c) = (b + c)a$ **48.** $a[b + (c + d)] = ab + a(c + d)$

49. $ab + a(c + d) = ab + ac + ad$ **50.** $ab + ac = ba + ac$

Which of the following sets are closed under the stated operation?

Example   {integers}, division

Solution   Not closed. The set is not closed, because the quotient of two integers is not always an integer; for example, the quotient $-2/3 \notin J$.

51. {even natural numbers}, addition   **52.** {odd natural numbers}, addition

53. {odd natural numbers}, multiplication

54. {even natural numbers}, multiplication

55. {0, 1}, multiplication   **56.** {0, 1}, addition

57. {natural numbers}, division   **58.** {natural numbers}, subtraction

1.4   *Field Properties*

The field postulates together with the postulates for equality imply other properties of the real numbers. Such implications are generally stated as **theorems**. A theorem is simply an assertion of a fact that follows logically from the postulates (or axioms) and other theorems. We shall list some of the theorems ordinarily encountered in lower-level algebra courses. We have numbered these theorems to provide an efficient way to refer to them later and have also named those that have commonly accepted names.

Addition law for equality

First, consider the following result, which reaffirms the uniqueness of the sum of two real numbers.

Theorem 1.1   *If $a, b, c \in R$ and $a = b$, then*

$$a + c = b + c \quad and \quad c + a = c + b.$$

This result follows directly from the reflexive and substitution laws of equality. Since $a = b$, and, by the reflexive law,

$$a + c = a + c \quad and \quad c + a = c + a,$$

we can substitute b for a in the right-hand member of each equation to obtain

$$a + c = b + c \quad and \quad c + a = c + b.$$

Both of these equations were included in the conclusion of Theorem 1.1, because applications to further results are sometimes immediate from one form and sometimes from the other.

Multiplication law for equality

A theorem closely analogous to Theorem 1.1 can be stated as follows.

Theorem 1.2   *If $a, b, c \in R$ and $a = b$, then*

$$ac = bc \quad and \quad ca = cb.$$

The proof of Theorem 1.2 exactly parallels that of Theorem 1.1, and is left as an exercise.

Additional properties

The next theorem asserts that the additive inverse of a real number is unique—that is, that a given real number has only one additive inverse.

Theorem 1.3   *If $a, b \in R$ and $a + b = 0$, then*

$$b = -a \quad and \quad a = -b.$$

To prove this, we begin by observing that if $a + b = 0$, then, by Theorem 1.1,

$$(-a) + (a + b) = (-a) + 0.$$

Next, the associative law of addition and the additive-inverse law imply

$$[(-a) + a] + b = (-a) + 0 \quad and \quad 0 + b = -a + 0.$$

Finally, by the additive-identity law, we have

$$b = -a.$$

A similar argument establishes also that $a = -b$.

In the last step of the foregoing proof, we did not state that $0 + b = b$ and $(-a) + 0 = -a$ and then argue that $b = -a$ by the substitution law; this amount of detail is superfluous. The degree of rigorous detail desirable in proofs of this kind is purely relative. If a line of argument is clear and valid, most reasonable persons will not insist on dotting all i's and crossing all t's. What is important is that you understand the argument.

A result analogous to Theorem 1.3 holds for the multiplicative inverse and is left as an exercise, as are the proofs of the succeeding four theorems, which follow directly from the previous results.

Theorem 1.4   *If $a, b, c \in R$ and $a + c = b + c$, then $a = b$.*

Theorem 1.5   *If $a, b, c \in R$, $c \neq 0$, and $ac = bc$, then $a = b$.*

Theorem 1.6   *For every $a \in R$, $a \cdot 0 = 0$.*

Theorem 1.7   *If $a, b \in R$ and $a \cdot b = 0$, then either $a = 0$ or $b = 0$ or both.*

Combining Theorems 1.6 and 1.7, we see that for $a, b \in R$ we have $ab = 0$ *if and only if* at least one of the factors is 0.

The following theorem concerns the familiar "laws of signs" for operating with real numbers.

Theorem 1.8   *If $a, b \in R$, then*

$$\text{I} \quad -(-a) = a, \qquad\qquad \text{II} \quad (-a) + (-b) = -(a + b),$$

$$\text{III} \quad (-a)(b) = -(ab), \qquad\qquad \text{IV} \quad (-a)(-b) = ab,$$

$$\text{V} \quad \frac{-a}{b} = \frac{a}{-b} = -\frac{a}{b} \quad (b \neq 0), \qquad \text{VI} \quad \frac{-a}{-b} = \frac{a}{b} \quad (b \neq 0).$$

Part I of this theorem is an immediate consequence of the uniqueness of the additive inverse. That is, the symbol " $-(-a)$ " denotes the additive inverse of $-a$, as does the symbol " a." Hence, $-(-a) = a$.

The proofs of the remaining parts of Theorem 1.8 are left as exercises, as are the proofs of Theorems 1.9 to 1.11, which are concerned with the familiar properties of quotients.

Theorem 1.9   *If $a, b, c \in R$, then*

$$\frac{a}{b} = \frac{c}{d} \quad (b, d \neq 0) \quad \text{if and only if} \quad ad = bc.$$

As a direct consequence of this characterization of equal quotients, we have a theorem that is sometimes referred to as the ***fundamental principle of fractions***.

Theorem 1.10   *If $a, b, c \in R$, then*

$$\frac{ac}{bc} = \frac{a}{b} \quad (b, c \neq 0).$$

Next, let us group a number of assertions about quotients into a single theorem.

Theorem 1.11   *If $a, b, c, d \in R$, then*

I $\dfrac{1}{a} \cdot \dfrac{1}{b} = \dfrac{1}{ab}$ $(a, b \neq 0)$,   II $\dfrac{a}{b} \cdot \dfrac{c}{d} = \dfrac{ac}{bd}$ $(b, d \neq 0)$,

III $\dfrac{a}{c} + \dfrac{b}{c} = \dfrac{a + b}{c}$ $(c \neq 0)$,   IV $\dfrac{a}{b} + \dfrac{c}{d} = \dfrac{ad + bc}{bd}$ $(b, d \neq 0)$,

V $\dfrac{a}{b} - \dfrac{c}{d} = \dfrac{ad - bc}{bd}$ $(b, d \neq 0)$,   VI $1 \div \dfrac{a}{b} = \dfrac{b}{a}$ $(a, b \neq 0)$,

VII $\dfrac{a}{b} \div \dfrac{c}{d} = \dfrac{ad}{bc}$ $(b, c, d \neq 0)$.

Exercise 1.4

In Problems 1–20, *each statement is justified by one part of Theorems* 1.1–1.11. *Cite an appropriate justification. All variables denote elements of the set R of real numbers.*

Example   If $p + q + 3 = 4 + 3$, then $p + q = 4$.

Solution   Theorem 1.4

1. If $x = 3$, then $x + 9 = 3 + 9$.

2. $-2 - q = -(2 + q)$

3. $\dfrac{-2}{5} = -\dfrac{2}{5}$

4. $\dfrac{4(x + y)}{6} = \dfrac{2(x + y)}{3}$

5. $\dfrac{x}{3} + \dfrac{y + z}{3} = \dfrac{x + y + z}{3}$

6. $\dfrac{p}{3} \cdot \dfrac{q}{4} = \dfrac{pq}{12}$

7. If $4p = 0$, then $p = 0$.

8. $(-5)(-6) = 5 \cdot 6$

9. If $3x = 2y$, then $\dfrac{3}{2} = \dfrac{y}{x}$.

10. $\dfrac{-x}{-3} = \dfrac{x}{3}$

11. $\dfrac{\frac{2}{5}}{\frac{3}{7}} = \dfrac{2 \cdot 7}{5 \cdot 3}$

12. $\dfrac{\frac{1}{2}}{\frac{3}{3}} = \dfrac{3}{2}$

13. $(-3)(p) = -(3p)$

14. $x - (-y) = x + y$

15. $\dfrac{x}{3} + \dfrac{y}{2} = \dfrac{2x + 3y}{3 \cdot 2}$

16. $\dfrac{x}{-3} = -\dfrac{x}{3}$

17. If $p + q = 7$, then $4(p + q) = 4 \cdot 7$.

18. If $(r + s) + 7 = 0$, then $r + s = -7$.

19. If $6(x - y) = 3z$, then $2(x - y) = z$.

20. $\dfrac{x + 1}{4} - \dfrac{y + 3}{3} = \dfrac{3(x + 1) - 4(y + 3)}{4 \cdot 3}$

Prove each of the following statements. All variables denote elements of the set R of real numbers.

21. If $a = b$, then $ac = bc$ and $ca = cb$.

22. $\dfrac{1}{a} \ (a \neq 0)$ is unique; that is, if $ab = 1$, then $b = \dfrac{1}{a}$.

23. 0 is unique; that is, if $a + b = a$, then $b = 0$.

24. 1 is unique; that is, if $a \neq 0$ and $ab = a$, then $b = 1$.

25. If $a + c = b + c$, then $a = b$.

26. If $ac = bc$ and $c \neq 0$, then $a = b$.

27. If $a \in R$, then $a \cdot 0 = 0$.

28. If $ab = 0$, then either $a = 0$ or $b = 0$ or both.

29. $(-a) + (-b) = -(a + b)$ **30.** $(-a)(b) = -(ab)$

31. $(-a)(-b) = ab$ **32.** $\dfrac{-a}{b} = \dfrac{a}{-b} = -\dfrac{a}{b}$ $(b \neq 0)$

33. $\dfrac{-a}{-b} = \dfrac{a}{b}$ $(b \neq 0)$ **34.** $\dfrac{a}{b} = \dfrac{c}{d}$ $(b, d \neq 0)$ if and only if $ad = bc$.

35. $\dfrac{ac}{bc} = \dfrac{a}{b}$ $(b, c \neq 0)$ **36.** $\dfrac{1}{a} \cdot \dfrac{1}{b} = \dfrac{1}{ab}$ $(a, b \neq 0)$

37. $\dfrac{a}{b} \cdot \dfrac{c}{d} = \dfrac{ac}{bd}$ $(b, d \neq 0)$ **38.** $\dfrac{a}{c} + \dfrac{b}{c} = \dfrac{a + b}{c}$ $(c \neq 0)$

39. $\dfrac{a}{b} + \dfrac{c}{d} = \dfrac{ad + bc}{bd}$ $(b, d \neq 0)$ **40.** $\dfrac{a}{b} - \dfrac{c}{d} = \dfrac{ad - bc}{bd}$ $(b, d \neq 0)$

41. $\dfrac{\frac{1}{a}}{\frac{a}{b}} = \dfrac{b}{a}$ $(a, b \neq 0)$ **42.** $\dfrac{\frac{a}{b}}{\frac{c}{d}} = \dfrac{ad}{bc}$ $(b, c, d \neq 0)$

43. $\dfrac{a}{b} = q$ if and only if $a = bq$ $(b \neq 0)$.

44. Derive the second equation,

$$(b + c)a = ba + ca,$$

in Field Postulate F-8 from the other postulates and the first equation,

$$a(b + c) = ab + ac.$$

1.5   *Order and Completeness*

As you probably recall from earlier courses, there is a one-to-one correspondence between the real numbers and the points on a geometric line (to each real number there corresponds one and only one point on the line, and vice versa). To illustrate this, we imagine the line scaled in convenient units, with the positive direction (from 0 toward 1) denoted by an arrowhead. The line is then called a **number line**; the real number corresponding to a point on the line is called the **coordinate** of the point, and the point is called the **graph** of the number. For example, a number-line representation of {1, 3, 5} is shown in Figure 1.6.

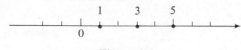

Figure 1.6

A horizontal number-line graph directed to the right can be used to illustrate the separation of the real numbers into three disjoint subsets: {negative real numbers}, {0}, {positive real numbers}. The point associated with 0 is called the **origin**. The set of numbers whose elements are associated with the points on the right-hand side of the origin belong to the set R_+ of **positive real numbers**, and the set whose elements are associated with the points on the other side belong to the set R_- of **negative real numbers**.

Notice that the word "negative" has now been used in two ways. In one case, we refer to the *negative of a number*, as in Postulate F-4, whereas, in the other, we refer to a *negative number*, which is the negative of a positive number.

Order
postulates

It is possible to categorize the set of positive real numbers without recourse to geometric considerations, though of course we shall continue to find it convenient to refer also to the number line. With this in mind, let us state two more postulates that apply to real numbers.

O-1   If a is a real number, then exactly one   *Trichotomy law.*
of the following statements is true: a is
positive, a is zero, or $-a$ is positive.

O-2   If a and b are positive real numbers, then   *Closure law for positive numbers.*
$a + b$ is positive and ab is positive.

The first of these postulates asserts that every real number belongs to one of the sets R_+, {0}, or R_-, but to only one of them. The second asserts that the set R_+ of positive real numbers is closed with respect to the binary operations of addition and multiplication.

Since the set R of real numbers satisfies Postulates O-1 and O-2 as well as Postulates F-1 through F-11, we say that R is an **ordered field**. Similarly, the set Q of rational numbers is an ordered field. But the set C of complex numbers is not an ordered field even though, as we shall see in Chapter 10, it is a field.

By Postulate O-2, if *a* and *b* are positive real numbers, then *ab* is a positive real number. This fact, together with Parts III and IV of Theorem 1.8, is sufficient to establish that the product of a positive real number and a negative real number is a negative real number, while the product of two negative real numbers is a positive real number.

Less than and
greater than
The addition of a positive real number *d* to a real number *a* can be visualized on a number-line graph as the process of locating the point corresponding to *a* on the line and then moving along the line *d* units to the right to arrive at the point corresponding to *a + d* (Figure 1.7). With this idea in mind, we define what is meant by "less than."

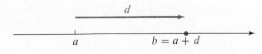

Figure 1.7

Definition 1.14   *If* $a, b \in R$, *then* *a* *is* **less than** *b* *if and only if there exists a positive real number* *d* *such that* $a + d = b$.

Since *d* is positive, this implies for any two real numbers *a* and *b*, that if the graph of *a* lies to the left of the graph of *b*, then *a* is less than *b*. The inequality symbol $<$ is used to denote the phrase "is less than," and $a < b$ is read "*a* is less than *b*." The inequality symbol $>$ means "is greater than." The statements $a < b$ and $b > a$ are taken as equivalent.

Definition 1.14 and the field postulates have the following implications, which we shall not prove.

Theorem 1.12   *For any* $a, b, c \in R$:

I   *If* $a < b$ *and* $b < c$, *then* $a < c$.   II   *If* $a < b$, *then* $a + c < b + c$.

III *If* $a < b$ *and* $c > 0$, *then* $ac < bc$.   IV   *If* $a < b$ *and* $c < 0$, *then* $ac > bc$.

As you can see, Theorem 1.12-I is comparable to the transitive law for equality. That is, we can say that "less than" is a transitive relationship.

Each of the symbols $\leq$ and $\geq$ (read "is less than or equal to" and "is greater than or equal to," respectively) is a contraction for two symbols, one of equality and one of inequality, connected by the word "or." For example, $x \leq 7$ is the statement that *x* is less than 7 *or* *x* equals 7.

We often write together two inequalities that express a transitive relationship. Thus $a < b$ and $b < c$ are written together as $a < b < c$ and read "*a* is less than *b* and *b* is less than *c*." Similarly, $0 < x$ and $x \leq 1$ are written together as $0 < x \leq 1$.

Absolute value
The graphs of the numbers *a* and $-a$ on a number line lie the same distance from the origin, but on opposite sides of it. If we wish to refer to the *distance* of the graph of a number from the origin, and not to the side of the origin on which it is located, then we use the term **absolute value**. Thus, the absolute value of *a* and the absolute value of $-a$ are the same nonnegative number. The symbol $|a|$ is used to denote the absolute value of *a*. We formalize the definition as follows:

Definition 1.15   *If $a \in R$, then the **absolute value** of a is given by*

$$|a| = \begin{cases} a, & \text{if } a \geq 0. \\ -a, & \text{if } a < 0. \end{cases}$$

For example, $|-3| = -(-3) = 3$, $|7| = 7$,   and $|0| = 0$.

Completeness postulate

On page 17 we mentioned in passing a property of the set R of real numbers that is of fundamental importance and is necessary for establishing the existence of irrational numbers. Let us formalize this by assuming the following **completeness property**.

O-3   There is a one-to-one correspondence between the set of real numbers and the set of points on a geometric line.

Although Postulate O-3 might be asserted more precisely, for our purposes this formulation is quite sufficient. Because of the completeness property, the set R of real numbers is said to be a **complete ordered field**. Note, however, that while the set Q of rational numbers is an ordered field, and between any two rational numbers there are infinitely many other rational numbers, Q is not a complete ordered field—because between any two rational numbers there are also infinitely many irrational numbers; and the set of points corresponding to Q does not completely "fill" the number line.

Exercise 1.5

For $x, y \in R$, justify each statement by citing one part of Theorem 1.12.

Example

If $2 < 3x$, then $-4 > -6x$.

Solution

Part IV. Each member of $2 < 3x$ is multiplied by -2.

1. If $x < 3$ and $y < x$, then $y < 3$.   2. If $x + 1 < 0$, then $x < -1$.
3. If $y < 8$, then $3y < 24$.   4. If $y < 4$, then $y + 2 < 6$.
5. If $x < 7$, then $x - 2 < 5$.   6. If $x < 9$, then $-2x > -18$.
7. If $-6x < 12$, then $x > -2$.   8. If $x - 3 < 5$, then $x < 8$.

Express each statement by means of symbols.

Examples

a.   5 is not greater than 7.   b. x is between 5 and 8.

Solutions

a. $5 \ngtr 7$, or $5 \leq 7$ b. $5 < x < 8$

9. 7 is greater than 3.   10. 2 is less than 5.
11. -4 is less than -3.   12. -4 is greater than -7.
13. x is between -1 and 1, inclusive.   14. x is negative.
15. x is positive.   16. x is nonpositive.
17. x is nonnegative.   18. $2x$ is less than or equal to 8.

Write each expression without using absolute-value notation.

Examples   a. $|-11|$

b. $|x - 6|$

Solutions   a. 11

b. $x - 6$, if $x - 6 \geq 0$;
$-(x - 6)$, if $x - 6 < 0$

19. $|-3|$ **20.** $|-5|$ **21.** $|7|$ **22.** $|4|$
23. $-|-2|$ **24.** $-|-5|$ **25.** $|3x|$ **26.** $|2y|$
27. $|x + 1|$ **28.** $|x + 2|$ **29.** $|y - 3|$ **30.** $|y - 4|$

Replace the comma with an appropriate order symbol to form a true statement.

31. $|-2|, |-5|$ **32.** $|3|, |-4|$ **33.** $-7, |-1|$
34. $5, |-2|$ **35.** $|-3|, 0$ **36.** $-|-4|, 0$

Write an equivalent relation without using the negation symbol, $/$.

37. $2 \not\geq 5$ **38.** $-1 \not\leq -2$ **39.** $7 \not\geq 8$
40. $|x| \not< 3$ **41.** $|x| \not\geq 3$ **42.** $x \not> |y|$

Chapter Review

[1.1]   *Replace the colored comma with either $\in$ or $\subset$ to make a true statement.*

1. $6, \{2, 4, 6, 8\}$ **2.** $\{2\}, \{1, 2, 3, 4\}$
3. $\emptyset, \{2, 4\}$ **4.** $25, \{5, 10, 15, \ldots\}$

5. List the subsets of $\{1, 2, 3, 4\}$ that have $\{1, 3\}$ as a subset.
6. Designate {multiples of 5} by using set-builder notation.

[1.2]   *Let $U = \{1, 2, 3, 4, 5, 6, 7, 8, 9\}$, $A = \{2, 4, 6, 8\}$, $B = \{5, 6, 7, 8, 9\}$, and $C = \{3, 5, 7, 9\}$. List the members of each of the following.*

7. $A \cup B$ **8.** $A \cap B$ **9.** $A \cap C$
10. $A' \cap B$ **11.** $(A \cap C)'$ **12.** $(B \cup C)'$

[1.3]   *Let $A = \{-8, -\dfrac{15}{7}, -\sqrt{5}, -1, 0, 3, \dfrac{13}{2}, \sqrt{50}, 21\}$. List the members of each of the following sets.*

13. {natural numbers in A}   **14.** {integers in A}
15. {rational numbers in A}   **16.** {irrational numbers in A}

For a, b, c, d ∈ R, justify each statement by citing the appropriate postulate.

17. $(c + d)a = (d + c)a$

18. $ca + bc = ca + cb$

19. $\dfrac{1}{a + b} \cdot (a + b) = 1 \quad (a \neq -b)$

20. $(c + d) \cdot \dfrac{1}{b} = c \cdot \dfrac{1}{b} + d \cdot \dfrac{1}{b} \ (b \neq 0)$

21. $(4 + a) + [-(4 + a)] = 0$

22. $c[a + (b + d)] = ca + c(b + d)$

[1.4] *For x, y ∈ R, justify each statement by citing one part of Theorems* 1.1–1.11.

23. If $x = 6$, then $x + 2 = 6 + 2$.

24. $(-5)(-x) = 5x$

25. $\dfrac{4}{5} \cdot \dfrac{x + 2}{3} = \dfrac{4(x + 2)}{5 \cdot 3}$

26. $\dfrac{4(x + 3)}{8} = \dfrac{x + 3}{2}$

27. $-\dfrac{-x}{-2} = -\dfrac{x}{2}$

28. $\dfrac{x}{3} + \dfrac{y}{5} = \dfrac{5x + 3y}{3 \cdot 5}$

29. $\dfrac{\frac{1}{3}}{\frac{3}{4}} = \dfrac{4}{3}$

30. $\dfrac{\frac{4}{7}}{\frac{2}{3}} = \dfrac{4}{7} \cdot \dfrac{3}{2}$

[1.5] *For x, y ∈ R, justify each statement by citing one part of Theorem* 1.12.

31. If $x < 3$, then $-2x > -6$.

32. If $x + 7 < 3$, then $x < -4$.

Express each statement by means of symbols.

33. -6 is greater than -9.

34. 4 is less than or equal to y.

35. y is nonpositive.

36. $y + 2$ is nonnegative.

Rewrite each expression without using absolute-value notation.

37. $|-7|$

38. $|x - 5|$

Write each statement without using the negation symbol.

39. $x \ngeq 4$

40. $|x| \nless 6$

2 *Polynomials*

2.1   *Definitions; Sums of Polynomials*

Any grouping of constants and variables generated by applying a finite number of the elementary operations—addition, subtraction, multiplication, division, or the extraction of roots—is called an **algebraic expression**. (Roots will be discussed in Chapter 3.) For example,

$$\frac{3x^2 + \sqrt{2x - 1}}{3x^3 + 7} \quad \text{and} \quad xy + 3x^2z - \sqrt[5]{z}$$

are algebraic expressions. If two expressions have equal values for all values of the variables for which both expressions are defined, then we say that on the set of these values the expressions are **equivalent**.

Natural-number powers

You should recall that an expression of the form a^n is called a **power** of a, where a is the **base** of the power and n is the **exponent** of the power.

Definition 2.1   *If $n \in N$ and $a \in R$, then*

$$a^n = \underbrace{a \cdot a \cdot a \cdot \cdots \cdot a.}_{n \text{ factors}}$$

Polynomial expressions

In any algebraic expression of the form $A + B + C + \cdots$, where $A, B, C, \ldots$ are algebraic expressions, $A, B, C, \ldots$ are called **terms** of the expression. For example, in $x + (y + 3)$ the terms are x and $(y + 3)$, but in $x + y + 3$ the terms are x, y, and 3. An algebraic expression that contains only nonnegative integral powers of a variable and contains no variable in a denominator is a **polynomial**. For example,

$$5x, \quad \frac{3x^2}{2} - \frac{7x}{2}, \quad 0, \quad 2x^2 - 3x + 4, \quad \text{and} \quad \frac{y}{4} - \frac{\sqrt{7}}{4}$$

are polynomials, whereas

$$\frac{3}{x}, \quad 3 + \sqrt{x}, \quad \text{and} \quad \frac{2\sqrt{x-1}}{2\sqrt{x+1}}$$

are algebraic expressions, but not polynomials, in the variable x.

Polynomials consisting of one, two, or three terms are also called **monomials, binomials**, and **trinomials**, respectively. Thus $3x^2y$ is a monomial, $x + 4x^2$ is a binomial, and $x + y + z$ is a trinomial.

The **degree** of a monomial is given by the exponent of the variable in the monomial. Thus, 5 is of degree zero (in Chapter 3 we shall define x^0 to be equal to 1 for all $x \in R$, $x \neq 0$), $2x$ is of the first degree, and $3x^4$ is of fourth degree; but no degree is assigned to the special monomial 0. If a monomial contains more than one variable, its degree is given by the sum of the exponents on the variables; $3x^2y^3z$ is of sixth degree in x, y, and z. It can also be described as being of second degree in x, third degree in y, fifth degree in x and y, and so on. The constant factor in the monomial, 3, is called the **coefficient** of the monomial. The **degree of a polynomial** is the same as the degree of its term of highest degree. Since no degree is assigned to the monomial 0, no degree is assigned to the polynomial 0, either.

Because $a - b$ is defined to be $a + (-b)$, we shall view the signs in any polynomial as signs denoting numbers or their negatives, and the operation involved to be addition. Thus

$$3x - 5y + 4z = (3x) + (-5y) + (4z),$$

and its terms are $3x$, $-5y$, and $4z$. Again, an expression such as

$$a - (bx + cx^2),$$

in which a set of parentheses is preceded by a negative sign, can be written

$$a + [-(bx + cx^2)],$$

or

$$a + [-bx - cx^2],$$

or, finally,

$$a + (-bx) + (-cx^2).$$

Also, since

$$\frac{a}{b} = a\left(\frac{1}{b}\right),$$

we can view division by a constant as multiplication by its reciprocal (multiplicative inverse), and, for example, write

$$\frac{3x^2}{4} + \frac{x}{2} \quad \text{as} \quad \frac{3}{4}x^2 + \frac{1}{2}x.$$

Accordingly, since a polynomial can be considered to involve only the operations of addition and multiplication, and since the set R of real numbers is closed with respect to these operations, it follows that, for any specific real value of x, a polynomial with real coefficients represents a real number. Therefore, the postulates for the real numbers are applicable to the terms in such polynomials and to the polynomials themselves.

Simplification
of
polynomials

By applying the commutative, associative, and distributive laws in various ways, we can frequently rewrite polynomials and sums of polynomials in what might be termed "simpler" forms.

Example

$$(2x^2 + 3x + 5) + 2x + (6x^2 + 7) = 2x^2 + 6x^2 + 3x + 2x + 5 + 7$$
$$= (2 + 6)x^2 + (3 + 2)x + 5 + 7$$
$$= 8x^2 + 5x + 12$$

Here we have reduced the number of terms from six to three, and the original expression is said to have been "simplified."

We shall often be concerned with polynomials in one variable. A polynomial of degree n, $n \geq 0$, in x can be represented—when its terms are rearranged, if need be—by an expression of the form

$$a_0 x^n + a_1 x^{n-1} + a_2 x^{n-2} + \cdots + a_{n-1}x + a_n \quad (a_0 \neq 0),$$

where it is understood that the a's are the (constant) coefficients of the powers of x in the polynomial.

The term $a_0 x^n$ is called the **leading term**, and the coefficient a_0 is called the **leading coefficient** in the polynomial. It is often convenient to have the leading term of the form x^n. In this case (that is, when $a_0 = 1$), the polynomial is said to be **monic**.

If the coefficients in a polynomial P are real numbers, then the polynomial is called a **polynomial over the real-number field**, or simply a **polynomial over** R. If the variable is restricted to represent only real numbers, then the polynomial is said to be a **polynomial in the real variable** x. If *both* the coefficients and the variable are restricted to real values, we say that the polynomial is a **real polynomial**.

Symbols for
polynomials

Polynomials are frequently represented by symbols such as

$$P(x), \quad D(y), \quad \text{and} \quad Q(z),$$

where the symbol in parentheses designates the variable. Thus, we might write

$$P(x) = 2x^3 - 3x + 2,$$
$$D(y) = y^6 - 2y^2 + 3y - 2,$$
$$Q(z) = 8z^4 + 3z^3 - 2z^2 + z - 1.$$

Value of a
polynomial

The notation $P(x)$ can be used to denote values of the polynomial for specific values of x. Thus, $P(2)$ means the value of the polynomial $P(x)$ when x is replaced by 2. For example, if

$$P(x) = x^2 - 2x + 1,$$

then

$$P(2) = 2^2 - 2(2) + 1 = 1,$$
$$P(3) = 3^2 - 2(3) + 1 = 4,$$
$$P(-4) = (-4)^2 - 2(-4) + 1 = 25.$$

In some applications, the notation $P(x)|_a^b$ denotes $P(b) - P(a)$. Thus if

$$P(x) = \frac{x^2}{2} - 4x,$$

then

$$P(x)\Big|_2^3 = P(3) - P(2) = \left[\frac{3^2}{2} - 4(3)\right] - \left[\frac{2^2}{2} - 4(2)\right] = -\frac{3}{2}.$$

Exercise 2.1

Simplify each expression.

Example

$(y^2 - 3y + 2) + (4y^2 - 7) - (2y^2 + 6y - 5)$

Solution

$y^2 - 3y + 2 + 4y^2 - 7 - 2y^2 - 6y + 5$
$(y^2 + 4y^2 - 2y^2) + (-3y - 6y) + (2 - 7 + 5)$
$3y^2 - 9y$

1. $(z^3 - 3z^2 + 2) + (z^3 + 5z - 3) - (2z^3 + 2z^2 - 3z + 1)$
2. $(x^2 - 3x + 5) - (x^2 + 3x - 5) - (2x^2 + 3x - 1)$
3. $(2y^4 - 3y^2 + 2) - (y^4 + 3y^3 - 2) + (-y^4 + 2y^3 - 3y^2 + y - 1)$
4. $(3z^4 + 2z^3 - 3z) + (5z^3 - 2z^2 + z) - (z^4 + 5z^3 - 2z^2 + 4z - 1)$
5. $(3x^2 - xy + 4y^2) - (x^2 + 2xy + y^2) + (-x^2 + 3xy - y^2)$
6. $(2y^2z - 3yz^2 + 5yz) + (y^2z + yz^2 - 4yz) - (3y^2z + yz^2 - yz)$
7. $(x^2 - 2x + 3) - [(2x^2 + x - 5) - (x^2 + 3x + 1)]$
8. $(2y^2 + 3y - 2) - [(y^2 + 5y - 3) - (-y^2 + 3y + 2)]$

Express $P(x) + Q(x)$ and $P(x) - Q(x)$ as polynomials.

9. $P(x) = 3x - 2, \quad Q(x) = 3 - x$
10. $P(x) = x^2 + 3x - 2, \quad Q(x) = 2x^2 + x - 2$
11. $P(x) = x^2 - 2x + 3, \quad Q(x) = 2x^2 - 2x - 1$
12. $P(x) = 2x^3 - 3x^2 + x - 1, \quad Q(x) = x^3 + 3x - 2$
13. $P(x) = x^3 + 4x^2 - 2x + 1, \quad Q(x) = x^2 - 2x + 3$
14. $P(x) = 2x^2 - 3x + 4, \quad Q(x) = 2x^3 + x^2 - x - 1$
15. $P(x) = 3x^4 - 2x^2 + 1, \quad Q(x) = x^5 - 3x^3 + 4x$
16. $P(x) = 4x^5 - x^3 + 3x, \quad Q(x) = 2x^4 + x^2 - 5$

For the polynomials

$$P(x) = 2x^2 - 3x + 2, \quad Q(x) = 3 - 2x + x^2, \quad \text{and} \quad S(x) = -2x^2 + 3x - 5,$$

write each given expression as an equivalent polynomial.

17. $P(x) + Q(x)$ **18.** $P(x) + [Q(x) - S(x)]$

19. $P(x) - [Q(x) + S(x)]$ **20.** $P(x) - [Q(x) - S(x)]$

21. $[Q(x) - P(x)] - S(x)$ **22.** $S(x) - [-P(x) - Q(x)]$

Example   Given $P(x) = 2x^2 - x + 3$, find $P(3), P(-3), P(0), P(a), P(x)|_0^3$.

Solution

$$P(3) = 2(3)^2 - (3) + 3 = 18$$
$$P(-3) = 2(-3)^2 - (-3) + 3 = 24$$
$$P(0) = 2(0)^2 - (0) + 3 = 3$$
$$P(a) = 2a^2 - a + 3$$
$$P(x)|_0^3 = P(3) - P(0) = 18 - 3 = 15$$

✦23. Given $P(x) = x^3 - 3x^2 + x + 1$, find $P(2), P(-2), P(0), P(x)|_0^2, P(x)|_{-2}^2$.

24. Given $P(x) = 2x^3 + x^2 - 3x + 4$, find $P(3), P(-3), P(0), P(x)|_0^3, P(x)|_{-3}^0$.

✗25. Given $P(x) = x^{12}$, find $P(1), P(-1), P(0), P(x)|_0^1, P(x)|_{-1}^1$.

26. Given $P(x) = x^{13}$, find $P(1), P(-1), P(0), P(x)|_0^1, P(x)|_{-1}^1$.

✗27. Given $Q(x) = 2x^4 - x^2 + 3$, find $Q(0), Q(h), Q(-h)$.

28. Given $R(x) = x^4 + 2x^3 - 5$, find $R(0), R(t), R(-t)$.

✦29. Given $P(x) = x^2 + 2x - 3$, find $P(x + h), P(x - h), P(x^2)$.

30. Given $N(x) = 2x^2 - x + 5$, find $N(x - h), N(x + h), N(x^3)$.

Example   Given $P(x) = x - 4$ and $Q(x) = x + 2$, find $P(Q(2))$.

Solution $Q(2) = 2 + 2 = 4$; hence

$$P(Q(2)) = P(4) = 4 - 4 = 0.$$

31. Given $P(x) = x + 2$ and $Q(x) = x - 3$, find $P(Q(2))$ and $Q(P(2))$.

32. Given $P(x) = 2x + 1$ and $Q(x) = \frac{1}{2}(x - 1)$, find $P(Q(-2))$ and $Q(P(-2))$.

✦33. Given $P(x) = x^2 + 3$ and $Q(x) = 6$, find $Q(P(2))$ and $P(Q(0))$.

34. Given $P(x) = 2x^3 - 3x$ and $Q(x) = x^2 + 1$, find $P(Q(0))$ and $Q(P(0))$.

35. If $P(x)$ is of degree n and $Q(x)$ is of degree $n - 2$, what is the degree of $P(x) + Q(x)$? Of $P(x) - Q(x)$?

36. If $P(x)$ and $Q(x)$ are polynomials, with $P(0) = 4$ and $Q(0) = 3$, what is the value of $P(x) + Q(x)$ for $x = 0$? Of $P(x) - Q(x)$ for $x = 0$?

2.2   *Products of Polynomials*

Laws of
exponents
for natural-
number
exponents

By Definition, 2.1 (page 22), for $a \in R$ and $n \in N$, we have

$$a^n = \underbrace{a \cdot a \cdot a \cdot \cdots \cdot a}_{n \text{ factors}}.$$

The product $a^m \cdot a^n$, where m and n are natural numbers, is then given by

$$a^m \cdot a^n = \underbrace{(a \cdot a \cdot a \cdot \cdots \cdot a)}_{m \text{ factors}} \underbrace{(a \cdot a \cdot a \cdot \cdots \cdot a)}_{n \text{ factors}}$$

$$= \underbrace{a \cdot a \cdot a \cdot \cdots \cdot a.}_{(m + n) \text{ factors}}$$

We state the result formally, together with two related properties whose proofs are similar to the one above and are omitted.

Theorem 2.1   *If $a \in R$ and $m, n \in N$, then*

$$\text{I} \quad a^m \cdot a^n = a^{m+n},$$
$$\text{II} \quad (a^m)^n = a^{mn},$$
$$\text{III} \quad (ab)^n = a^n b^n.$$

This theorem justifies writing the product of two natural-number powers of the same base as a power (of the same base) with an exponent equal to the sum of the two exponents. For example,

$$x^2 x^3 = x^5,$$
$$y^3 y^4 y^2 = y^9.$$

In rewriting the product of two monomials, say

$$(3x^2 y)(2xy^2),$$

we use the commutative and associative laws and Theorem 2.1 to write

$$6x^3 y^3.$$

Products of polynomials

The **generalized distributive law**,

$$a(b_1 + b_2 + \cdots + b_n) = ab_1 + ab_2 + \cdots + ab_n,$$

can be applied to write as a polynomial the product of a monomial and a polynomial containing more than one term. For example,

$$3x(x + y + z) = 3xx + 3xy + 3xz = 3x^2 + 3xy + 3xz.$$

In its complete form, the generalized distributive law requires mathematical induction for its proof, but its validity as applied in our example here—or in any similar example—can readily be verified; thus we have

$$3x(x + y + z) = 3x[x + (y + z)]$$
$$= 3x(x) + 3x(y + z)$$
$$= 3x^2 + 3xy + 3xz.$$

The distributive law can be applied successively to the products of polynomials containing more than one term.

Example

$$(3x + 2y)(x - y) = 3x(x - y) + 2y(x - y)$$
$$= 3x^2 - 3xy + 2xy - 2y^2$$
$$= 3x^2 - xy - 2y^2$$

Of course, products of polynomials should ordinarily be simplified mentally if it is convenient to do so. The following binomial products are types so frequently encountered that you should learn to recognize them on sight:

$$(x + a)(x + b) = x^2 + (a + b)x + ab$$
$$(x + a)^2 = x^2 + 2ax + a^2$$
$$(x + a)(x - a) = x^2 - a^2$$

Exercise 2.2

Simplify by writing each product in equivalent polynomial form, in each term combining all constants and all powers of each variable.

Examples   a. $(-2x^3)(3xy)(y^2)$ b. $a^n \cdot a^{n+1}$

Solutions   a. $-6x^4y^3$ b. $a^{n+(n+1)}$
                                                                    a^{2n+1}

1. $(-3x^2)(-2xy)(-y^3)$ 2. $(a^3)(-2ab^2)(-b^3)$
3. $(4x^2y)(2xy^3)(-xy)$ ⅄4. $(2ab^3)(-3a^3b^2)(-a^2b^2)$
5. $x^{n+2} \cdot x^{2n-2}$ ⅄6. $z^{n-3} \cdot z^3$
7. $3^{n+1} \cdot 3^{1-n}$ ⅄8. $4^{2n+1} \cdot 4^{2-2n}$

Examples   a. $2(x^2 - x - 1)$ b. $(x - 3)(x + 5)$

Solutions   a. $2x^2 - 2x - 2$ b. $x^2 + 5x - 3x - 15$
                                                                    $x^2 + 2x - 15$

9. $abc(a - b + 2c)$ 10. $-ab(2a - b + 3c)$
11. $(x + 2)(x + 5)$ Ⅹ12. $(x - 3)(x + 2)$
13. $(x - 2y)^2$ 14. $(2x - y)^2$
15. $(5x + 1)(2x + 3)$ Ⅹ16. $(2x + 3)(x - 5)$
17. $2(3a + 2b)(3a - 2b)$ 18. $4(5x - y)(5x + y)$
19. $-(2a - b)(c - 3d)$ 20. $-(3a - b)(c + 3d)$

Example $(y - 3)(y^2 + 3y - 1)$

Solution
$$(y - 3)(y^2 + 3y - 1) = y(y^2 + 3y - 1) - 3(y^2 + 3y - 1)$$
$$= y^3 + 3y^2 - y - 3y^2 - 9y + 3$$
$$= y^3 - 10y + 3$$

Alternate format
$$
\begin{array}{r}
y^2 + 3y - 1 \\
y - 3 \\
\hline
y^3 + 3y^2 - y \\
-3y^2 - 9y + 3 \\
\hline
y^3 - 10y + 3
\end{array}
$$

21. $(x + 4)(x^2 + 2x - 1)$ **22.** $(x - 2)(x^2 - x + 3)$
23. $(3x - 1)(2x^2 + 3x - 1)$ **24.** $(2x - 3)(3x^2 - 2x + 5)$
25. $a(a - b)(a^2 + ab + b^2)$ **26.** $b(a + b)(a^2 - ab + b^2)$

Example $5\{2x - [x - 3(x + 2) + 3] - 2\}$

Solution   Eliminate the inner grouping symbols first and work toward the outside grouping symbols, removing them last:

$$5\{2x - [x - 3(x + 2) + 3] - 2\} = 5\{2x - [x - 3x - 6 + 3] - 2\}$$
$$= 5\{2x - [-2x - 3] - 2\}$$
$$= 5\{2x + 2x + 3 - 2\}$$
$$= 5\{4x + 1\}$$
$$= 20x + 5$$

27. $2\{a - [a - 2(a + 1) + 1] + 1\} + 1\}$
28. $-\{4 - [3 - 2(a - 1) + a] + a\}$
29. $2x\{x + 3[2(2x - 1) - (x - 1)] + 5\}$
30. $-x\{4 - 2[(x + 1) - 3(x + 2)] - x\}$
31. $5y^2 - 3[y^2 - (y - 2)^2 + 1] - 4$
32. $2x^2 + 5[x - (x + 3)^2 - x^2] + 2x^2$
33. $2\{z^2 + 2[2z - (5 + z) + 2]^2 - 3\} + 2$
34. $-1\{3z^2 - 2[3 - (3z + 1) + 2]^2 - z^2\} + 2z$
35. Given $P(x) = x^2 - 3x + 7$, find $P(x - 1)$ and $P(2 - x)$.
36. Given $P(x) = x^2 + 2x + 1$, find $P(x + h)$ and $P(x - h)$.
37. Given $P(x) = 3 - x^2$, find $[P(3)]^2$ and $P(3^2)$.
38. Given $P(x) = x^2 - 3x$, find $[P(-x)]^2$ and $P(-x^2)$.
39. Simplify the expression $(a + b)^2 - (a^2 + b^2)$. What are the conditions on a and b for $a^2 + b^2$ to be greater than $(a + b)^2$? For $a^2 + b^2$ to be less than $(a + b)^2$?
40. If $P(x)$ and $Q(x)$ are polynomials of degree m and n, respectively, what is the degree of $P(x) \cdot Q(x)$?

2.3   Factoring Polynomials

What do we mean when we say that we have _factored_ an integer or a polynomial? It is true, for example, that

$$2 = 4\left(\frac{1}{2}\right),$$

but we would not ordinarily say that 4 and $1/2$ are factors of 2. On the other hand, since

$$10 = (2)(5),$$

we do say that 2 and 5 are factors of 10 in the domain J of integers.

Now consider the polynomial

$$2x^2 - 10,$$

which is completely factored as

$$2x^2 - 10 = 2(x^2 - 5)$$

if we are limited to **integral** coefficients—that is, coefficients that are integers; but if we consider polynomials in x having real numbers as coefficients, its complete factorization is

$$2x^2 - 10 = 2(x - \sqrt{5})(x + \sqrt{5}).$$

Thus the result depends in part on the coefficients we consider permissible. In this book, we are primarily concerned with polynomials whose coefficients are integers.

In factoring polynomials having integer coefficients, we consider as factors only polynomials having integer coefficients with no common integer factor other than 1 or -1, and we say that such a polynomial is **prime** if it is not the product of two polynomials of this sort and its leading coefficient is positive.

We say that a polynomial other than 0 or 1 is **completely factored** if it is written equivalently as a product of prime polynomials or as such a product times -1.

Example

$$
\begin{aligned}
6x^4 - 96 &= 6(x^4 - 16) \\
&= 2 \cdot 3(x^2 - 4)(x^2 + 4) \\
&= (2)(3)(x - 2)(x + 2)(x^2 + 4)
\end{aligned}
$$

Factors of quadratics

One very common type of factoring is that involving quadratic (second-degree) binomials or trinomials with integer coefficients. From Section 2.2, we recall that

$$(x + a)(x + b) = x^2 + (a + b)x + ab, \tag{1}$$

$$(x + a)^2 = x^2 + 2ax + a^2, \tag{2}$$

$$(x + a)(x - a) = x^2 - a^2. \tag{3}$$

These three forms, each of which involves prime polynomial factors, are those most commonly encountered in the chapters that follow. In this section, we are interested in viewing these relationships from right to left—that is, from polynomial to factored form.

A few other polynomials occur frequently enough to justify a study of their factorization. In particular, the forms

$$(a + b)(x + y) = ax + ay + bx + by, \tag{4}$$

$$(x + a)(x^2 - ax + a^2) = x^3 + a^3, \tag{5}$$

$$(x - a)(x^2 + ax + a^2) = x^3 - a^3 \tag{6}$$

are often encountered in one or another part of mathematics. We are again interested in viewing these relationships from right to left. Expressions such as the right-hand member of form (4) are factorable by grouping. For example, to factor

$$3x^2y + 2y + 3xy^2 + 2x,$$

we write it in the form

$$3x^2y + 2x + 3xy^2 + 2y$$

and factor the common monomial x from the first group of two terms and y from the second group of two terms, obtaining

$$x(3xy + 2) + y(3xy + 2).$$

If we now factor the common binomial $(3xy + 2)$ from each term, we have

$$(3xy + 2)(x + y),$$

in which both factors are prime.

The application of forms (5) and (6) is direct.

Example $a^3 - 1 = (a - 1)(a^2 + a + 1)$

Example $8a^3 + b^3 = (2a)^3 + b^3$
$$= (2a + b)[(2a)^2 - 2ab + b^2]$$
$$= (2a + b)[4a^2 - 2ab + b^2]$$

Exercise 2.3

Factor completely into products of polynomials with integer coefficients. (Assume that all variables in exponents represent natural numbers.)

Examples   a. $18x^2y - 24xy^2 + 6xy$ b. $x^{2n} + x^n$ c. $4a^3 - 5a^2 + a$

Solutions   a. $(2)(3)xy(3x - 4y + 1)$ b. $x^n(x^n + 1)$ c. $a(4a^2 - 5a + 1)$
$$a(4a - 1)(a - 1)$$

1. $9x^5y - 3x^4y + 6x^3y$ 2. $x^2y^2z^2 + 2xyz - xz$
3. $x^2 - 3x - 4$ 4. $y^2 + 5y - 6$
5. $x^2 - 8x + 12$ 6. $z^2 + 4z - 12$
7. $y^2 + 4y + 4$ 8. $t^2 - 6t + 9$
9. $x^2 - 25$ 10. $z^2 - 36$
11. $2x^2 + 5x + 3$ 12. $2n^2 - n - 3$
13. $1 + 5a + 6a^2$ 14. $3 - 7z + 2z^2$
15. $6z^2 + 8z + 2$ 16. $6n^3 + 21n^2 + 9n$
17. $x^4y^2 - x^2y^2$ 18. $3xy^2 - 12xy^2$
19. $x^{2n} - 1$ 20. $x^{4n} - x^{2n}y^{2n}$
21. $x^{n+2} - x^{n+1} + 2x^n$ 22. $x^{n-2} - 3x^{n-1} + x^n$

Examples   a. $by - ay + bx - ax$ b. $8x^3 - y^3$

Solutions   a. $y(b - a) + x(b - a)$ b. $(2x)^3 - y^3$
                       $(b - a)(y + x)$ $(2x - y)(4x^2 + 2xy + y^2)$

✗ 23. $y^4 + 3y^2 + 2$ **24.** $x^4 - 5x^2 + 4$
✗ 25. $2a^4 - a^2 - 1$ **26.** $3z^4 - 11z^2 - 4$
27. $x^4 - (y - 2x)^4$ **28.** $ax^2 + x + ax + 1$
✗ 29. $x^2 + ax + xy + ay$ **30.** $3x + y - 6x^2 - 2xy$
✗ 31. $a^3 + 2ab^2 - 4b^3 - 2a^2b$ **32.** $6x^3 - 4x^2 + 3x - 2$
✗ 33. $y^3 - 27x^3$ **34.** $8 + x^3y^3$
✗ 35. $x^3 + (x - y)^3$ **36.** $(x + y)^3 - z^3$

Examples   a. $a^{2n} - 9$ b. $x^{4n} - 3x^{2n} - 4$

Solutions   a. $(a^n - 3)(a^n + 3)$ b. $(x^{2n} - 4)(x^{2n} + 1)$
                                                                  $(x^n - 2)(x^n + 2)(x^{2n} + 1)$

37. $a^{2n} - 4$ **38.** $x^{2n} - y^{2n}$
39. $x^{4n} - y^{4n}$ **40.** $x^{4n} - 2x^{2n} + 1$
41. $3x^{4n} - 10x^{2n} + 3$ **42.** $6y^{2n} + 30y^n - 900$
43. $2y^{2n} - 12y^n - 1440$ **44.** $2x^{2n} - 23x^ny^n - 39y^{2n}$

45. Show that $ac - ad + bd - bc$ can be factored both as $(a - b)(c - d)$ and as $(b - a)(d - c)$.

46. Show that $a^2 - b^2 - c^2 + 2bc$ can be factored as $(a - b + c)(a + b - c)$.

47. Consider the polynomial $x^4 + x^2y^2 + 25y^4$. If $9x^2y^2$ is both added to and subtracted from this expression, we have

$$x^4 + x^2y^2 + 25y^4 + 9x^2y^2 - 9x^2y^2,$$
$$(x^4 + 10x^2y^2 + 25y^4) - 9x^2y^2,$$
$$(x^2 + 5y^2)^2 - (3xy)^2,$$
$$[(x^2 + 5y^2) - 3xy][(x^2 + 5y^2) + 3xy],$$
$$(x^2 - 3xy + 5y^2)(x^2 + 3xy + 5y^2).$$

By adding and subtracting an appropriate monomial, factor $x^4 + x^2y^2 + y^4$.

48. Use the method of Problem 47 to factor $x^4 - 3x^2y^2 + y^4$.

2.4   *Quotients of Polynomials*

It is easy to show that the set of integers is not closed with respect to division; $2/3$, $1/5$, $8/9$, and $-3/4$ are all examples of noninteger quotients of integers. Similarly, we can see that the set of polynomials is not closed with respect to division, because $1/x$ is a counterexample. That is, $1/x$ is the quotient of the polynomials 1 and x but is not, itself, a polynomial. In Exercise 1.4, Problem 43, you were asked to show that if a, b, and q are real numbers, with $b \neq 0$, then

$$\frac{a}{b} = q \quad \text{if and only if} \quad a = bq.$$

When we observe that the closure laws guarantee the applicability of the real-number postulates to real polynomials for all values of the variables, the result stated in this problem suffices to validate the following theorem.

Theorem 2.2   *If A, D, and Q are real polynomials, then for values of the variables for which $D \neq 0$,*

$$\frac{A}{D} = Q \quad \text{if and only if} \quad A = DQ.$$

If a polynomial Q exists such that $A = DQ$, then A is said to be **exactly divisible** by D, and we write $A/D = Q$. If A is not exactly divisible by D, then the quotient A/D cannot be written as a polynomial.

Let us examine some ways in which we can rewrite quotients of polynomials, even if the resulting expressions are not always, themselves, polynomials. We begin with the simplest case, that in which A and D are monomials and their quotient is a polynomial. Consider

$$\frac{a^m}{a^n} \quad (a \neq 0, \, m, n \in N, \text{ and } m > n).$$

We have

$$\frac{a^m}{a^n} = a^m \cdot \frac{1}{a^n}$$

$$= (a^{m-n} \cdot a^n) \cdot \frac{1}{a^n}$$

$$= a^{m-n} \cdot \left(a^n \cdot \frac{1}{a^n} \right)$$

$$= a^{m-n} \cdot 1,$$

$$\frac{a^m}{a^n} = a^{m-n}.$$

Laws of exponents for quotients

This establishes the first part of the following result. Proof of the second part is similar to the proof of Theorem 2.1 and is omitted.

Theorem 2.3   *If $a, b \in R \, (a, b \neq 0)$, $m, n \in N$, then*

$$\text{I} \quad \frac{a^m}{a^n} = a^{m-n}, \quad m > n,$$

$$\text{II} \quad \left(\frac{a}{b} \right)^m = \frac{a^m}{b^m}.$$

This theorem enables us, for example, to write

$$\frac{12a^5b^3}{4a^2b^2} = \frac{12}{4} \cdot a^{5-2}b^{3-2} = 3a^3b \quad (a, b \neq 0).$$

Note that a and b are not permitted to take the value 0, because if they were, $3a^3b$ would represent a real number, 0, although $(12a^5b^3)/(4a^2b^2)$ would not be defined.

Quotients of
polynomials

Theorem 1.11-III (together with the closure laws) permits us to rewrite quotients of polynomials of the form $(A + B)/C$, $C \neq 0$, in the form $A/C + B/C$, and then to do such further rewriting as seems indicated. For example,

$$\frac{2x^3 + 4x^2 + 8x}{2x} = \frac{2x^3}{2x} + \frac{4x^2}{2x} + \frac{8x}{2x},$$

and, by an application of Theorem 2.3-I, the right-hand member can then be denoted by the expression

$$x^2 + 2x + \frac{8x}{2x} \quad (x \neq 0).$$

Though Theorem 2.3-I is not applicable to the variable factors in expressions such as $(8x)/(2x)$, by Theorem 1.11-II and the fact that $x/x = 1$ for all $x \neq 0$, we have

$$\frac{8x}{2x} = \frac{8}{2} \cdot \frac{x}{x} = \frac{8}{2} \cdot 1 = 4,$$

for every $x \neq 0$. Thus,

$$\frac{2x^3 + 4x^2 + 8x}{2x} = x^2 + 2x + 4 \quad (x \neq 0).$$

As mentioned at the start of this section, however, quotients of polynomials cannot always be represented by polynomials. For example, we have

$$\frac{2x^3 + 4x + 1}{x} = \frac{2x^3}{x} + \frac{4x}{x} + \frac{1}{x}$$

$$= 2x^2 + 4 + \frac{1}{x} \quad (x \neq 0),$$

where the resulting expression is not a polynomial.

If the divisor of a quotient contains more than one term, the familiar **long-division algorithm** involving successive subtractions can be used to rewrite the quotient. For example, the computation

$$
\begin{array}{r}
x - 3 \\
x^2 + 2x - 1 \overline{\smash{)}\, x^3 - x^2 - 7x + 3} \\
\underline{x^3 + 2x^2 - x} \\
-3x^2 - 6x + 3 \\
\underline{-3x^2 - 6x + 3} \\
0
\end{array}
$$

shows that, for $x^2 + 2x - 1 \neq 0$,

$$\frac{x^3 - x^2 - 7x + 3}{x^2 + 2x - 1} = x - 3.$$

It is most convenient to arrange the dividend and the divisor in descending powers of the variable before using the division algorithm, and to leave an appropriate space for any missing terms (terms with coefficient 0) in the dividend.

When the divisor is not a factor of the dividend, the division process will produce a nonzero remainder. For example, from

$$
\begin{array}{r}
x^3 - 3x^2 + 10x - 28 \\
x + 3 \overline{\smash{\big)}\ x^4 \qquad\quad +\quad x^2 +\ 2x - 1} \\
\underline{x^4 + 3x^3} \\
-3x^3 +\quad x^2 \\
\underline{-3x^3 -\ 9x^2} \\
10x^2 +\ 2x \\
\underline{10x^2 + 30x} \\
-28x -\ 1 \\
\underline{-28x - 84} \\
83 \text{ (remainder)}
\end{array}
$$

we see that

$$
\frac{x^4 + x^2 + 2x - 1}{x + 3} = x^3 - 3x^2 + 10x - 28 + \frac{83}{x + 3} \quad (x \neq -3).
$$

Observe in the foregoing example that a space is left in the dividend for a term involving x^3, even though the dividend contains no such term.

Exercise 2.4

Write each quotient as a polynomial. (Assume that all variables in exponents represent natural numbers.)

Examples

a. $\dfrac{6x^2 y^3}{2xy}$ b. $\dfrac{x^{2n+3}}{x^{n+1}}$ c. $\dfrac{2y^3 - 6y^2 + y}{y}$

Solutions

a. $\dfrac{6}{2} \cdot x^{2-1} y^{3-1}$ b. $x^{2n+3-(n+1)}$ c. $\dfrac{2y^3}{y} - \dfrac{6y^2}{y} + \dfrac{y}{y}$

   $3xy^2 \quad (x, y \neq 0)$ $x^{n+2} \quad (x \neq 0)$ $2y^2 - 6y + 1 \quad (y \neq 0)$

1. $\dfrac{8a^3 y^5}{2a^2 y^3}$ 2. $\dfrac{27x^4 b^2}{3x^2 b}$ 3. $\dfrac{38a^3 b^5}{19ab^3}$ 4. $\dfrac{121y^2 x^5 z}{11y^2 xz}$

5. $\dfrac{x^{2n}}{x^n}$ 6. $\dfrac{a^{2n+3}}{a^{n-4}}$ 7. $\dfrac{x^{2n} y^{n+1}}{x^n y}$ 8. $\dfrac{r^{2n} s^{n+5}}{r^{n-1} s^{n+3}}$

9. $\dfrac{8a^2 + 4a + 4}{2}$ 10. $\dfrac{12x^3 - 8x^2 + 36x}{4x}$

11. $\dfrac{x^3 - 4x^2 - 3x}{x}$ 12. $\dfrac{8a^2 x^2 - 4ax^2 + ax}{ax}$

Write each quotient, P/D, either in the form Q or in the form $Q + R/D$, where Q and R are polynomials and the degree of R is less than that of D.

Examples   a. $\dfrac{2y^3 - 6y^2 + 2y - 4}{y}$ b. $\dfrac{y^3 + y^2 - 5y + 2}{y^2 - 2y}$

Solutions   a. $\dfrac{2y^3}{y} - \dfrac{6y^2}{y} + \dfrac{2y}{y} - \dfrac{4}{y}$

$\qquad\qquad 2y^2 - 6y + 2 - \dfrac{4}{y} \quad (y \neq 0)$

b.

$$\begin{array}{r}
y\ + 3 \\
y^2 - 2y\ \overline{)\ y^3 +\ \ y^2 - 5y + 2} \\
\underline{y^3 - 2y^2} \\
3y^2 - 5y \\
\underline{3y^2 - 6y} \\
y + 2
\end{array}$$

$$y + 3 + \dfrac{y + 2}{y^2 - 2y} \quad (y \neq 0, 2)$$

13. $\dfrac{15x^3 - 10x^2 + 3}{5x}$ 14. $\dfrac{38a^2 - 19a + 3}{19a}$

15. $\dfrac{2x^2 + x - 15}{x + 3}$ 16. $\dfrac{4y^2 - 4y - 5}{2y + 1}$

17. $\dfrac{4y^3 + 12y + 5}{2y + 1}$ 18. $\dfrac{3x^3 + 11x^2 + 11x + 15}{x + 3}$

19. $\dfrac{2x^3 - 5x^2 + 8x + 3}{2x - 1}$ 20. $\dfrac{2x^4 + 13x^3 - 7}{2x - 1}$

21. $\dfrac{4y^5 - 4y^2 - 5y + 1}{2y^2 + y + 1}$ 22. $\dfrac{2x^3 - 3x^2 - 15x - 1}{x^2 + 5}$

23. $\dfrac{x^4 - 4x^3 + 10x^2 - 12x + 9}{x^2 - 2x + 3}$ 24. $\dfrac{z^4 + 4z^3 - 2z^2 - 4z + 1}{z^2 + 2z + 1}$

2.5   Synthetic Division

If the divisor is of the form $x + c$, then the process of dividing one polynomial by another can be simplified by a process called **synthetic division**. Consider the example on page 35, where $x^4 + x^2 + 2x - 1$ is divided by $x + 3$. If we omit the variables, writing only the coefficients of the terms, and use zero for the coefficient of any missing power, we have

$$\begin{array}{r}
1 - 3 + 10 - 28 \\
1 + 3\ \overline{)\ 1 + 0 +\ \ 1 +\ \ 2 -\ \ 1} \\
\underline{1 + 3} \\
-3 + (1) \\
\underline{-3 -\ \ 9} \\
10 + (2) \\
\underline{10 + 30} \\
-28 - (1) \\
\underline{-28 - 84} \\
83\ \text{(remainder).}
\end{array}$$

Now, observe that the numerals shown in color are repetitions of the numerals written immediately above and are also repetitions of the coefficients of the associated variables in the quotient; the numerals in parentheses, ( ), are repetitions of the coefficients of the dividend. Therefore, the whole process can be written in compact form as

$$
\begin{array}{rl}
(1) & \underline{3\,|\,1\quad\ \ 0\quad\ \ 1\quad\ \ 2\quad -1} \\
(2) & \phantom{\underline{3\,|\,1}}\ \ 3\quad -9\quad 30\quad -84 \\
(3) & \ \ 1\quad -3\quad 10\quad -28\quad\ \ 83 \quad (\text{remainder}: 83)
\end{array}
$$

where the repetitions are omitted and where 1, the coefficient of x in the divisor, has also been omitted.

The entries in line (3), which are the coefficients of the variables in the quotient and the remainder, have been obtained by *subtracting* the **detached coefficients** in line (2) from the detached coefficients of terms of the same degree in line (1). We could obtain the same result by replacing 3 with -3 in the divisor and *adding* instead of subtracting at each step, and this is what is done in the *synthetic-division* process. The final form then appears:

$$
\begin{array}{rl}
(1) & \underline{-3\,|\,1\quad\ \ 0\quad\ \ 1\quad\ \ 2\quad -1} \\
(2) & \phantom{\underline{-3\,|\,1}}\ -3\quad 9\quad -30\quad 84 \\
(3) & \ 1\quad -3\quad 10\quad -28\quad\ \ 83 \quad (\text{remainder}: 83).
\end{array}
$$

Comparing the results of using synthetic division with those obtained by the same process using long division, on page 35, we observe that the entries in line (3) are the coefficients of the polynomial $x^3 - 3x^2 + 10x - 28$, and that there is a remainder of 83.

Example   Write $\dfrac{3x^3 - 4x - 1}{x - 2}$ in the form $Q + \dfrac{r}{D}$, where Q and D are polynomials and r is a constant.

Solution   Using synthetic division, we write

$$
2\,|\ 3\quad\ \ 0\quad -4\quad -1,
$$

where 0 has been inserted in the position that would be occupied by the coefficient of a second-degree term if such a term were present in the dividend. Our divisor is the negative of -2, or 2.

$$
\begin{array}{rl}
(1) & \underline{2\,|\,3\quad\ \ 0\quad -4\quad -\ 1} \\
(2) & \phantom{\underline{2\,|\,3}}\ \ 6\quad\ \ 12\quad\ \ 16 \\
(3) & \ 3\quad\ \ 6\quad\ \ 8\quad\ \ 15 \quad (\text{remainder}: 15).
\end{array}
$$

This process employs these steps:

1. 3 is "brought down" from line (1) to line (3).
2. 6, the product of 2 and 3, is written in the next position on line (2).
3. 6, the sum of 0 and 6, is written on line (3).
4. 12, the product of 2 and 6, is written in the next position on line (2).
5. 8, the sum of -4 and 12, is written on line (3).
6. 16, the product of 2 and 8, is written in the next position on line (2).
7. 15, the sum of -1 and 16, is written on line (3).

We can use the first three entries on line (3) as coefficients to write a polynomial of degree one less than the degree of the dividend. This polynomial is the quotient lacking the remainder. The last number is the remainder. Thus, for $x - 2 \neq 0$, the quotient when $3x^3 - 4x - 1$ is divided by $x - 2$ is $3x^2 + 6x + 8$ with a remainder of 15; that is,

$$\frac{3x^3 - 4x - 1}{x - 2} = 3x^2 + 6x + 8 + \frac{15}{x - 2} \quad (x \neq 2).$$

The foregoing example illustrates a theorem that we shall state without proof.

Theorem 2.4   *If $P(x)$ is a real polynomial and c is any real number, then there exists a unique real polynomial $Q(x)$ and a real number r, such that*

$$P(x) = (x - c)Q(x) + r.$$

Although this theorem does not directly involve the quotient $\dfrac{P(x)}{x - c}$, it does assure us that, for $x \neq c$,

$$\frac{P(x)}{x - c} = Q(x) + \frac{r}{x - c}.$$

Exercise 2.5

Use synthetic division to write each quotient $P(x)/D(x)$ either in the form $Q(x)$ or in the form $Q(x) + r/D(x)$, where $Q(x)$ is a polynomial and r is a constant.

Examples   a. $\dfrac{2x^4 + x^3 - 1}{x + 2}$ b. $\dfrac{x^3 - 1}{x - 1}$

Solutions   a.
$$\begin{array}{r|rrrrr} -2 & 2 & 1 & 0 & 0 & -1 \\ & & -4 & 6 & -12 & 24 \\ \hline & 2 & -3 & 6 & -12 & 23 \end{array}$$

b.
$$\begin{array}{r|rrrr} 1 & 1 & 0 & 0 & -1 \\ & & 1 & 1 & 1 \\ \hline & 1 & 1 & 1 & 0 \end{array}$$

$2x^3 - 3x^2 + 6x - 12 + \dfrac{23}{x + 2} \quad (x \neq -2)$ $x^2 + x + 1 \quad (x \neq 1)$

1. $\dfrac{2x^2 + x - 15}{x + 3}$ 2. $\dfrac{z^2 + 5z - 14}{z - 2}$ 3. $\dfrac{z^2 - 36}{z + 6}$

4. $\dfrac{z^2 - 25}{z - 5}$ 5. $\dfrac{3z^3 + 14z^2 - 5z + 4}{z - 4}$

6. $\dfrac{2y^3 + 15y^2 + 22y - 15}{y + 5}$ 7. $\dfrac{x^4 - 3x^3 + 2x^2 - 1}{x - 2}$

8. $\dfrac{x^4 + 2x^2 - 3x + 5}{x - 3}$ 9. $\dfrac{2x^3 + x - 5}{x + 1}$ 10. $\dfrac{3x^3 + x^2 - 7}{x + 2}$

11. $\dfrac{2x^4 - x + 6}{x - 5}$ **12.** $\dfrac{3x^4 - x^2 + 1}{x - 4}$ **13.** $\dfrac{x^3 + 4x^2 + x - 2}{x + 2}$

14. $\dfrac{x^3 - 7x^2 - x + 3}{x + 3}$ **15.** $\dfrac{x^6 + x^4 - x}{x - 1}$ **16.** $\dfrac{x^6 + 3x^3 - 2x - 1}{x - 2}$

17. $\dfrac{x^5 - 1}{x - 1}$ **18.** $\dfrac{x^5 + 1}{x + 1}$

19. Use synthetic division to show that the first three terms in the expansion of the quotient

$$\frac{x^n - 1}{x - 1} \quad (x \neq 1, n \in N)$$

are given by the polynomial $x^{n-1} + x^{n-2} + x^{n-3}$, and then argue that the quotient is $x^{n-1} + x^{n-2} + x^{n-3} + \cdots + x + 1$ and the remainder is 0.

2.6   *Equivalent Fractions*

Rational expressions

A fraction is an expression denoting a quotient. If the numerator (dividend) and the denominator (divisor) are polynomials, then the fraction is said to be a **rational expression**. Trivially, any polynomial can be considered as being a rational expression, since it is the quotient of itself and 1. For each replacement of the variable(s) for which the numerator and denominator of a fraction represent real numbers and for which the denominator is not zero, the fraction represents a real number. Of course, for any value of the variable(s) for which the denominator vanishes (is equal to zero), the fraction does not represent a real number and its value is undefined.

Since for each replacement of the variable(s) for which its denominator is not zero, a rational expression represents a real number, some theorems for rational expressions follow directly from Theorems 1.8, 1.9, and 1.10 and the laws of closure.

Theorem 2.5   *If A, B, C, and D represent polynomials, then for values of the variables for which the denominators do not vanish,*

$$\text{I} \quad \frac{A}{B} = \frac{C}{D} \quad \textit{if and only if} \quad AD = BC,$$

$$\text{II} \quad -\frac{A}{B} = \frac{-A}{B} = \frac{A}{-B} = -\frac{-A}{-B},$$

$$\text{III} \quad \frac{A}{B} = \frac{-A}{-B} = -\frac{-A}{B} = -\frac{A}{-B},$$

$$\text{IV} \quad \frac{AC}{BC} = \frac{A}{B} \quad (\textit{fundamental principle of fractions}).$$

Reducing
fractions

A fraction is said to be in **lowest terms** when the numerator and denominator do not contain prescribed types of factors in common. The arithmetic fraction a/b, where a and b are integers and $b \neq 0$, is in lowest terms provided a and b are relatively prime—that is, provided they contain no common positive integral factors other than 1. If the numerator and denominator of a fraction are polynomials with integral coefficients, then the fraction is said to be in lowest terms if the numerator and denominator cannot be expressed as products of polynomials with integral coefficients having a common factor other than ± 1.

To express a given fraction in lowest terms (called **reducing** the fraction), we can factor the numerator and denominator and apply the fundamental principle of fractions.

Example

$$\frac{y}{y^2} = \frac{1 \cdot y}{y \cdot y} = \frac{1}{y} \quad (y \neq 0).$$

Diagonal lines are sometimes used to abbreviate the procedure of reducing a fraction. Thus in the above example, we may write

$$\frac{y}{y^2} = \frac{\overset{1}{\cancel{y}}}{\underset{y}{\cancel{y^2}}} = \frac{1}{y} \quad (y \neq 0).$$

Reducing a fraction to lowest terms should be accomplished mentally whenever convenient.

To reduce fractions with polynomial numerators and denominators, you should, when possible, write them in factored form. Common factors are then evident by inspection.

Example

$$\frac{2x^2 + x - 15}{2x + 6} = \frac{(2x - 5)(x + 3)}{2(x + 3)} = \frac{2x - 5}{2} \quad (x \neq -3).$$

Building
fractions

We can also change fractions to equivalent fractions in higher terms by applying the fundamental principle in the form

$$\frac{A}{B} = \frac{AC}{BC} \quad (B, C \neq 0).$$

We might want to do this, for instance, in order to express two given fractions A/B and D/C as equivalent fractions with the same denominator, BC.

In general, to change a fraction A/B to an equivalent fraction with BC as a denominator, we can determine the factor C by inspection, and then can multiply the numerator and the denominator of the original fraction by this factor.

Exercise 2.6

Reduce fractions to lowest terms where possible. Specify restrictions on the variable or variables for which the reduction is not valid.

Examples   a. $\dfrac{6x^2y^3}{2xy^4}$ b. $\dfrac{a-b}{b^2-a^2}$ c. $\dfrac{a+b}{b}$

Solutions   a. $\dfrac{3x(2xy^3)}{y(2xy^3)}$ b. $\dfrac{-(b-a)}{(b-a)(b+a)}$ c. Expression is in lowest terms $(b \neq 0)$.

$\dfrac{3x}{y}$ $(x, y \neq 0)$ $\dfrac{-1}{b+a}$ $(b \neq a, b \neq -a)$

1. $\dfrac{a^2bc}{ab^2c}$

2. $\dfrac{24x^4y^2z}{16x^3y^2z}$

3. $\dfrac{2x+2y}{x+y}$

4. $\dfrac{x^2+x}{x+1}$

5. $\dfrac{a-b}{b-a}$

6. $\dfrac{x^2-xy}{y-x}$

7. $\dfrac{x^2-1}{1-x}$

8. $\dfrac{x^2-16}{4-x}$

9. $\dfrac{2x^3-4x^2-3x}{2x}$

10. $\dfrac{8a^2x^2-4ax^2+ax}{2ax}$

11. $\dfrac{y^2+5y-14}{y-2}$

12. $\dfrac{x^2+5x+6}{x+3}$

13. $\dfrac{4y^2+8y-5}{1-2y}$

14. $\dfrac{2x^2+13x-7}{1-2x}$

15. $\dfrac{x^2+6x+9}{x^2+2x-3}$

16. $\dfrac{x^2+5x+6}{x^2+6x+9}$

17. $\dfrac{y^2-2y+1}{y^2-1}$

18. $\dfrac{4-9z^2}{9z^2+12z+4}$

19. $\dfrac{n^2-8n+15}{n^2+n-12}$

20. $\dfrac{t^2-2t-8}{t^2-t-6}$

21. $\dfrac{3x^2-27}{x^2-11x+24}$

22. $\dfrac{5s^2-45s+90}{180-5s^2}$

23. $\dfrac{x^3-y^3}{x^2-y^2}$

24. $\dfrac{x^4-y^4}{x^2+y^2}$

25. $\dfrac{y^4-16}{y^4-y^2-12}$

26. $\dfrac{a^3-a^2+b^2+b^3}{3a^2+6ab+3b^2}$

Express each of the given fractions as an equivalent fraction with the given denominator. Specify values for the variable or variables for which the fractions are equivalent.

27. $\dfrac{3}{4}; \dfrac{}{12}$ 28. $\dfrac{1}{5}; \dfrac{}{10}$ 29. $\dfrac{b}{a}; \dfrac{}{a^2b}$ 30. $\dfrac{b}{2a}; \dfrac{}{6a^3b^2}$

31. $\dfrac{3}{y+2}; \dfrac{}{y^2-y-6}$ 32. $\dfrac{2}{x+3}; \dfrac{}{x^2+x-6}$

33. $\dfrac{3}{a+3}; \dfrac{}{a^3+27}$ 34. $\dfrac{-2}{x^2+y^2}; \dfrac{}{x^4-y^4}$

35. Is the value of the fraction $\dfrac{x(1-x)}{x^2-3x+2}$ equal to that of $\dfrac{x}{2-x}$ for all values of x? If not, for what value(s) of x does the equality fail to hold?

36. Write three equivalent forms of the fraction $\dfrac{1}{a-b}$ $(a \neq b)$ by changing the sign or signs of the numerator, denominator, or fraction itself.

37. What is the condition on a and b for the fraction $\dfrac{-1}{a-b}$ to represent a positive number? A negative number?

38. Is the fraction $\dfrac{x-2}{1+x^2}$ defined for all values of $x \in R$? For what value(s) of x does the fraction equal zero?

2.7   *Sums of Rational Expressions*

Since rational expressions represent real numbers for each replacement of the variable(s) for which the denominators are not zero, the following theorem is a direct consequence of Theorem 1.11-III and the laws of closure.

Theorem 2.6   *If A, B, and C are polynomials, then for values of the variables for which* $C \neq 0$,

$$\frac{A}{C} + \frac{B}{C} = \frac{A+B}{C}.$$

Examples   a. $\dfrac{3x^2}{5} + \dfrac{x}{5} = \dfrac{3x^2+x}{5}$ b. $\dfrac{3}{y} + \dfrac{x}{y} = \dfrac{3+x}{y}$ $(y \neq 0)$.

This principle, of course, extends to any number of fractions.

If the fractions in a sum have unlike denominators, we can replace the fractions with equivalent fractions having common denominators and then write the sum as a single fraction. Thus,

$$\frac{A}{B} + \frac{C}{D} = \frac{AD}{BD} + \frac{CB}{DB}$$

$$= \frac{AD+CB}{BD} \quad (B, D \neq 0).$$

The difference

$$\frac{A}{B} - \frac{C}{D}$$

may be viewed as the sum

$$\frac{A}{B} + \left(-\frac{C}{D}\right)$$

and then written as

$$\frac{AD}{BD} + \left(\frac{-CB}{DB}\right) = \frac{AD-CB}{BD} \quad (B, D \neq 0).$$

Least
common
multiple

In rewriting fractions in a sum so that they share a common denominator, any such denominator may be used. If the **least common multiple** of the denominators (called the **least common denominator**) is used, however, the resulting fraction will be in simpler form than if any other common denominator is employed. The least common multiple of two or more natural numbers is the least natural number that is exactly divisible (that is, each quotient is a natural number) by each of the given numbers.

The notion of a least common multiple among several polynomial expressions is, in general, meaningless without further specification of what is desired. We can, however, define the least common multiple of a set of polynomials with integer coefficients to be the polynomial of lowest degree with integer coefficients yielding a polynomial quotient upon division by each of the given polynomials, and to be, among all such polynomials, the one having the least possible positive leading coefficient.

Very often, the least common multiple of a set of natural numbers or polynomials can be determined by inspection. When inspection fails us, however, we can find the least common multiple of a set of polynomials with integer coefficients as follows:

1. Express each polynomial in completely factored form.
2. Write as factors of a product each *different* factor occurring in any of the polynomials, including each factor the greatest number of times it occurs in any one of the given polynomials.

Example

Find the least common multiple of 12, 15, and 18.

Solution

$$12 \qquad 15 \qquad 18$$
$$2\cdot 2\cdot 3 \qquad 3\cdot 5 \qquad 3\cdot 3\cdot 2$$

The least common multiple is $2^2\cdot 3^2\cdot 5$, or 180.

Example

Find the least common multiple of x^2, $x^2 - 9$, and $x^3 - x^2 - 6x$.

Solution

$$x^2 \qquad x^2 - 9 \qquad x^3 - x^2 - 6x$$
$$x\cdot x \qquad (x-3)(x+3) \qquad x(x-3)(x+2)$$

The least common multiple is $x^2(x+2)(x-3)(x+3)$.

To simplify sums of fractions having different denominators, we can ascertain the least common denominator of the fractions, determine the factor necessary to express each of the fractions as a fraction having this common denominator, write the fractions accordingly, and then express the sum as a single fraction.

Example

Write $\dfrac{3}{x} + \dfrac{2}{x^2} + \dfrac{3}{xy}$ as a single fraction.

(Solution overleaf)

Solution

The least common denominator of the fraction is x^2y. We have

$$\frac{3}{x} + \frac{2}{x^2} + \frac{3}{xy} = \frac{3(xy)}{x(xy)} + \frac{2(y)}{x^2(y)} + \frac{3(x)}{xy(x)}$$

$$= \frac{3xy}{x^2y} + \frac{2y}{x^2y} + \frac{3x}{x^2y}$$

$$= \frac{3xy + 2y + 3x}{x^2y} \quad (x, y \neq 0).$$

Exercise 2.7

Write each sum or difference as a single fraction in lowest terms. Assume that no variable in a denominator takes a value for which the denominator vanishes.

1. $\dfrac{x-1}{2y} + \dfrac{x}{2y}$

2. $\dfrac{y+1}{x} + \dfrac{y-1}{x}$

3. $\dfrac{2a-b}{a} - \dfrac{a-b}{a}$

4. $\dfrac{3a-1}{b} - \dfrac{2-a}{b}$

5. $\dfrac{a+2}{3} - \dfrac{a-3}{9}$

6. $\dfrac{a-2}{9} - \dfrac{a+1}{3}$

7. $\dfrac{2}{a+b} + \dfrac{1}{2a+2b}$

8. $\dfrac{7}{5x-10} + \dfrac{5}{3x-6}$

9. $\dfrac{2}{3-x} - \dfrac{1}{x-3}$

10. $\dfrac{7}{y-3} + \dfrac{3}{3-y}$

11. $\dfrac{a+1}{a+2} - \dfrac{a+2}{a+3}$

12. $\dfrac{5x-y}{3x+y} - \dfrac{6x-5y}{2x-y}$

13. $\dfrac{x+2y}{2x-y} - \dfrac{2x+y}{x-2y}$

14. $\dfrac{x-2y}{x+y} - \dfrac{2x-y}{x-y}$

15. $\dfrac{1}{y^2-y-2} + \dfrac{1}{y^2+2y+1}$

16. $\dfrac{2}{z^2-z-6} + \dfrac{3}{z^2-9}$

17. $\dfrac{3n}{n^2+3n-10} - \dfrac{2n}{n^2+n-6}$

18. $\dfrac{5u}{u^2+3u+2} - \dfrac{3u-6}{u^2+4u+4}$

19. $\dfrac{y}{y^2-16} - \dfrac{y+1}{y^2-5y+4} + \dfrac{1}{y+4}$

20. $\dfrac{1}{b^2-1} - \dfrac{1}{b^2+2b+1} + \dfrac{1}{b+1}$

21. $x + \dfrac{1}{x-1} - \dfrac{1}{(x-1)^2}$

22. $y - \dfrac{2y}{y^2-1} + \dfrac{3}{y+1}$

23. $x - 1 + \dfrac{3}{2x-1} - \dfrac{x}{4x^2-1}$

24. $2y - 3 - \dfrac{1}{y^2+2y+1} + \dfrac{3}{y+1}$

25. $\dfrac{y+3}{3y^2+7y+4} - \dfrac{y-7}{3y^2+13y+12}$

26. $\dfrac{z+4}{2z^2-5z-3} + \dfrac{2z-1}{2z^2+3z+1}$

27. $\dfrac{xy}{(z-x)(x-y)} + \dfrac{yz}{(z-y)(x-z)} + \dfrac{xz}{(y-x)(y-z)}$

28. $\dfrac{a+b}{a^2+2ab-3b^2} - \dfrac{a-2b}{a^2-b^2} + \dfrac{2a+b}{a^2+4ab+3b^2}$

29. $\dfrac{1}{(a-b)(b-c)} + \dfrac{1}{(b-c)(c-a)} + \dfrac{1}{(c-a)(a-b)}$

30. Any set of fractions has an infinite number of common denominators. Why is it convenient to use the *least* common denominator in finding sums or differences of fractions?

2.8   *Products and Quotients of Rational Expressions*

By Parts II and VII of Theorem 1.11 and the laws of closure, we have the following result.

Theorem 2.7   *If A, B, C, and D are real polynomials, then for values of the variables for which the denominators do not vanish,*

$$\text{I} \quad \frac{A}{B} \cdot \frac{C}{D} = \frac{AC}{BD},$$

$$\text{II} \quad \frac{A}{B} \div \frac{C}{D} = \frac{AD}{BC}.$$

The quotient $\dfrac{A}{B} \div \dfrac{C}{D}$ can also be denoted by

$$\frac{\dfrac{A}{B}}{\dfrac{C}{D}}.$$

This latter form is called a **complex fraction**; that is, it is a fraction containing a fraction in either the numerator or denominator or both.

We can use Theorem 2.7 to rewrite a product or quotient of fractions as a single fraction in lowest terms.

Example

$$\frac{x^2 - 2x + 1}{x^2 + 2x - 3} \cdot \frac{x^2 + 3x}{x^2 + 2x} = \frac{(x-1)(x-1)}{(x+3)(x-1)} \cdot \frac{x(x+3)}{x(x+2)}$$

$$= \frac{(x-1)[(x-1)(x+3)x]}{(x+2)[(x-1)(x+3)x]}$$

$$= \frac{x-1}{x+2} \quad (x \neq -3, -2, 0, 1)$$

Since the factors of the numerator and denominator of the product of two fractions are just the factors of the numerators and denominators, respectively, of the fractions, we can divide common factors out of the numerators and denominators before writing the product as a single fraction. Thus, in the example above, we could write

$$\frac{x^2 - 2x + 1}{x^2 + 2x - 3} \cdot \frac{x^2 + 3x}{x^2 + 2x} = \frac{\overset{1}{(x-1)}(x-1)}{(x+3)(x-1)} \cdot \frac{\overset{1}{x}\overset{1}{(x+3)}}{x(x+2)}$$

$$= \frac{x - 1}{x + 2} \quad (x \neq -3, -2, 0, 1).$$

Example

$$\frac{x^2 - x - 2}{x^2 + x - 2} \div \frac{x + 1}{x - 1} = \frac{(x-2)\overset{1}{(x+1)}}{(x+2)(x-1)} \cdot \frac{\overset{1}{(x-1)}}{(x+1)}$$

$$= \frac{x - 2}{x + 2} \quad (x \neq -2, -1, 1)$$

Reduction of complex fractions

When the quotient of two fractions is given in the form of a complex fraction, we have a choice of procedures available to us for writing the quotient in the form of a simple (not complex) fraction.

Example

Write $\dfrac{x + \dfrac{3}{4}}{x - \dfrac{1}{2}}$ as a simple fraction in lowest terms.

Solution 1

We can apply the fundamental principle of fractions to multiply numerator and denominator by the least common denominator of the simple fractions involved. Thus, we have

$$\frac{\left(x + \dfrac{3}{4}\right)4}{\left(x - \dfrac{1}{2}\right)4} = \frac{4x + 3}{4x - 2} \quad \left(x \neq \frac{1}{2}\right).$$

Solution 2

Alternatively, we can rewrite the complex fraction as follows:

$$\frac{x + \dfrac{3}{4}}{x - \dfrac{1}{2}} = \frac{\dfrac{4x + 3}{4}}{\dfrac{2x - 1}{2}} = \frac{4x + 3}{4} \cdot \frac{2}{2x - 1} = \frac{4x + 3}{4x - 2} \quad \left(x \neq \frac{1}{2}\right).$$

In the event we have a more complicated expression involving a complex fraction, we can rewrite the expression by simplifying small parts of it at a time.

Example

Write $\dfrac{1}{x + \dfrac{1}{x + \dfrac{1}{x}}}$ as a simple fraction in lowest terms.

Solution   We can begin by concentrating on the lower right-hand expression, $\dfrac{1}{x+\dfrac{1}{x}}$.

We have

$$\frac{1}{x+\dfrac{1}{x}} = \frac{(1)x}{\left(x+\dfrac{1}{x}\right)x} = \frac{x}{x^2+1}.$$

Thus,

$$\frac{1}{x+\dfrac{1}{x+\dfrac{1}{x}}} = \frac{1}{x+\dfrac{x}{x^2+1}}.$$

From this point, we can apply either of the methods shown in the previous example to the right-hand member above. Using the first method, we have

$$\frac{1}{x+\dfrac{1}{x+\dfrac{1}{x}}} = \frac{1(x^2+1)}{\left(x+\dfrac{x}{x^2+1}\right)(x^2+1)} = \frac{x^2+1}{x^3+x+x} = \frac{x^2+1}{x^3+2x} \quad (x \neq 0).$$

Exercise 2.8

Write each product or quotient as a single fraction in lowest terms. Assume that no denominator vanishes.

1. $\dfrac{-12a^2b}{5c} \cdot \dfrac{10b^2c}{24a^3b}$

2. $\dfrac{a^2}{xy} \cdot \dfrac{3x^3y}{4a}$

3. $\dfrac{xy}{a^2b} \div \dfrac{x^3y^2}{ab}$

4. $\dfrac{24a^3b}{-6xy^2} \div \dfrac{3a^2b}{12x}$

5. $\dfrac{x^2-x-20}{x^2+7x+12} \cdot \dfrac{2x^2+6x}{x^2-25}$

6. $\dfrac{4x^2+8x+3}{2x^2-5x+3} \cdot \dfrac{6x^2-9x}{1-4x^2}$

7. $\dfrac{25a^2b^2-16}{4ab+1} \div \dfrac{5ab+4}{16a^2b^2+16ab+3}$

8. $\dfrac{a^2-25}{a^2-16} \div \dfrac{a^2+2a-15}{a^2+a-12}$

9. $\dfrac{x^2-y^2}{x^2} \cdot \dfrac{x^2-xy+y^2}{x^2} \div \dfrac{x^3+y^3}{x^4}$

10. $\dfrac{a^3+8b^3}{a^2-64b^2} \cdot \dfrac{a^3+512b^3}{a^2-ab-6b^2} \div \dfrac{a^4+4a^2b^2+16b^4}{a^2+11ab+24b^2}$

11. $\left(1+\dfrac{1}{x}\right) \cdot \left(1-\dfrac{1}{x}\right)$

12. $\left(x-\dfrac{1}{x}\right) \div \left(x+\dfrac{1}{x}\right)$

13. $\left[\dfrac{3}{x-1} - \dfrac{2}{x+1}\right] \cdot \dfrac{x-1}{x}$

14. $\left[\dfrac{x}{x^2-9} + \dfrac{2}{x-3}\right] \cdot \dfrac{x-1}{x}$

15. $\left[\dfrac{2y}{2y-1} - \dfrac{3}{y}\right] \div \dfrac{3}{2y^2-y}$

16. $\left[\dfrac{y}{y^2-1} - \dfrac{y}{y^2-2y+1}\right] \div \dfrac{y}{y-1}$

17. $\dfrac{\dfrac{2}{a} + \dfrac{3}{2a}}{5 + \dfrac{1}{a}}$

18. $\dfrac{1 + \dfrac{1}{x}}{1 - \dfrac{1}{x}}$

19. $a - \dfrac{a}{a + \dfrac{1}{4}}$

20. $x - \dfrac{x}{1 - \dfrac{x}{1 - x}}$

21. $1 - \dfrac{1}{1 - \dfrac{1}{y - 2}}$

22. $2y + \dfrac{3}{3 - \dfrac{2y}{y - 1}}$

23. $\dfrac{1 + \dfrac{1}{1 - \dfrac{a}{b}}}{1 - \dfrac{3}{1 - \dfrac{a}{b}}}$

24. $\dfrac{1 - \dfrac{1}{\dfrac{a}{b} + 2}}{1 + \dfrac{3}{\dfrac{a}{2b} + 1}}$

25. $\dfrac{a + 2 - \dfrac{12}{a + 3}}{a - 5 + \dfrac{16}{a + 3}}$

26. $\dfrac{a + 4 - \dfrac{7}{a - 2}}{a - 1 + \dfrac{2}{a - 2}}$

27. $\dfrac{\dfrac{x^2 - 2xy + y^2 - z^2}{x^2 + 2xy + y^2 - z^2}}{\dfrac{x - y + z}{x + y - z}}$

28. $\dfrac{\dfrac{9x^2 - 6x}{6x^2 - 7x + 2} \cdot \dfrac{2x^2 + 13x - 7}{2x^2 + 6x}}{\dfrac{4x^2 - 8x + 3}{2x^2 + 3x - 9}}$

29. $\left[\dfrac{a^2 b^2 + 2b^4}{27a^6} \div \dfrac{a^2 - 10ab + 25b^2}{3a^5 + 6a^3 b^2} \right] \cdot \dfrac{15a^3 b - 3a^4}{a^4 + 4a^2 b^2 + 4b^4}$

30. $\left(\dfrac{x^2 - xy + y^2}{x^2 + xy + y^2} \cdot \dfrac{x^3 - y^3}{x^3 + y^3} \right) \div \dfrac{(x - y)^2}{(x + y)^2}$

2.9   *Partial Fractions*

Theorems 2.5-IV and 2.6 justify writing the sum

$$\frac{A}{B} + \frac{C}{D}$$

as the single fraction

$$\frac{AD + BC}{BD}.$$

It is frequently useful to be able to reverse this process in the case of certain kinds of fractions—in particular, for rational expressions in which the numerator is a polynomial of lesser degree than the denominator. Such rational expressions are customarily referred to as "proper" fractions.

Notice first that by the long-division algorithm, any rational expression can be written as a polynomial or the sum of a polynomial and a proper fraction.

For example, by the long-division algorithm we find that

$$\frac{x^3 + 2x^2 + 2x + 3}{x^2 - 1} = x + 2 + \frac{3x + 5}{x^2 - 1}.$$

Theorem 2.8   *If* $a, b, r_1, r_2 \in R$ *are given, with* $r_1 \neq r_2$, *then there exist constants* $c_1, c_2 \in R$ *such that for* $x \neq r_1, r_2$,

$$\frac{ax + b}{(x - r_1)(x - r_2)} = \frac{c_1}{x - r_1} + \frac{c_2}{x - r_2}.$$

Proof Suppose first that there are such constants $c_1, c_2 \in R$. Multiplying each member of the equation by $(x - r_1)(x - r_2)$, we get

$$ax + b = c_1(x - r_2) + c_2(x - r_1).$$

Since we wish this equation to hold for *every value of* x other than r_1 and r_2, it must hold also for $x = r_1$ and $x = r_2$. Substituting r_1 and r_2 for x in turn, we get

$$ar_1 + b = c_1(r_1 - r_2) + c_2(r_1 - r_1) \quad \text{and} \quad ar_2 + b = c_1(r_2 - r_2) + c_2(r_2 - r_1),$$

from which, since $r_1 \neq r_2$,

$$c_1 = \frac{ar_1 + b}{r_1 - r_2} \quad \text{and} \quad c_2 = \frac{ar_2 + b}{r_2 - r_1}.$$

That these values satisfy the given fractional equation can be verified by direct substitution. Thus not only do c_1 and c_2 exist, but they have the values shown.

While proof of the foregoing theorem produces formulas for c_1 and c_2, it is preferable from the standpoint of efficiency to rely on the method used in the proof to obtain the numbers c_1 and c_2 in any particular problem. The technique is known as the method of **partial fractions**.

Example   Apply the method of partial fractions to express $\dfrac{3x - 2}{x^2 - x}$ as a sum of fractions.

Solution   We first observe that

$$\frac{3x - 2}{x^2 - x} = \frac{3x - 2}{x(x - 1)}. \tag{1}$$

Then by Theorem 2.8, we know there exist numbers $c_1, c_2 \in R$, such that

$$\frac{3x - 2}{x(x - 1)} = \frac{c_1}{x} + \frac{c_2}{x - 1}. \tag{2}$$

Multiplying each member of the equation by $x(x - 1)$, we obtain

$$3x - 2 = c_1(x - 1) + c_2(x). \tag{3}$$

We wish the members of Equation (2) to be equal for all values of x except 0 and 1. However, if this is to be the fact, then the members of (3) must be equal for all values of x, *including* 0 and 1. Hence, while we can arbitrarily select *any* values for x and solve for c_1 and c_2 in (3), let us take 0 and 1 because these values of x yield the values of c_1 and c_2 by inspection. Thus, if $x = 0$, we have from Equation (3)

$$3(0) - 2 = c_1(0 - 1) + c_2(0),$$

(*Solution continued overleaf*)

from which

$$c_1 = 2.$$

If $x = 1$, Equation (3) becomes

$$3(1) - 2 = c_1(1 - 1) + c_2(1),$$

$$c_2 = 1.$$

Therefore,

$$\frac{3x - 2}{x(x - 1)} = \frac{2}{x} + \frac{1}{x - 1}.$$

Theorem 2.8 generalizes to the following form, although the proof is omitted.

Theorem 2.9   *If $P(x)$ and $Q(x)$ are real polynomials, with $P(x)$ of degree less than $Q(x)$, and if $Q(x) = (x - r_1)(x - r_2) \cdots (x - r_n)$, where no two factors are identical, then there exist constants $c_1, c_2, \ldots, c_n \in R$ such that*

$$\frac{P(x)}{Q(x)} = \frac{c_1}{x - r_1} + \frac{c_2}{x - r_2} + \cdots + \frac{c_n}{x - r_n}.$$

If the denominator of a proper fraction can be factored into *equal* linear factors, then we need the following result, which is stated without proof.

Theorem 2.10   *If $P(x)$ and $Q(x)$ are real polynomials with $P(x)$ of degree less than $Q(x)$, and if $Q(x) = (x - r_1)^n$, then there exist constants $c_1, c_2, \ldots, c_n \in R$ such that*

$$\frac{P(x)}{Q(x)} = \frac{c_1}{(x - r_1)} + \frac{c_2}{(x - r_1)^2} + \cdots + \frac{c_n}{(x - r_1)^n}.$$

A further extension of Theorems 2.9 and 2.10 can be used to apply the method of partial fractions to rational expressions in which the denominator has one or more repeated factors. Each repeated factor contributes a sum such as the one in Theorem 2.10.

Example   Express $\dfrac{5x^2 + 2}{x^3 - 2x^2 + x}$ as a sum of fractions.

Solution   We first observe that $\dfrac{5x^2 + 2}{x^3 - 2x^2 + x} = \dfrac{5x^2 + 2}{x(x - 1)(x - 1)}.$

Then we write

$$\frac{5x^2 + 2}{x(x - 1)(x - 1)} = \frac{c_1}{x} + \frac{c_2}{x - 1} + \frac{c_3}{(x - 1)^2}.$$

Multiplying each member of the equation by $x(x - 1)^2$, we obtain

$$5x^2 + 2 = c_1(x - 1)^2 + c_2 x(x - 1) + c_3 x.$$

Arbitrarily selecting any values of x, we can solve for c_1, c_2, and c_3. Let us use 0, 1, and 2 because these numbers yield simple computations. Substituting 0 for x gives

$$5(0)^2 + 2 = c_1(-1)^2 + 0 + 0,$$
$$c_1 = 2.$$

Substituting 1 for x gives

$$5(1)^2 + 2 = 0 + 0 + c_3(1),$$
$$c_3 = 7.$$

Substituting 2 for x gives

$$5(2)^2 + 2 = c_1(1)^2 + c_2(2)(1) + c_3(2).$$

Now, since $c_1 = 2$ and $c_3 = 7$, we have

$$22 = 2 + 2c_2 + 14,$$
$$2c_2 = 6,$$
$$c_2 = 3.$$

Hence,

$$\frac{5x^2 + 2}{x^3 - 2x^2 + x} = \frac{2}{x} + \frac{3}{x - 1} + \frac{7}{(x - 1)^2}.$$

Note that Theorems 2.9 and 2.10 apply to fractions whose denominators can be factored into linear factors. Similar theorems apply to fractions whose denominators cannot be factored completely into linear factors. In particular, the numerator of each term of the partial-fraction expansion whose denominator is a nonfactorable quadratic expression will be a linear expression of the form $c_1 x + c_2$. However, we shall not consider such fractions here.

Exercise 2.9

Use the method of partial fractions to express each fraction as a sum of fractions with linear polynomials for denominators.

1. $\dfrac{2}{x^2 + x}$ 2. $\dfrac{-6}{y^2 - 2y}$ 3. $\dfrac{6x + 2}{x^2 + x}$

4. $\dfrac{4z + 12}{z^2 + 4z}$ 5. $\dfrac{1}{y^2 + y - 2}$ 6. $\dfrac{1}{z^2 + z - 12}$

7. $\dfrac{3z - 2}{z^2 + 3z - 4}$ 8. $\dfrac{2x}{4x^2 - 7x + 3}$

Example $\dfrac{x + 3}{x^3 - 3x^2 + 2x}$

Solution   Factoring the denominator leads to

$$\frac{x + 3}{x(x - 1)(x - 2)} = \frac{c_1}{x} + \frac{c_2}{x - 1} + \frac{c_3}{x - 2}.$$

(*Solution continued overleaf*)

Multiplying each member by $x(x - 1)(x - 2)$, we have

$$x + 3 = c_1(x - 1)(x - 2) + c_2(x)(x - 2) + c_3(x)(x - 1).$$

Substituting, in turn $x = 0$, $x = 1$, and $x = 2$ produces

$$x = 0: \quad 0 + 3 = c_1(0 - 1)(0 - 2) + c_2(0)(0 - 2) + c_3(0)(0 - 1),$$

$$3 = 2c_1, \quad \text{or} \quad c_1 = \frac{3}{2};$$

$$x = 1: \quad 1 + 3 = c_1(1 - 1)(1 - 2) + c_2(1)(1 - 2) + c_3(1)(1 - 1),$$

$$4 = -c_2, \quad \text{or} \quad c_2 = -4;$$

$$x = 2: \quad 2 + 3 = c_1(2 - 1)(2 - 2) + c_2(2)(2 - 2) + c_3(2)(2 - 1),$$

$$5 = 2c_3, \quad \text{or} \quad c_3 = \frac{5}{2}.$$

Therefore,

$$\frac{x + 3}{x^3 - 3x^2 + 2x} = \frac{3/2}{x} - \frac{4}{x - 1} + \frac{5/2}{x - 2}.$$

9. $\dfrac{7x - 4}{x^3 + x^2 - 2x}$

10. $\dfrac{1}{z^3 - z}$

11. $\dfrac{3x^2 - 7x - 17}{(x + 2)(x + 1)(x - 1)}$

12. $\dfrac{5y^2 + 16y - 12}{y^3 + y^2 - 6y}$

13. $\dfrac{1}{x^3 + 5x^2 + 4x}$

14. $\dfrac{2t^2 - 14}{(t + 1)(t - 2)(t - 3)}$

Exercises 15–20 involve repeated linear factors.

15. $\dfrac{1}{x^3 + x^2}$

16. $\dfrac{1}{x^3 - x^2}$

17. $\dfrac{x - 8}{x^3 - 4x^2 + 4x}$

18. $\dfrac{5x - 1}{(x - 2)(x + 1)^2}$

19. $\dfrac{1}{y^4 - y^2}$

20. $\dfrac{1}{y^4 - 2y^2 + 1}$

Use the long-division algorithm and the method of partial fractions to express each fraction as a sum of a polynomial and fractions with linear polynomials for denominators.

21. $\dfrac{2x^3}{x^2 - 1}$

22. $\dfrac{x^2}{x^2 - 4}$

23. $\dfrac{x^3 - x^2}{(x - 1)^2}$

24. $\dfrac{x^3 + 2x^2 - x + 1}{x^2 - x - 2}$

Chapter Review

[2.1]   *Simplify each expression.*

1. $(2y^2 - 3y + 4) - (y^2 - 3y + 5)$

2. $2[3(x^2 + 2x - 1) + 4x] + (3 - 4x - 6x^2)$

Given $P(x) = 3x^2 - 2x + 1$, find:

3. $P(-2)$

4. $P(x)\big|_{-2}^{4}$

[2.2] *Simplify.*

5. $2(3x - 1)(x + 2)$

6. $2\{z^2 - 2[z - 3(z^2 - 1) + 1] + 3z^2\}$

[2.3] *Factor completely.*

7. $z^2 - 7z + 12$

8. $u^5 - 2u^3 + u$

9. $4y^{2n} - 1$

10. $8t^3 - 27$

[2.4] *Write each quotient as a polynomial.*

11. $\dfrac{48x^5y^2z}{6x^4yz}$

12. $\dfrac{24z^3 - 16z^2 + 72z}{8z}$

13. $\dfrac{2n^2 - 5n - 3}{n - 3}$

14. $\dfrac{2r^2 + 3r + 1}{2r + 1}$

[2.5] *Use synthetic division to write each quotient $P(x)/D(x)$ either in the form $Q(x)$ or in the form $Q(x) + \dfrac{r}{D(x)}$, where $Q(x)$ is a polynomial and r is a constant.*

15. $\dfrac{2x^3 - 11x^2 + 13x - 4}{x - 4}$

16. $\dfrac{3y^3 + 7y^2 - 3y + 10}{y + 3}$

[2.6] *Reduce to lowest terms where possible. State all restrictions on the variable or variables.*

17. $\dfrac{8r^2s^2 - 4rs^2 + 12rs}{2rs}$

18. $\dfrac{2x^2 - 2}{x + 1}$

19. $\dfrac{x^2 - 4x + 4}{x^2 - 4}$

20. $\dfrac{8n^2 + 40n + 32}{32 - 2n^2}$

[2.7] *Simplify each expression.*

21. $\dfrac{x + 1}{4} + \dfrac{x + 2}{8}$

22. $\dfrac{3}{6x - 3} - \dfrac{2}{2x - 1}$

23. $\dfrac{2x - 5}{2 - x} + \dfrac{x}{2x - 4}$

24. $\dfrac{4y - 2}{y^2 - 3y + 2} - \dfrac{3}{y - 1}$

[2.8] *Simplify each expression.*

25. $\dfrac{2x^2y^2}{9a^2b^2} \cdot \dfrac{45ab}{14xy}$

26. $\dfrac{x^2 + 3x + 2}{x^2 - 1} \cdot \dfrac{x^2 + 2x - 3}{x^2 + 5x + 6}$

27. $\dfrac{13t^2}{20r^2} \div \dfrac{39t^3}{5r}$

28. $\dfrac{x^2 - 4}{ax - 2a} \div \dfrac{x^2 + 4x + 4}{x^2 + 2x}$

29. $\dfrac{\dfrac{a-b}{b}}{\dfrac{a+b}{3b}}$

30. $\dfrac{\dfrac{1}{xy} - \dfrac{1}{y}}{\dfrac{1}{y} - \dfrac{1}{xy}}$

[2.9] *Use the method of partial fractions to express each fraction as a sum of fractions with linear polynomials for denominators.*

31. $\dfrac{7}{2x^2 + 5x - 3}$

32. $\dfrac{2x^2 + 3x - 1}{x^3 - x}$

3 Rational Exponents
—Radicals

Powers with Integral Exponents

In Chapter 2, powers of real numbers were defined for natural-number exponents, and some simple properties of products and quotients of powers were examined. Powers with integral and rational exponents can be defined in a manner consistent with these properties.

Reason for defining a^0 to be 1

We saw that for $a \in R$, $a \neq 0$, and $m, n \in N$, $m > n$, we have

$$\frac{a^m}{a^n} = a^{m-n}.$$ (1)

If (1) is to hold also for $m = n$, then we must have

$$\frac{a^n}{a^n} = a^{n-n} = a^0.$$

Since $a^n/a^n = 1$ for $a \neq 0$, we can state the following definition.

Definition 3.1   *If $a \in R$, $a \neq 0$, then*

$$a^0 = 1.$$

Reason for defining a^{-n} to be $1/a^n$

In a similar way, if (1) is to hold for $m = 0$, then we must have

$$\frac{a^0}{a^n} = a^{0-n} = a^{-n}.$$

Since $a^0 = 1$, we can make the following definition.

Definition 3.2   *If $a \in R$, $a \neq 0$, and $n \in N$, then*

$$a^{-n} = \frac{1}{a^n}.$$

*Laws of
exponents
for integral
exponents*

Now for m and n nonnegative integers, the equation in Theorem 2.1-I, with $-n$ in place of n, is

$$a^m \cdot a^{-n} = a^{m-n} \quad (a \in R, \, a \neq 0),$$

which is true by Definition 3.2 and Theorem 2.3-I (extended to include $m = 0$ or $n = 0$ or both; for a complete discussion, several cases must be considered). Again,

$$a^{-m} \cdot a^{-n} = \frac{1}{a^m} \cdot \frac{1}{a^n} = \frac{1}{a^{m+n}} = a^{-(m+n)} = a^{-m-n}.$$

Thus the equation in Theorem 2.1-I is true for all $a \in R$, $a \neq 0$, and all integers m, n. By Definitions 1.13 and 3.2 and Theorem 2.1-I, we can show that, for $m, n \in J$,

$$\frac{a^m}{a^n} = a^m \cdot \frac{1}{a^n} = a^m \cdot a^{-n} = a^{m-n},$$

which extends Theorem 2.3-I to m, $n \in J$. The other parts of Theorem 2.1 and 2.3, appropriately reworded for integers, also follow from Definitions 3.1 and 3.2.

We now restate these theorems for integral exponents m, $n \in J$.

Theorem 3.1   *If a, $b \in R$ (a, $b \neq 0$) and m, $n \in J$, then*

$$\text{I} \quad a^m \cdot a^n = a^{m+n},$$

$$\text{II} \quad \frac{a^m}{a^n} = a^{m-n},$$

$$\text{III} \quad (a^m)^n = a^{mn},$$

$$\text{IV} \quad (ab)^n = a^n b^n,$$

$$\text{V} \quad \left(\frac{a}{b}\right)^n = \frac{a^n}{b^n}.$$

Some examples of applications of this theorem should prove enlightening.

Examples

a. $(x^2)^3 = x^{2 \cdot 3} = x^6$ By Theorem 3.1-III.

b. $(xy)^3 = x^3 y^3$ By Theorem 3.1-IV.

c. $\left(\dfrac{x^2}{y}\right)^{-4} = \dfrac{(x^2)^{-4}}{(y)^{-4}}$ By Theorem 3.1-V.

$= \dfrac{x^{-8}}{y^{-4}}$ By Theorem 3.1-III.

$= \dfrac{1/x^8}{1/y^4}$ By Definition 3.2.

$= \dfrac{y^4}{x^8} \quad (x, \, y \neq 0)$ By Theorem 1.11-VII.

Although we shall not do anything about proving it here, Parts III, IV, and V of Theorem 3.1 can be shown to apply to expressions involving more than two factors.

Examples

a. $\left(\dfrac{x^2z^4}{5y}\right)^{-2} = \dfrac{x^{-4}z^{-8}}{5^{-2}y^{-2}}$

$\qquad\qquad = \dfrac{5^2y^2}{x^4z^8}$

$\qquad\qquad = \dfrac{25y^2}{x^4z^8}$

b. $\dfrac{xy^{-2}}{x^{-1}+y} = \dfrac{x\cdot\dfrac{1}{y^2}}{\dfrac{1}{x}+y} = \dfrac{\dfrac{x}{y^2}}{\dfrac{1+xy}{x}}$

$\qquad\qquad = \dfrac{x}{y^2}\cdot\dfrac{x}{1+xy} = \dfrac{x^2}{y^2(1+xy)}$

In order to avoid the necessity of constantly noting exceptions, we shall assume that in the exercises in this chapter the variables are restricted so that no denominator vanishes.

Exercise 3.1

For each of the following expressions, write an equivalent basic numeral—that is, a numeral with exponent 1.

Examples

a. $3\cdot5^{-2}$ b. $\dfrac{3}{2^{-3}}$ c. $4^2+4^{-2}-4^0$

Solutions

a. $3\cdot\dfrac{1}{5^2}$ b. $\dfrac{3}{\dfrac{1}{2^3}}$ c. $16+\dfrac{1}{16}-1$

$\quad 3\cdot\dfrac{1}{25}$ $3\cdot2^3$ $15+\dfrac{1}{16}$

$\quad \dfrac{3}{25}$ 24 $\dfrac{241}{16}$

1. 5^{-1} 2. 4^{-1} 3. 3^{-2} 4. 2^{-3}

5. $\dfrac{1}{3^{-3}}$ 6. $\dfrac{1}{4^{-2}}$ 7. $\dfrac{2^{-1}\cdot3^0}{5}$ 8. $\dfrac{2}{3^{-2}\cdot4^0}$

9. $\left(\dfrac{3}{5}\right)^{-1}$ 10. $\left(\dfrac{1}{3}\right)^{-2}$ 11. $\dfrac{5^{-1}}{3^{-2}}$ 12. $\dfrac{3^{-3}}{6^{-2}}$

13. $3^{-2}+3^2$ 14. $5^{-1}+25^0$ 15. $4^{-1}-4^{-2}$ 16. $8^{-2}-2^0$

Write each expression as a product or quotient in which each variable occurs at most once in the expression and involves positive exponents only.

Examples

a. $x^{-3}x^5$ b. $(x^2y^{-3})^{-1}$ c. $\left(\dfrac{x^{-1}y^2z^0}{x^3y^{-4}z^2}\right)^{-1}$

(Solutions overleaf)

Solutions   a. x^{-3+5} b. $x^{-2}y^3$ c. $\dfrac{xy^{-2}z^0}{x^{-3}y^4z^{-2}}$

$\quad\quad\quad\quad x^2$

$\quad\quad\quad\quad\quad\quad\quad\quad\quad\quad \dfrac{1}{x^2}\cdot y^3$ $\quad\quad\quad\quad\quad \dfrac{x\cdot x^3\cdot 1\cdot z^2}{y^4\cdot y^2}$

$\quad\quad\quad\quad\quad\quad\quad\quad\quad\quad \dfrac{y^3}{x^2}$ $\quad\quad\quad\quad\quad\quad\quad \dfrac{x^4z^2}{y^6}$

17. $\dfrac{x^2}{y^{-3}}$ $\quad$ **18.** $\dfrac{x^3}{y^{-2}}$ $\quad$ **19.** $(x^3y^2)^2$ $\quad$ **20.** $(xy^2)^3$

21. $\left(\dfrac{3x}{y^3}\right)^2$ $\quad$ **22.** $\left(\dfrac{x^2}{2y}\right)^3$ $\quad$ ✓**23.** $\left(\dfrac{x^2}{y}\right)^3\left(\dfrac{2y}{x}\right)^2$ $\quad$ **24.** $\left(\dfrac{3x}{y^2}\right)^2\left(\dfrac{2y^3}{x}\right)^2$

25. $x^{-3}x^7$ $\quad$ **26.** $\dfrac{x^3}{x^{-2}}$ $\quad$ **27.** $(x^{-2}y^0)^3$ $\quad$ **28.** $(x^{-2}y^3)^0$

29. $\dfrac{x^{-1}}{y^{-1}}$ $\quad$ **30.** $\dfrac{x^{-3}}{y^{-2}}$ $\quad$ **31.** $\dfrac{8^{-1}x^0y^{-3}}{(2xy)^{-5}}$ $\quad$ **32.** $\left(\dfrac{x^{-1}y^3}{2x^0y^{-5}}\right)^{-2}$

Represent each expression as a single fraction involving positive exponents only.

Examples   a. $x^{-1}+y^{-2}$ b. $(x^{-1}+x^{-2})^{-1}$ c. $\dfrac{x^{-1}}{x^{-1}+y^{-1}}$

Solutions   a. $\dfrac{1}{x}+\dfrac{1}{y^2}$ b. $\left(\dfrac{1}{x}+\dfrac{1}{x^2}\right)^{-1}$ c. $\dfrac{\dfrac{1}{x}\ (xy)}{\left(\dfrac{1}{x}+\dfrac{1}{y}\right)(xy)}$

$\quad\quad\quad\quad \dfrac{(y^2)1}{(y^2)x}+\dfrac{1(x)}{y^2(x)}$ $\quad\quad\quad \left(\dfrac{x+1}{x^2}\right)^{-1}$ $\quad\quad\quad\quad \dfrac{y}{y+x}$

$\quad\quad\quad\quad \dfrac{y^2+x}{xy^2}$ $\quad\quad\quad\quad\quad\quad \dfrac{x^2}{x+1}$

✓**33.** $x^{-1}-y^{-2}$ $\quad$ **34.** $x^{-2}+y$ $\quad$ **35.** $x^{-1}y+xy^{-1}$

36. $\dfrac{x}{y^{-1}}+\dfrac{x^{-1}}{y}$ $\quad$ **37.** $x(x-y)^{-1}$ $\quad$ **38.** $y(x+y)^{-2}$

✓**39.** $xy^{-1}+x^{-1}y$ $\quad$ **40.** $x^{-1}y-xy^{-1}$ $\quad$ **41.** $\dfrac{x^{-1}+y^{-1}}{(xy)^{-1}}$

42. $\dfrac{x}{y^{-1}}+\left(\dfrac{x}{y}\right)^{-1}$ $\quad$ **43.** $(x^{-1}-y^{-1})^{-1}$ $\quad$ **44.** $\dfrac{x^{-1}+y^{-1}}{x^{-1}-y^{-1}}$

For each expression, write an equivalent product in which each variable occurs only once.

Examples   a. $\dfrac{x^nx^{n+1}}{x^{n-1}}$ b. $(y^{n-1})^{-3}$ c. $\dfrac{(y^{n-1})^2}{y^{n-2}}$

Solutions   a. $x^{n+(n+1)-(n-1)}$ b. y^{-3n+3} c. $\dfrac{y^{(2n-2)-(n-2)}}{y^n}$
                     x^{n+2}

45. $x^n x^{n-1}$ **46.** $x^{n-1} x^{2n}$ **47.** $\dfrac{x^{n+1} x^{2n-1}}{x^{3n}}$

48. $\dfrac{y^{n+1} y^{n+2}}{y^n}$ ✓**49.** $\dfrac{(y^{n+1} y)^2}{y^{2n}}$ **50.** $\left(\dfrac{y^{2n-1} y^n}{y^{2n}}\right)^3$

✓**51.** $\left(\dfrac{x^n}{x^{n-3}}\right)^2$ **52.** $\left(\dfrac{x^n}{x^{n-1}}\right)^3$ **53.** $\dfrac{x^n y^{n+1}}{x^{2n-1} y^n}$

54. $\dfrac{x^n y^{2n-1}}{x^{n+1} y^{2n}}$ ✓**55.** $\left(\dfrac{x^{2n}}{x^{n+1}}\right)^{-2}$ **56.** $\left(\dfrac{x^{2n} y^{n-1}}{x^{n-1} y}\right)^{-2}$

Write each number as the product of a number between 1 and 10 and a power of 10. (This exponential form is called **scientific notation.**)

Examples   a. 680,000   b. 0.032   c. 0.0000431

Solutions   Factor.

a. $6.8 \times 100{,}000$ b. $3.2 \times \dfrac{1}{100}$ c. $4.31 \times \dfrac{1}{100{,}000}$

Write in exponential form.

6.8×10^5 3.2×10^{-2} 4.31×10^{-5}

57. 2,540   **58.** 38,421   **59.** 642,000   **60.** 2,541,000

61. 0.0014   **62.** 0.0000006   **63.** 0.0000230   **64.** 0.5020

3.2 Powers with Rational Exponents

In Section 3.1, we defined powers of real numbers with 0 and negative-integer exponents so that Theorem 3.1-I would be consistent with Theorem 2.1-I. It followed that the remaining laws of exponents hold for these exponents also. Now we want to give meaning to powers of real numbers with rational numbers as exponents. As before, we shall want any such definition to be consistent with all of the laws of exponents for integers.

We observed in Chapter 2 that for $a \in R$ and $m, n \in N$,

$$(a^m)^n = a^{mn}. \tag{1}$$

If (1) is to hold for $m = 1/n$, and $a^{1/n}$ is a real number, then we must have

$$(a^{1/n})^n = a^{(1/n)(n)} = a^{n/n} = a^1 = a,$$

so that the *n*th power of $a^{1/n}$ must be *a*. A number having *a* as its *n*th power is called an **nth root** of *a*. In particular, for *n* equal to 2 or 3, respectively, an *n*th root is called a **square root** or a **cube root**.

Number of
roots

For n odd, each $a \in R$ has just one real nth root. Thus

$$(-2)^3 = -8 \quad \text{and} \quad 2^3 = 8,$$

so that -2 is the cube root of -8, and 2 is the cube root of 8.

For n even and $a > 0$, a has two real nth roots. Thus,

$$(-2)^4 = 16 \quad \text{and} \quad 2^4 = 16,$$

so that -2 and 2 are both fourth roots of 16.

For n even and $a < 0$, a has no real nth root. Thus -1 has no real square root, since the square of each real number is nonnegative.

If $a = 0$, then a has exactly one real nth root, namely 0.

We therefore make the following definition.

Definition 3.3   *If $a \in R$, $n \in N$, then $a^{1/n}$ is the real number if one exists, and is the positive real number if two exist, such that*

$$(a^{1/n})^n = a.$$

Examples

a. $25^{1/2} = 5$ b. $-25^{1/2} = -5$ c. $(-25)^{1/2}$ is not a real number.

d. $27^{1/3} = 3$ e. $-27^{1/3} = -3$ f. $(-27)^{1/3} = -3$

Notice in parts b and e that $-25^{1/2}$ and $-27^{1/3}$ denote $-(25^{1/2})$ and $-(27^{1/3})$, respectively.

To generalize from rational exponents of the form $1/n$, for $n \in N$, to rational exponents of the form m/n, for $m \in J$, $n \in N$, we need the results expressed in the following two theorems.

Theorem 3.2   *If $a^{1/n} \in R$, $m \in J$, and $n \in N$, with $a \neq 0$ for $m \leq 0$, then*

$$(a^{1/n})^m = (a^m)^{1/n}.$$

Proof Since $n \in N$, by Theorem 3.1-III we have

$$[(a^{1/n})^m]^n = (a^{1/n})^{mn} = (a^{1/n})^{nm} = [(a^{1/n})^n]^m = a^m.$$

Thus, since $[(a^{1/n})^m]^n = a^m$, it follows that $(a^{1/n})^m$ is an nth root of a^m, and by Definition 3.3 this root is denoted by $(a^m)^{1/n}$. Hence $(a^{1/n})^m = (a^m)^{1/n}$.

Observe that Theorem 3.2 requires that $a^{1/n}$ be a real number. This requirement is not satisfied if n is even and a is negative. If, for instance, $a = -3$, $m = 2$, and $n = 2$, then $a^{1/n} = (-3)^{1/2}$ is not a real number.

Theorem 3.3   *If $a^{1/np} \in R$, $m \in J$, and $n, p \in N$, with $a \neq 0$ for $m \leq 0$, then*

$$(a^{1/np})^{mp} = (a^{1/n})^m.$$

The proof of the foregoing result is similar to that of Theorem 3.2 and will be omitted.

Theorems 3.2 and 3.3 show that if $a^{1/np}$ is a real number, then $(a^{1/n})^m$, $(a^m)^{1/n}$, $(a^{1/np})^{mp}$, and $(a^{mp})^{1/np}$ all represent the same number.

Examples   a. $(16^{1/2})^3 = 4^3 = 64$ b. $(16^3)^{1/2} = 4096^{1/2} = 64$

c. $(16^{1/4})^6 = 2^6 = 64$ d. $(16^6)^{1/4} = 16{,}777{,}216^{1/4} = 64$

We make the following definition.

Definition 3.4   *If $a^{1/n} \in R$, $m \in J$, and $n \in N$, then*

$$a^{m/n} = (a^{1/n})^m.$$

Observe that requiring $n \in N$ does not alter the fact that m/n can represent every rational number, since all that is done is to restrict the denominator of the fraction representing the rational number to be positive, which is always possible in light of Theorems 1.8-V and 1.8-VI.

Examples   a. $8^{2/3} = (8^{1/3})^2 = 2^2 = 4$

b. $16^{-3/4} = (16^{1/4})^{-3} = 2^{-3} = \dfrac{1}{8}$

Because we define $a^{1/n}$ to be the positive nth root of a for a positive and n an even natural number, and since a^m is positive for a negative and m an even natural number, it follows that, for m and n *even* natural numbers and a *any* real number,

$$(a^m)^{1/n} = |a|^{m/n}.$$

For the special case $m = n$ (m and n even),

$$(a^n)^{1/n} = |a|.$$

Properties of rational powers   Powers with rational exponents have the same fundamental properties as powers with integral exponents, as long as the powers are real numbers. Theorem 3.1, appropriately reworded for rational-number exponents, can be invoked to rewrite exponential expressions involving rational exponents.

Examples   a. $\dfrac{x^{2/3}}{x^{1/3}} = x^{2/3 - 1/3}$ b. $\left(\dfrac{a^3 b^6}{c^{12}}\right)^{2/3} = \dfrac{(a^3)^{2/3}(b^6)^{2/3}}{(c^{12})^{2/3}}$ c. $(a^6)^{1/2} = |a|^{6/2}$

$= x^{1/3}$ $(x \neq 0)$ $= \dfrac{a^2 b^4}{c^8}$ $(c \neq 0)$ $= |a|^3$

Observe that in the last example it was necessary to use absolute-value notation because the expression has been defined to be positive for m and n even. For example, with $a = -2$, we have

$$[(-2)^6]^{1/2} = |-2|^{6/2} = |-2|^3 = 8,$$

whereas

$$(-2)^3 = -8.$$

Exercise 3.2

For each of the following expressions, write an equivalent basic numeral—that is, a numeral with exponent 1.

Examples

a. $64^{1/2}$

b. $\left(\dfrac{8}{27}\right)^{-2/3}$

c. $(-27)^{4/3}$

Solutions

a. 8

b. $\left[\left(\dfrac{8}{27}\right)^{1/3}\right]^{-2}$

$\left(\dfrac{2}{3}\right)^{-2}$

$\dfrac{9}{4}$

c. $[(-27)^{1/3}]^4$

$(-3)^4$

81

1. $16^{1/2}$ 2. $27^{1/3}$ 3. $27^{2/3}$ 4. $27^{4/3}$

5. $16^{-1/2}$ 6. $8^{-1/3}$ 7. $16^{-3/4}$ 8. $27^{-2/3}$

9. $(-8)^{-2/3}$ 10. $(-8)^{-4/3}$ 11. $\left(\dfrac{4}{9}\right)^{3/2}$ 12. $\left(\dfrac{4}{9}\right)^{-3/2}$

Write each of the following expressions as a product or quotient of powers in which each variable occurs but once, and all exponents are positive. Assume all variable bases are positive and all variable exponents are natural numbers.

Examples

a. $\dfrac{x^{5/6}}{x^{2/3}}$

b. $\dfrac{(x^{1/2}y^2)^2}{(x^{2/3}y)^3}$

c. $(y^{2n} \cdot y^{n/2})^4$

Solutions

a. $x^{5/6-2/3}$

$x^{5/6-4/6}$

$x^{1/6}$

b. $\dfrac{xy^4}{x^2y^3}$

$\dfrac{y}{x}$

c. $y^{8n} \cdot y^{2n}$

y^{10n}

13. $x^{2/3}x^{4/3}$ 14. $x^{1/4}x^{5/4}$ 15. $y^{1/2}y^{3/4}$

16. $y^{5/6}y^{1/3}$ 17. $\dfrac{x^{5/6}}{x^{1/2}}$ 18. $\dfrac{x^{3/4}}{x^{1/2}}$

19. $\dfrac{y^{2/3}}{y^{1/2}}$ 20. $\dfrac{y^{3/5}}{y^{1/2}}$ 21. $\left(\dfrac{x^{1/2}}{y^2}\right)^2 \left(\dfrac{y^4}{x^2}\right)^{1/2}$

22. $\left(\dfrac{y^{2/3}}{x}\right)^3 \left(\dfrac{x^2}{y^{1/2}}\right)^2$ 23. $\left(\dfrac{x^5y^8}{y^{13}}\right)^{1/4}$ 24. $\left(\dfrac{125x^3y^4}{27x^{-6}y}\right)^{1/3}$

25. $(x^2)^{n/2}(y^{2n})^{2/n}$ 26. $(x^{n/2})^2(y^n)^{5/n}$ 27. $\dfrac{x^{2n}}{x^{n/2}}$

28. $\left(\dfrac{a^n}{b}\right)^{1/2} \cdot \left(\dfrac{b}{a^{2n}}\right)^{3/2}$ **29.** $\dfrac{x^{3n}y^{2m-1}}{(x^n y^m)^{1/2}}$ **30.** $\left(\dfrac{x^{2n^2}}{x^{4n}}\right)^{1/n}$

31. $\left(\dfrac{xy^{-m}}{x^n y}\right)^{-1}$ **32.** $\left(\dfrac{x^{2n}y^n}{x^n y^{3n}}\right)^{-1/n}$

Apply the distributive law to write each product as a sum.

Examples   **a.** $x^{1/4}(x^{3/4} - x)$ **b.** $(x^{1/2} - x)(x^{1/2} + x)$

Solutions   **a.** $x^{1/4}x^{3/4} - x^{1/4}x$ **b.** $x^{1/2}x^{1/2} - x^2$

                           $x - x^{5/4}$ $x - x^2$

33. $x^{1/2}(x^{1/2} - 1)$ **34.** $x^{1/3}(x^{1/3} + 2)$

35. $y^{2/3}(y - y^{1/3})$ **36.** $y^{1/2}(y^2 - y^{1/2})$

37. $(x^{1/2} - y^{1/2})(x^{1/2} + y^{1/2})$ **38.** $(x^{-1/2} + y^{1/2})(x^{-1/2} - y^{1/2})$

39. $(x + y)^{1/2}[(x + y)^{1/2} - (x + y)]$ **40.** $(x - y)^{2/3}[(x - y)^{-1/3} + (x - y)]$

Factor as indicated.

Examples   **a.** $y^{-1/2} + y^{1/2} = y^{-1/2}(?)$ **b.** $x^{3/2} - x^{-1/2} = x^{-1/2}(?)$

Solutions   **a.** $y^{-1/2}(1 + y)$ **b.** $x^{-1/2}(x^2 - 1)$

41. $x^{3/2} + x = x(?)$ **42.** $y - y^{2/3} = y^{1/3}(?)$

43. $x^{-3/2} + x^{-1/2} = x^{-1/2}(?)$ **44.** $y^{3/4} - y^{-1/4} = y^{-1/4}(?)$

45. $(x + 1)^{1/2} - (x + 1)^{-1/2} = (x + 1)^{-1/2}(?)$

46. $(y + 2)^{1/5} - (y + 2)^{-4/5} = (y + 2)^{-4/5}(?)$

In the previous problems, the variables were restricted to represent positive numbers. In Problems 47–52, consider each variable base to be any *element of the set of real numbers and simplify.*

Examples   **a.** $[(-3)^2]^{1/2}$ **b.** $[u^2(u + 5)]^{1/2}$

Solutions   **a.** $[(-3)^2]^{1/2} = |-3| = 3$ **b.** $[u^2(u + 5)]^{1/2} = |u|(u + 5)^{1/2}$

47. $[(-5)^2]^{1/2}$ **48.** $[(-3)^{12}]^{1/4}$ **49.** $[4x^2]^{1/2}$

50. $[x^2(x - 1)]^{1/2}$ **51.** $\dfrac{2}{[x^2(x + 1)]^{1/2}}$ **52.** $\left[\dfrac{9}{x^6(x^2 + 1)}\right]^{1/2}$

3.3   *Radical Expressions*

Powers of real numbers with rational numbers for exponents are frequently denoted by symbols involving the use of the radical sign, $\sqrt{}$.

Definition 3.5   *If $a^{1/n} \in R$ and $n \in N, n \neq 1$, then*

$$\sqrt[n]{a} = a^{1/n}.$$

Naturally, the radical expression on the left is not defined if the power on the right is not. In the symbolism $\sqrt[n]{a}$, a is called the **radicand** and n the **index** of the radical, and the expression is called a **radical expression of order** n. If no index is shown with a radical expression, as, for example, in the case $\sqrt{a}$, then the index 2 is understood to apply. The symbol $\sqrt{a}$ denotes the nonnegative square root of a, where, of course, a cannot be negative. The symbol $\sqrt{x^2}$, where $x \in R$, therefore provides us with an alternative means of writing $|x|$. That is, $\sqrt{x^2} = \lceil x \rceil$.

Properties of radicals   An immediate consequence of the foregoing definition and the theorems pertaining to exponents is the following.

Theorem 3.4   *For real values of a and b for which all the radical expressions in the equation denote real numbers,*

$$\text{I}_A \quad \sqrt[n]{a^n} = a \quad (n \text{ an odd natural number}),$$

$$\text{I}_B \quad \sqrt[n]{a^n} = |a| \quad (n \text{ an even natural number}),$$

$$\text{II} \quad \sqrt[n]{a^m} = (\sqrt[n]{a})^m \quad (n \in N, m \in J),$$

$$\text{III} \quad \sqrt[n]{a} \cdot \sqrt[n]{b} = \sqrt[n]{ab} \quad (n \in N),$$

$$\text{IV} \quad \frac{\sqrt[n]{a}}{\sqrt[n]{b}} = \sqrt[n]{\frac{a}{b}} \quad (b \neq 0, n \in N),$$

$$\text{V} \quad \sqrt[cn]{a^{cm}} = \sqrt[n]{a^m} \quad (m \in J, n, c \in N).$$

It might be noted in III and IV that if $a, b < 0$ and n is even, then the radical in the left-hand member is not defined, even though the radical in the right-hand member is. Notice also that $n \neq 1$ is implied, since $\sqrt[1]{a}$ is not defined.

The several parts of this theorem can be used to rewrite radical expressions in various ways, and, in particular, to write them in what is called **standard form**. A radical expression is said to be in standard form if

a. the radicand contains no polynomial factor raised to a power equal to or greater than the index of the radical,

b. the radicand contains no fractions,

c. no radical expressions are contained in denominators of fractions, and

d. the index of the radical is as small as possible.

Examples

a. $\sqrt[3]{24x^3y^2} = \sqrt[3]{8x^3}\sqrt[3]{3y^2} = 2x\sqrt[3]{3y^2}$ By Theorem 3.4-III and Definition 3.5.

b. $\sqrt[6]{49} = \sqrt[3\cdot2]{7^{1\cdot2}} = \sqrt[3]{7}$ By Theorem 3.4-V.

c. $\dfrac{\sqrt[3]{4a^2}}{\sqrt[3]{b}} = \dfrac{\sqrt[3]{4a^2}\,\sqrt[3]{b^2}}{\sqrt[3]{b}\,\sqrt[3]{b^2}}$ By Theorem 1.10 (the fundamental principle of fractions).

$= \dfrac{\sqrt[3]{4a^2b^2}}{\sqrt[3]{b^3}} = \dfrac{\sqrt[3]{4a^2b^2}}{b}$ $(b \neq 0)$ By Theorem 3.4-III and Definition 3.5.

The process employed in simplifying the expression in part c in the foregoing example is called "rationalizing the denominator," because the result is a fraction with denominator free of radicals. This does not exclude the possibility that the denominator is an irrational number.

It should be noted, though, that it is not *always* preferable, in working with fractions, to have their denominators rationalized. Sometimes, in fact, it is desirable to rationalize the numerator. Thus, for example,

$$\frac{\sqrt[3]{2}}{\sqrt{3}} = \frac{\sqrt[3]{2}\sqrt[3]{4}}{\sqrt{3}\sqrt[3]{4}} = \frac{2}{\sqrt{3}\sqrt[3]{4}}.$$

Exercise 3.3

(*Assume that all variables represent positive real numbers and that all radicands are positive.*)

Write in radical form.

Examples

a. $5^{1/2}$ b. $xy^{2/3}$ c. $(x - y^2)^{-1/2}$

Solutions

a. $\sqrt{5}$ b. $x\sqrt[3]{y^2}$ c. $\dfrac{1}{\sqrt{x - y^2}}$

1. $2^{1/3}$ 2. $3^{1/2}$ 3. $3x^{1/3}$ 4. $2y^{1/4}$
5. $(2y)^{1/2}$ 6. $(3x)^{1/3}$ 7. $x^{2/3}$ 8. $y^{3/4}$
9. $xy^{2/3}$ 10. $x^{1/2}y^{2/3}$ 11. $(x^3y)^{1/4}$ 12. $(xy)^{2/3}$
13. $(x - y)^{1/2}$ 14. $(x^2 - y)^{1/3}$ 15. $(x^2 + y)^{-2/3}$ 16. $(2x + y)^{-3/4}$

Write an equivalent expression using positive fractional exponents in lowest terms.

Examples

a. $\sqrt{x^3}$ b. $7\sqrt[3]{a^2}$ c. $\sqrt[4]{a - b}$

Solutions

a. $x^{3/2}$ b. $7a^{2/3}$ c. $(a - b)^{1/4}$

17. $\sqrt[3]{y^2}$ 18. $\sqrt[4]{y^3}$ 19. $\sqrt[4]{2xy^2}$ 20. $\sqrt[5]{x^2y^3}$
21. $3\sqrt[4]{x^3y}$ 22. $\sqrt[7]{x^5y^2}$ 23. $\sqrt[3]{a + b^2}$ 24. $\sqrt{a^2 - b}$

Find the root indicated.

Examples   a. $\sqrt[5]{-32}$ b. $\sqrt[3]{x^6 y^3}$ c. $\sqrt{x^2 y^6}$

Solutions   a. -2 b. $x^2 y$ c. xy^3

    25. $\sqrt{81}$ **26.** $\sqrt{64}$ **27.** $\sqrt[3]{-27}$ **28.** $\sqrt[3]{-8}$

    29. $\sqrt{x^4 y^2}$ **30.** $\sqrt[3]{x^3 y^6}$ **31.** $\sqrt{\dfrac{4}{9} x^6 y^{10}}$ **32.** $\sqrt[3]{\dfrac{-8x^3 y^3}{125}}$

Write in simplest form.

Examples   a. $\sqrt[3]{2x^7 y^3}$ b. $\sqrt{2xy}\sqrt{8x}$ c. $\dfrac{\sqrt{6a}\sqrt{5a}}{\sqrt{15}}$

Solutions   a. $\sqrt[3]{x^6 y^3}\,\sqrt[3]{2x}$ b. $\sqrt{16x^2}\sqrt{y}$ c. $\sqrt{\dfrac{15}{15} \cdot 2a^2}$

           $x^2 y \sqrt[3]{2x}$ $4x\sqrt{y}$ $a\sqrt{2}$

    33. $\sqrt{4x^5}$ **34.** $\sqrt{16y^3}$ **35.** $\sqrt[4]{3x^5 y^5}$

    36. $\sqrt[3]{-8x^6}$ **37.** $\sqrt{3xy}\sqrt{6y}$ **38.** $\sqrt[3]{y^4}\sqrt[3]{y^7}$

    39. $\dfrac{\sqrt{8y}}{\sqrt{2y}}$ **40.** $\dfrac{\sqrt{x^3 y^3}}{\sqrt{xy^2}}$ **41.** $\dfrac{\sqrt[3]{2a^2 b^7}}{\sqrt[3]{ab^2}}$

    42. $\dfrac{\sqrt[3]{24a^5 b^2}}{\sqrt[3]{6a^2 b}}$ **43.** $\dfrac{\sqrt{ab}\sqrt{ab^4}}{\sqrt{b}}$ **44.** $\dfrac{\sqrt[3]{4ab^2}\sqrt[3]{2a}}{\sqrt[3]{4ab}}$

Rationalize the denominator of each expression.

Examples   a. $\dfrac{1}{\sqrt{2}}$ b. $\sqrt{\dfrac{3x}{7y}}$ c. $\sqrt[3]{\dfrac{2}{y}}$

Solutions   a. $\dfrac{1}{\sqrt{2}}\dfrac{\sqrt{2}}{\sqrt{2}}$ b. $\dfrac{\sqrt{3x}\sqrt{7y}}{\sqrt{7y}\sqrt{7y}}$ c. $\dfrac{\sqrt[3]{2}\sqrt[3]{y}\sqrt[3]{y}}{\sqrt[3]{y}\sqrt[3]{y}\sqrt[3]{y}}$

           $\dfrac{\sqrt{2}}{2}$ $\dfrac{\sqrt{21xy}}{7y}$ $\dfrac{\sqrt[3]{2y^2}}{y}$

    45. $\dfrac{2}{\sqrt{3}}$ **46.** $\dfrac{4}{\sqrt{5}}$ ✓**47.** $\sqrt{\dfrac{x}{2y}}$ **48.** $\sqrt{\dfrac{5x}{y}}$

    ✓**49.** $\sqrt[3]{\dfrac{2}{x^2}}$ **50.** $\sqrt[3]{\dfrac{3}{y^2}}$ **51.** $\sqrt[3]{\dfrac{3}{y}}$ **52.** $\sqrt[3]{\dfrac{5}{x}}$

Rationalize the numerator of each expression.

53. $\dfrac{\sqrt{3}}{3}$ **54.** $\dfrac{\sqrt{2}}{4}$ **55.** $\dfrac{\sqrt{x}}{\sqrt{y}}$ **56.** $\dfrac{\sqrt{xy}}{x}$

Reduce the order of each radical.

Examples   a. $\sqrt[4]{5^2}$ b. $\sqrt[12]{81}$ c. $\sqrt[4]{x^2 y^2}$

Solutions   a. $\sqrt[2\cdot2]{5^{1\cdot2}} = \sqrt{5}$ b. $\sqrt[3\cdot4]{3^{1\cdot4}} = \sqrt[3]{3}$ c. $\sqrt[2\cdot2]{x^{1\cdot2}y^{1\cdot2}} = \sqrt{xy}$

✓**57.** $\sqrt[6]{81}$ **58.** $\sqrt[10]{32}$ ✓**59.** $\sqrt[4]{16x^2}$ **60.** $\sqrt[9]{8a^3}$

✓**61.** $\sqrt[6]{8x^3}$ **62.** $\sqrt[6]{125z^3}$ **63.** $\sqrt{(x-1)^4}$ **64.** $\sqrt[12]{(x-2y)^4}$

3.4   Operations on Radical Expressions

We have defined our radical expressions so that they represent real numbers. Hence the properties of the real numbers can be applied. For example, the distributive law permits us to express certain sums as products.

Examples   a. $2\sqrt{5} + 4\sqrt{5} = (2+4)\sqrt{5} = 6\sqrt{5}$

b. $5\sqrt{2} - 9\sqrt{2} = (5-9)\sqrt{2} = -4\sqrt{2}$

c. $5\sqrt{2} + \sqrt{75} = 5\sqrt{2} + 5\sqrt{3} = 5(\sqrt{2} + \sqrt{3})$

The distributive law in the form $c(a + b) = ca + cb$ permits us to write certain products as sums or differences.

Examples   a. $2(\sqrt{2} - 7) = 2\sqrt{2} - 14$

b. $(\sqrt{x} - 3)(\sqrt{x} + 3) = \sqrt{x}(\sqrt{x} + 3) - 3(\sqrt{x} + 3)$
$$= x - 9$$

c. $(\sqrt{x} + 2)(\sqrt{x} - 1) = \sqrt{x}(\sqrt{x} - 1) + 2(\sqrt{x} - 1)$
$$= x - \sqrt{x} + 2\sqrt{x} - 2$$
$$= x + \sqrt{x} - 2$$

Binomial   The distributive law also provides us with a means of rationalizing denominators
denominators   of fractions in which radicals occur in one or both of two terms. To accomplish this, we first observe that

$$(a - \sqrt{b})(a + \sqrt{b}) = a^2 - b,$$

where the expression in the right-hand member contains no radical term. Each of the two factors of a product exhibiting this property is said to be the **conjugate** of the other. Now consider a fraction of the form

$$\frac{a}{b + \sqrt{c}},$$

where c is positive and $b \neq -\sqrt{c}$. If we multiply the numerator and denominator of this fraction by the conjugate of the denominator, then the denominator of the resulting fraction will contain no term involving $\sqrt{c}$ and hence will be free of radicals. That is,

$$\frac{a}{b + \sqrt{c}} = \frac{a(b - \sqrt{c})}{(b + \sqrt{c})(b - \sqrt{c})} = \frac{ab - a\sqrt{c}}{b^2 - c},$$

where the denominator has been rationalized. This process is equally applicable to radical fractions of the form

$$\frac{a}{\sqrt{b} + \sqrt{c}},$$

since

$$\frac{a}{\sqrt{b} + \sqrt{c}} = \frac{a(\sqrt{b} - \sqrt{c})}{(\sqrt{b} + \sqrt{c})(\sqrt{b} - \sqrt{c})} = \frac{a\sqrt{b} - a\sqrt{c}}{b - c}.$$

As noted on page 65, it is sometimes preferable, in working with fractions, to have their numerators rationalized. Thus, for example,

$$\frac{\sqrt{b} + \sqrt{c}}{a} = \frac{(\sqrt{b} + \sqrt{c})(\sqrt{b} - \sqrt{c})}{a(\sqrt{b} - \sqrt{c})} = \frac{b - c}{a(\sqrt{b} - \sqrt{c})}.$$

Exercise 3.4

Write each sum as a product.

Examples

a. $4\sqrt{2} + 3\sqrt{2} - \sqrt{2}$

b. $2\sqrt{3} + 4\sqrt{12}$

Solutions

a. $(4 + 3 - 1)\sqrt{2}$

    $6\sqrt{2}$

b. $2\sqrt{3} + 4 \cdot 2\sqrt{3}$

    $10\sqrt{3}$

1. $5\sqrt{2} + 3\sqrt{2}$
2. $8\sqrt{3} - 3\sqrt{3}$
3. $4\sqrt{5} + \sqrt{5} - 6\sqrt{5}$
4. $\sqrt{2} - 5\sqrt{2} + 2\sqrt{2}$
5. $4\sqrt{3} + 2\sqrt{12}$
6. $5\sqrt{5} - \sqrt{20}$
7. $\sqrt{8} - \sqrt{50} - \sqrt{2}$
8. $\sqrt{50} + 2\sqrt{32} - \sqrt{2}$
9. $3\sqrt[3]{16} - \sqrt[3]{2}$
10. $\sqrt[3]{54} + 2\sqrt[3]{128}$
11. $\sqrt[4]{48} + 5\sqrt[4]{3}$
12. $3\sqrt[4]{32} - \sqrt[4]{162}$

Multiply factors and write all radicals in the result in simplest form. Assume that all variables represent positive real numbers and that all radicands are positive.

Examples   a. $\sqrt{x}\left(\sqrt{2x}-\sqrt{x}\right)$ b. $\left(\sqrt{x}-2\sqrt{y}\right)\left(2\sqrt{x}+\sqrt{y}\right)$

Solutions   a. $x\sqrt{2}-x$ b. $2x-2y-3\sqrt{xy}$

13. $\sqrt{2}\left(2-\sqrt{2}\right)$ 14. $\sqrt{3}\left(4+\sqrt{3}\right)$
15. $\sqrt{3}\left(2+\sqrt{2}\right)$ 16. $\sqrt{2}\left(3-\sqrt{5}\right)$
17. $\left(3+\sqrt{5}\right)\left(2-\sqrt{5}\right)$ 18. $\left(\sqrt{3}-5\right)\left(\sqrt{3}+5\right)$
19. $\left(\sqrt{2}-1\right)\left(\sqrt{2}+1\right)$ 20. $\left(2-\sqrt{3}\right)\left(2+\sqrt{3}\right)$
21. $\left(\sqrt{x}-2\sqrt{3}\right)\left(\sqrt{x}+\sqrt{3}\right)$ 22. $\left(\sqrt{3}-2\sqrt{x}\right)\left(\sqrt{3}+\sqrt{x}\right)$
23. $\left(\sqrt{2}-\sqrt{x}\right)\left(\sqrt{2}+\sqrt{x}\right)$ 24. $\left(2\sqrt{x}-\sqrt{3}\right)\left(2\sqrt{x}+\sqrt{3}\right)$

Rationalize denominators.

Examples   a. $\dfrac{3}{\sqrt{2}-1}$ b. $\dfrac{1}{\sqrt{x}-\sqrt{y}}$

Solutions   a. $\dfrac{3}{\left(\sqrt{2}-1\right)}\dfrac{\left(\sqrt{2}+1\right)}{\left(\sqrt{2}+1\right)}$ b. $\dfrac{1}{\left(\sqrt{x}-\sqrt{y}\right)}\dfrac{\left(\sqrt{x}+\sqrt{y}\right)}{\left(\sqrt{x}+\sqrt{y}\right)}$

$\dfrac{3\sqrt{2}+3}{2-1}$ $\dfrac{\sqrt{x}+\sqrt{y}}{x-y}$

$3\sqrt{2}+3$

25. $\dfrac{-4}{1+\sqrt{3}}$ 26. $\dfrac{1}{2-\sqrt{2}}$ 27. $\dfrac{x}{\sqrt{x}-3}$

28. $\dfrac{\sqrt{x}}{\sqrt{x}-\sqrt{y}}$ 29. $\dfrac{4\sqrt{2}-\sqrt{5}}{4\sqrt{2}+\sqrt{5}}$ 30. $\dfrac{4\sqrt{a}-\sqrt{3a+1}}{\sqrt{a}+\sqrt{3a+1}}$

Rationalize numerators.

31. $\dfrac{1-\sqrt{2}}{2}$ 32. $\dfrac{\sqrt{3}-2}{3}$ 33. $\dfrac{1-\sqrt{x+a}}{\sqrt{x+a}}$ 34. $\dfrac{\sqrt{x-a}+1}{\sqrt{x-a}}$

35. Find the value of y^2-3y+1, for $y=3-\sqrt{2}$.

36. Find the value of $3x^2-4x-2$, for $x=\dfrac{2-\sqrt{10}}{3}$.

In the previous problems, the variables were restricted to represent positive numbers. In Problems 37–40, consider each variable base to denote any element of the set of real numbers and simplify. Hint: See Problems 47–52 on page 63.

37. $\sqrt{4x^2}$ 38. $\sqrt{9x^2y^4}$ 39. $\sqrt{9(x^3-x^2)}$ 40. $\sqrt{18(x^6-x^7)}$

41. Prove Part I of Theorem 3.4.

3.5   *Approximations for Irrational Numbers*

Terminating and periodic decimals

Any rational number can be expressed either as a terminating decimal numeral or as a periodic (repeating) decimal numeral. A periodic decimal numeral is one in which a group (one or more) of successive digits repeats endlessly, without intervening digits, in the numeral. For example, $0.33\overline{3}$, $5.28\overline{5714}$, and $1.13\overline{6}36$ are repeating decimal numerals, where the bar above a digit or group of digits indicates that this group repeats without end. These particular numerals correspond to $\frac{1}{3}$, $5\frac{2}{7}$, and $1\frac{3}{22}$, respectively.

To see why every quotient of two integers a/b $(b \neq 0)$ can be expressed as a terminating decimal numeral or as a repeating decimal numeral, consider the quotient $5/22$. If we wished to use the division algorithm to find a decimal numeral equivalent to $5/22$, we would begin by writing

$$22 \overline{)5.0000 \ldots} \, ,$$

where we envisage an unending procession of zeros in the numeral for the dividend. Now, the first step in the division process will be to determine the initial digit in the quotient, which is 2, and then subtract the product, 2×22, from the first two digits in $5.000 \ldots$. The remainder must be one of the nonnegative integers less than 22. If the remainder is 0, the decimal numeral for the quotient terminates. As it turns out, the difference is 6:

$$
\begin{array}{r}
.2 \\
22 \overline{)5.0000 \ldots} \\
4\,4 \\
\hline
6
\end{array}
$$

The next step, then, is to determine the next digit in the numeral for the quotient. This is also 2, so we subtract 2×22 from 60, and obtain one of the nonnegative integers less than 22 for a difference. Again, if the difference is 0, the decimal numeral for the quotient terminates. If the difference is 6, as before, then we can expect this difference to recur endlessly, since each step hereafter is a repetition of the previous one. As it turns out, the difference is 16:

$$
\begin{array}{r}
.22 \\
22 \overline{)5.0000 \ldots} \\
4\,4 \\
\hline
60 \\
44 \\
\hline
16
\end{array}
$$

This process is then repeated until one of the differences from the set of positive integers less than 22 repeats itself or 0 occurs. This must happen within 22 steps. Actually, we have

$$
\begin{array}{r}
.227 \\
22 \overline{)5.0000} \\
4\,4 \\
\hline
60 \\
44 \\
\hline
160 \\
154 \\
\hline
6
\end{array}
$$

and 6 has repeated itself as a remainder. It follows, then, that the block of digits 27 will recur indefinitely, and

$$\frac{22}{5} = 0.2\overline{27}.$$

A general argument can be made to show that an analogous situation holds for every quotient a/b, where $a, b \in J$ and $b \neq 0$. In Chapter 12, conversely, we shall show a general means of finding a common fraction equivalent to any repeating decimal numeral. What we have, then, is the following assertion.

Theorem 3.5   *Every terminating and every repeating decimal numeral names a rational number, and, conversely, every rational number can be named by either a terminating or a repeating decimal numeral.*

Decimal numerals

Since, by this theorem, every terminating or repeating decimal numeral names a rational number, it follows that, if irrational numbers have decimal representations, these decimal numerals must be nonterminating and nonrepeating. Since we wish to be able to use numerals in a practical way, we find a rational number as close as need be to any given irrational number and then use this rational number as an approximation to the irrational number.

The most practical means for finding an approximation for $\sqrt[n]{a}$, where $a \in N$, is to use a table of roots such as Table IV at the end of this text. Tables of greater accuracy than the one that appears in this text, as well as tables for roots of higher index, can be found in other publications.

When suitable tables are not available, an intelligent guess can be made and adjusted. For example, to approximate $\sqrt[7]{2}$, we perhaps might guess 1.1. But a computation yields $(1.1)^7 = 1.94 \ldots$. Since this is just a bit too small, we next try 1.11, etc.

Approximations for other kinds of irrational numbers, such as π or values of trigonometric or logarithmic functions, for example, are also available in tables.

Given such a table, one can find decimal numerals for approximating sums, differences, products, and quotients involving irrational numbers. The properties of radicals can be used to simplify the calculations involved.

Example

Find a decimal numeral for a rational number approximating $\dfrac{2 - \sqrt{3}}{5 - \sqrt{3}}$ to the nearest hundredth.

Solution

To make the work less laborious, we first rationalize the denominator of this expression. Thus we have

$$\frac{(2 - \sqrt{3})(5 + \sqrt{3})}{(5 - \sqrt{3})(5 + \sqrt{3})} = \frac{10 - 3\sqrt{3} - 3}{25 - 3} = \frac{7 - 3\sqrt{3}}{22}.$$

From Table IV, $\sqrt{3} \approx 1.732$, so that

$$\frac{7 - 3\sqrt{3}}{22} \approx \frac{7 - 3(1.732)}{22} = \frac{7 - 5.196}{22} = \frac{1.804}{22} = 0.082 \approx 0.08.$$

The symbol $\approx$ is read "is approximately equal to."

Exercise 3.5

Find a terminating or repeating decimal equivalent to the given fraction.

1. $\dfrac{2}{7}$ 2. $\dfrac{209}{700}$ 3. $\dfrac{15}{16}$

4. $\dfrac{71}{32}$ 5. $\dfrac{3}{15}$ 6. $\dfrac{5}{14}$

Find a decimal numeral approximation for each of the following radical expressions to the nearest hundredth. Use Table IV on page 350.

7. $\sqrt{3} - \sqrt{37}$ 8. $\sqrt{72} - 2\sqrt{2}$ 9. $\dfrac{3\sqrt{17} - 1}{3}$

10. $\dfrac{27 - 3\sqrt{17}}{5}$ 11. $\dfrac{1}{\sqrt{3}}$ 12. $\dfrac{5 + \sqrt{2}}{\sqrt{2}}$

13. $\dfrac{4}{1 - \sqrt{2}}$ 14. $\dfrac{\sqrt{3}}{\sqrt{3} + 2}$ 15. $\dfrac{\sqrt{7} - \sqrt{3}}{3\sqrt{7} + \sqrt{3}}$

16. $\dfrac{2\sqrt{5} - \sqrt{6}}{\sqrt{5} - \sqrt{6}}$ 17. $\dfrac{\sqrt{7} - 2\sqrt{3}}{1 - \sqrt{3}}$ 18. $\dfrac{\sqrt{31} - \sqrt{23}}{2\sqrt{7} + 1}$

19. Find a positive rational number in $\{x \mid x^2 \leq 10\}$ that approximates $\sqrt{10}$ to within 1/1000. Similarly, find a positive rational number in $\{x \mid x^2 \geq 10\}$ that approximates $\sqrt{10}$ to within 1/1000. *Hint:* Modify tabular values in Table IV as necessary.

20. Find a positive rational number in $\{x \mid x^2 < 127\}$ which approximates $\sqrt{127}$ to within 1/1000. Similarly, find such a rational number in $\{x \mid x^2 > 127\}$.

21. A numeral for an irrational number between the rational number 1.41 and the irrational number $\sqrt{2}$ can be constructed by writing 1.413, which names a rational number in the desired interval, and then assigning additional digits in such fashion as to guarantee that no group ever repeats. An example is 1.413010010001 ..., where an additional zero is used each time to separate the digits 1 in the rest of the numeral. Since $1.414 < \sqrt{2}$, our number is less than $\sqrt{2}$. Use a similar method to find an irrational number in the interval $\{x \mid 1.73 < x < \sqrt{3}\}$.

22. See Problem 21. Find an irrational number in $\{x \mid 5929 \leq x^2 \leq 6084\}$.

Chapter Review

[3.1]   *Write each expression as a product or quotient in which each variable occurs at most once in the expression and all exponents are positive.*

1. $(x^2 y^4)^3$ 2. $\dfrac{x^2 y^{-3}}{x^{-1} y}$ 3. $\dfrac{x^0 y^{-2}}{(xy)^{-2}}$ 4. $\left(\dfrac{x^{-2} y^0}{y^{-3}} \right)^{-1}$

Represent each expression as a single fraction involving positive exponents only.

5. $x^{-1}y^{-1} + \dfrac{x}{y^{-1}}$

6. $\dfrac{x^{-2} - y^{-1}}{(xy)^{-1}}$

Write each expression as an equivalent expression in which each variable occurs only once.

7. $\dfrac{x^{2n}x^n}{x^{n+1}}$

8. $\left(\dfrac{x^n y^{n+1}}{xy^{n-1}}\right)^2$

Write each number as the product of a number between 1 and 10 and a power of 10.

9. 47300

10. 0.000045

[3.2] *Write each expression as a product or quotient of powers in which each variable occurs only once and all exponents are positive. Assume that all variable bases are positive and all variables in exponents are natural numbers.*

11. $x^{4/3} \cdot x^{1/2}$

12. $\left(\dfrac{x^2}{y^3}\right)^{-1/6}$

13. $(x^n y^{2n})^{-1/2}$

14. $\left(\dfrac{y^n}{x^n y^{2n}}\right)^{-1/n}$

Apply the distributive law to write each product as a sum.

15. $y^{1/2}(y^{1/2} - y)$

16. $(y - y^{1/2})(y + y^{1/2})$

Factor as indicated.

17. $x^{-1/4} + x^{1/2} = x^{-1/4}(?)$

18. $x^{2/3} - x^{-2/3} = x^{-2/3}(?)$

[3.3] *Write in simplest form.*

19. $\sqrt{9x^3 y^5}$

20. $\sqrt{2xy}\,\sqrt{8x}$

21. $\dfrac{\sqrt{3y}\,\sqrt{2xy}}{\sqrt{3}}$

22. $\dfrac{\sqrt[3]{6x}\,\sqrt[3]{9x^2 y}}{\sqrt[3]{2y}}$

23. Rationalize the denominator: $\sqrt[3]{\dfrac{x}{y}}$

24. Rationalize the numerator: $\dfrac{\sqrt{3y}}{6y}$

Reduce the order of each radical.

25. $\sqrt[4]{9}$

26. $\sqrt[6]{8x^3 y^3}$

[3.4] *Write each sum as a product.*

27. $3\sqrt{18} - 2\sqrt{18} + \sqrt{2}$

28. $4\sqrt[3]{24} + 2\sqrt[3]{3}$

Multiply factors and write all radicals in the result in simplest form.

29. $(\sqrt{5} - \sqrt{2})(\sqrt{5} + \sqrt{2})$

30. $(3\sqrt{x} - 4)(\sqrt{x} + 2)$

Rationalize denominators.

31. $\dfrac{1}{2 - \sqrt{3}}$

32. $\dfrac{2\sqrt{x} - 1}{\sqrt{x} + 1}$

Rationalize numerators.

33. $\dfrac{\sqrt{2} + 3}{2}$

34. $\dfrac{\sqrt{y} - 3}{\sqrt{y} + 1}$

[3.5] *Use Table IV to find a decimal numeral approximation for the value of the given expression to the nearest hundredth.*

35. $14 - 2\sqrt{5}$

36. $\dfrac{4 + \sqrt{6}}{4 - \sqrt{6}}$

4

Equations and Inequalities in One Variable

Equations and inequalities that involve variables are called **open sentences**. Equations and inequalities involving only constants are referred to as **statements**. For example,

$$x + 2 = 7, \quad x^2 - y \geq 4, \quad \text{and} \quad |x - 3| < 2$$

are open sentences, whereas

$$3 + 2 = 5, \quad 7 < 10 - 1, \quad \text{and} \quad |3 + 2| > 0$$

are statements. Although a statement can be adjudged true or false, no such judgment is possible in the case of an open sentence. Thus, $3 - 2 = 1$ is a true statement, and $3 - 2 = 2$ is false, but we cannot assert that $x - 2 = 1$ is either true or false until we know something more about x. Open sentences can be looked upon as set selectors. Given any set of numbers, an equation such as $x - 2 = 1$ or an inequality such as $x - 2 > 1$ will serve to select certain numbers from the universal set U and reject others, depending on whether the numbers make the resulting statement true or false.

In this chapter, we shall be concerned with open sentences over the real numbers (both equations and inequalities) in one variable.

4.1 Equivalent Equations; First-Degree Equations

If we replace the variable x in $P(x) = Q(x)$ with an element from its replacement set U, and if the resulting statement is true, the element is a **solution**, or **root**, of the equation and is said to **satisfy** the equation. Thus, if $x \in J$, then 2 is a solution of $x + 3 = 5$, because $2 + 3 = 5$ is a true statement. On the other hand, 3 is not a solution of the equation, because $3 + 3 = 5$ is false. The subset of U consisting of all solutions of an equation is said to be the **solution set** of the equation. In the example we have been using here, $x + 3 = 5$ and $x \in J$, the solution set is $\{2\}$.

Equations that have the same solution set are called **equivalent equations**. For example, if $x \in J$, the equations

$$x + 3 = -3 \quad \text{and} \quad x = -6$$

are equivalent, because the solution set of each is $\{-6\}$.

Elementary transformations

To solve an equation over a given set, we usually either determine the members of the solution set by inspection or else generate a sequence of equivalent equations until we arrive at one with an obvious solution set. The following theorem, which is a direct consequence of the addition and multiplication laws for real numbers, is frequently used in generating equivalent equations over the set of real numbers.

Theorem 4.1   *If $P(x)$, $Q(x)$, and $R(x)$ are expressions, then for all values of x for which $P(x)$, $Q(x)$, and $R(x)$ are real numbers, the sentence*

$$P(x) = Q(x)$$

is equivalent to each of the following sentences:

I $P(x) + R(x) = Q(x) + R(x)$,

II $P(x) - R(x) = Q(x) - R(x)$,†

III $P(x) \cdot R(x) = Q(x) \cdot R(x)$

IV $\dfrac{P(x)}{R(x)} = \dfrac{Q(x)†}{R(x)}$

$\left. \right\}$ for $x \in \{x \mid R(x) \neq 0\}$.

Any application of any part of Theorem 4.1 is called an **elementary transformation**. An elementary transformation *always* produces an equivalent equation. Care must be exercised in the application of Parts III and IV, however, for we have specifically excluded multiplication or division by zero. For example, to solve the equation

$$\frac{x}{x-3} = \frac{3}{x-3} + 2 \qquad (x \in R), \tag{1}$$

we might first multiply each member by $(x - 3)$ to find an equation that is free of fractions. We have

$$(x-3)\frac{x}{x-3} = (x-3)\frac{3}{x-3} + (x-3)2,$$

or

$$x = 3 + 2x - 6, \tag{2}$$

from which

$$x = 3.$$

Thus 3 is a solution of Equation (2). But, upon substituting 3 for x in Equation (1), we have

$$\frac{3}{3-3} = \frac{3}{3-3} + 2 \quad \text{or} \quad \frac{3}{0} = \frac{3}{0} + 2,$$

and neither member is defined. In obtaining Equation (2), each member of Equation (1) was multiplied by $(x - 3)$; but if x is 3, then $(x - 3)$ is zero, and Theorem 4.1-III is not applicable. Equation (2) is *not* equivalent to Equation (1), and in fact Equation (1) has no solution.

† Parts II and IV can be considered special cases of Parts I and III, respectively.

We can always ascertain whether what we think is a solution of an equation is such in reality by substituting the suggested solution in the original equation and determining whether or not the resulting statement is true. If each equation in a sequence is obtained by means of an elementary transformation, the sole purpose for such checking is to detect arithmetic errors. We shall dispense with checking solution sets in the examples that follow unless we apply what may be a nonelementary transformation—that is, unless we multiply or divide by an expression that vanishes for some value or values of the variable.

Solution of a linear equation

The equation

$$ax + b = 0, \tag{3}$$

where $a, b \in R$ and $a \neq 0$, is a **first-degree**, or **linear**, **equation**. Any equation that can be reduced to this form by elementary transformations, therefore, is equivalent to a first-degree equation. We can show that such an equation always has one and only one solution. By Theorem 4.1-I,

$$ax = -b \tag{4}$$

is equivalent to Equation (3); further, by Theorem 4.1-III, Equation (4) is equivalent to

$$x = -\frac{b}{a}, \tag{5}$$

which, of course, has the unique solution $-b/a$. Since (3), (4), and (5) are equivalent, Equation (3) has the unique solution $-b/a$.

An equation containing more than one variable, or containing symbols such as a, b, and c, representing constants, can often be solved for one of the symbols in terms of the remaining symbols by applying elementary transformations until the desired symbol is obtained by itself as one member of an equation.

Example

Solve $ay = b + y$, for y.

Solution

We generate the following sequence of equivalent equations.

$$ay - y = b$$

$$y(a - 1) = b$$

$$y = \frac{b}{a - 1} \quad (a \neq 1)$$

Exercise 4.1

Solve. Consider R to be the replacement set of the variable.

1. $6y - 2(2y + 5) = 6(5 + y)$
2. $3(7 - 2z) = 30 - 7(z + 1)$
3. $-3[x - (2x + 3) - 2x] = -9$
4. $-2[x - (x - 1)] = -3(x + 1)$
5. $4 - (x - 3)(x + 2) = 10 - x^2$
6. $6 + 3x - x^2 = 4 - (x - 2)(x + 3)$

7. $\dfrac{5x}{2} - 1 = x - \dfrac{1}{2}$

8. $\dfrac{x}{5} - \dfrac{x}{2} = 9$

9. $\dfrac{2x - 1}{5} = \dfrac{x + 1}{2}$

10. $\dfrac{2y}{3} - \dfrac{2y + 5}{6} = \dfrac{1}{2}$

11. $\dfrac{2}{x - 9} = \dfrac{9}{x + 12}$

12. $\dfrac{x}{x - 2} = \dfrac{2}{x - 2} + 7$

13. $\dfrac{5}{x - 3} = \dfrac{x + 2}{x - 3} + 3$

14. $\dfrac{2}{y + 1} + \dfrac{1}{3y + 3} = \dfrac{1}{6}$

15. $\dfrac{y}{y + 2} - \dfrac{3}{y - 2} = \dfrac{y^2 + 8}{y^2 - 4}$

16. $\dfrac{4}{2x - 3} + \dfrac{4x}{4x^2 - 9} = \dfrac{1}{2x + 3}$

Solve. Assume that all constants are real numbers and that the replacement set of all variables is R. Leave the results in the form of an equation equivalent to the given equation. Indicate any restrictions on the variables.

Example   Solve $l = a + (n - 1)d$, for d.

Solution   Add $-a$ to each member.

$$l - a = (n - 1)d$$

Multiply each member by $\dfrac{1}{n - 1}$ $(n \neq 1)$.

$$\dfrac{l - a}{n - 1} = d \quad (n \neq 1)$$

$$d = \dfrac{l - a}{n - 1} \quad (n \neq 1)$$

17. $v = k + gt$, for k

18. $v = k + gt$, for t

19. $A = \dfrac{h}{2}(b + c)$, for c

20. $S = \dfrac{a}{1 - r}$, for r

21. $l = a + (n - 1)d$, for n

22. $\dfrac{1}{r} = \dfrac{1}{r_1} + \dfrac{1}{r_2}$, for r

23. $x^2 y' - 3x - 2y^3 y' = 1$, for y'

24. $2xy' - 3y' + x^2 = 0$, for y'

25. $x_1 x_2 - 2x_1 x_3 = x_4$, for x_1

26. $3x_1 x_3 + x_1 x_2 = x_4$, for x_1

27. $\dfrac{y - y_1}{x - x_1} = 6$, for y

28. $\dfrac{y - y_1}{x - x_1} = 2$, for x

29. For what value of k does the equation $2x - 3 = \dfrac{4 + x}{k}$ have as its solution set $\{-1\}$?

30. Find a value of k so that the equation $3x - 1 = k$ is equivalent to the equation $2x + 5 = 1$.

4.2   *Second-Degree Equations*

The equation

$$ax^2 + bx + c = 0 \quad (a \neq 0)$$

is a **second-degree**, or **quadratic**, **equation**. Any equation that can be reduced to this form by elementary transformations is therefore equivalent to a quadratic equation. We shall designate the form shown above as the **standard form** for such equations.

Solution by factoring

The following theorem, which is a consequence of Theorems 1.6 and 1.7, will prove useful to us in solving quadratic equations.

> **Theorem 4.2**   *If $a, b \in R$, then $ab = 0$ if and only if*
>
> $$a = 0 \quad or \quad b = 0 \quad or\ both.$$

Example

Find the solution set in R of $x^2 + 2x - 15 = 0$.

Solution

The equation

$$x^2 + 2x - 15 = 0$$

is equivalent to

$$(x + 5)(x - 3) = 0.$$

Since $(x + 5)(x - 3) = 0$ is true if and only if

$$x + 5 = 0 \quad or \quad x - 3 = 0,$$

we can see by inspection that the only values of x that satisfy the original equation are -5 and 3. Hence the solution set is $\{-5, 3\}$.

Number of solutions

In general, the solution set of a quadratic equation *over the real numbers* may contain two, one, or no real numbers as elements (see page 83). The equation in the foregoing example has two real solutions. Now, consider the equation

$$x^2 - 2x + 1 = 0.$$

Since $x^2 - 2x + 1 = 0$ is equivalent to

$$(x - 1)^2 = 0,$$

and since the only value of x for which $(x - 1)^2 = 0$ is 1, this is the only member of the solution set of $x^2 - 2x + 1 = 0$. For reasons of convenience, in Chapter 11 we shall wish to consider every quadratic equation to have two roots. Accordingly, we say that the solution of any quadratic equation having only one solution is of **multiplicity two**; that is, we shall count it twice as a solution.

Before exhibiting an example of a quadratic equation over the real numbers in the real variable x having an empty solution set, let us consider the special case of a quadratic equation,

$$x^2 - a = 0, \quad \text{where } a > 0.$$

Since $x^2 - a = 0$ is equivalent to $x^2 = a$, and since $x^2 = a$ implies that x must be a square root of a, we have as the solution set $\{\sqrt{a}, -\sqrt{a}\}$. This method of solving a quadratic equation is sometimes called **extraction of roots**.

A quadratic equation over R having no real solutions is, for example, $x^2 + 1 = 0$, since there exists no number $x \in R$ such that its square is negative. The solution set in R of the given equation is $\emptyset$. In Chapter 10, when we turn our attention to the set C of complex numbers, you will see that *every* quadratic equation over this set has a nonempty solution set in C.

General solution

Quadratic equations of the form
$$(x - a)^2 = b,$$
where $b \geq 0$, can be solved by observing that $x - a$ must be one of the square roots of b. That is, either
$$x - a = \sqrt{b} \quad \text{or} \quad x - a = -\sqrt{b},$$
and conversely. Thus it is evident that for $b \geq 0$ the solution set of $(x - a)^2 = b$ is $\{a + \sqrt{b}, a - \sqrt{b}\}$.

Being able to find solution sets for quadratic equations of the form $(x - a)^2 = b$ enables us to find the solution set of any quadratic equation. Let us first consider the general quadratic equation in standard form
$$ax^2 + bx + c = 0,$$
for the special case in which $a = 1$, that is,
$$x^2 + bx + c = 0. \tag{1}$$
We can write this equation in the equivalent form
$$(x - p)^2 = q,$$
which we can solve as above. We begin the process by adding $-c$ to each member of Equation (1), which yields
$$x^2 + bx \qquad = -c. \tag{2}$$
If we then add $(b/2)^2$ to each member of Equation (2), we obtain
$$x^2 + bx + \left(\frac{b}{2}\right)^2 = -c + \left(\frac{b}{2}\right)^2, \tag{3}$$
in which the left-hand member is equal to $(x + b/2)^2$, and we have
$$\left(x + \frac{b}{2}\right)^2 = -c + \frac{b^2}{4}. \tag{4}$$

Since we have performed only elementary transformations, Equation (4) is equivalent to (2), and we can solve (4) by the method of the preceding section, provided
$$-c + \frac{b^2}{4} \geq 0.$$

The technique used to obtain Equations (3) and (4) is called **completing the square**. We can determine the term necessary to complete the square in (2) by dividing the coefficient b of the first-degree term by the number 2 and squaring the result. The expression obtained, $x^2 + bx + (b/2)^2$, is called a **perfect square** and may be written in the form $(x + b/2)^2$.

Exercise 4.2

Solve by factoring.

Example

$x^2 + x = 30$

Solution

Write an equivalent equation in standard form, and then factor the left-hand member.

$$x^2 + x - 30 = 0$$
$$(x + 6)(x - 5) = 0$$

Determine solutions by inspection, or set each factor equal to zero, and solve each linear equation.

$$x + 6 = 0 \qquad x - 5 = 0$$
$$x = -6 \qquad x = 5$$

The solution set is $\{-6, 5\}$.

1. $x^2 + 2x = 0$
2. $x^2 - x = 5x$
3. $x^2 + 5x - 14 = 0$
4. $3x^2 - 6x = -3$
5. $2x^2 - 5x = 3$
6. $2y^2 = -(3y + 1)$
7. $6x^2 = 5x + 4$
8. $35 = t + 6t^2$
9. $x(2x - 3) = -1$
10. $(x - 2)(x + 1) = 4$
11. $3 = \dfrac{10}{x^2} - \dfrac{7}{x}$
12. $\dfrac{2}{x - 3} - \dfrac{6}{x - 8} = -1$

Solve by the extraction of roots.

Example

$(x + 3)^2 = 7$

Solution

Set $x + 3$ equal to each square root of 7.

$$x + 3 = \sqrt{7} \qquad x + 3 = -\sqrt{7}$$
$$x = -3 + \sqrt{7} \qquad x = -3 - \sqrt{7}$$

The solution set is $\{-3 + \sqrt{7}, -3 - \sqrt{7}\}$.

13. $x^2 = 4$
14. $9x^2 - 100 = 0$
15. $x^2 = 5$
16. $(x - 1)^2 = 4$
17. $(x - 6)^2 = 5$
18. $(x - a)^2 = 4$

Solve by completing the square.

Example

$2x^2 + x - 1 = 0$

Solution

Write an equivalent equation with the constant term as the right-hand member and the coefficient of x^2 equal to 1.

$$x^2 + \frac{1}{2}x \qquad = \frac{1}{2}$$

(Solution continued overleaf)

Add the square of one-half of the coefficient of the first-degree term to each member.

$$x^2 + \frac{1}{2}x + \frac{1}{16} = \frac{1}{2} + \frac{1}{16}$$

Rewrite the left-hand member as the square of an expression.

$$\left(x + \frac{1}{4}\right)^2 = \frac{9}{16}$$

Set $x + \frac{1}{4}$ equal to each square root of $\frac{9}{16}$.

$$x + \frac{1}{4} = \frac{3}{4} \qquad x + \frac{1}{4} = -\frac{3}{4}$$

$$x = \frac{1}{2} \qquad x = -1$$

The solution set is $\left\{\frac{1}{2}, -1\right\}$.

19. $x^2 + 4x - 12 = 0$	**20.** $x^2 - 2x + 1 = 0$	**21.** $x^2 + 9x + 20 = 0$
22. $x^2 - 2x - 1 = 0$	**23.** $2x^2 = 2 - 3x$	**24.** $2x^2 + 4x = -1$

Reduce each equation to an equivalent equation of the form $(x - h)^2 + (y - k)^2 = r^2$ by completing the squares in x and y.

Example $x^2 + y^2 - 4x + 6y = 5$

Solution   Write an equivalent equation in the form

$$[x^2 - 4x + (\quad)] + [y^2 + 6y + (\quad)] = 5 + (\quad) + (\quad).$$

Complete the squares in x and y.

$$[x^2 - 4x + 4] + [y^2 + 6y + 9] = 5 + 4 + 9$$

Write each expression in brackets as the square of a binomial.

$$(x - 2)^2 + (y + 3)^2 = 18 \quad \text{or} \quad (x - 2)^2 + [(y - (-3)]^2 = (\sqrt{18})^2$$

25. $x^2 + y^2 - 4x - 4y - 17 = 0$	**26.** $x^2 + y^2 + 6x - 6y + 18 = 0$
27. $x^2 + y^2 + 6x - 2y + 6 = 0$	**28.** $x^2 + y^2 - 2x + 4y + 2 = 0$
29. $4x^2 + 4y^2 - 4x + 8y = 11$	**30.** $16x^2 + 16y^2 - 8x + 16y = 59$

Reduce each of the following equations to the form $y = (x - a)^2 + b$, where a and b are constants, by completing the square in x.

Example $y = x^2 + 4x + 3$

Solution   Write the equation in the indicated form.

$$y = [x^2 + 4x + (\quad)] + 3$$

Complete the square in x; add $4 + (-4)$ to the right-hand member.

$$y = [x^2 + 4x + (4)] + 3 + (-4)$$
$$y = (x + 2)^2 - 1, \quad \text{or} \quad y = [x - (-2)]^2 + (-1)$$

31. $y = x^2 - 2x + 5$ **32.** $y = x^2 + 6x + 3$ **33.** $y = x^2 + 8x + 4$

34. $y = x^2 - 4x + 7$ **35.** $y = x^2 - 3x + 5$ **36.** $y = x^2 + 5x - 2$

Since r_1 and r_2 are solutions of the quadratic equation $(x - r_1)(x - r_2) = 0$, it follows that $(x - r_1)(x - r_2) = 0$, or $x^2 - (r_1 + r_2)x + r_1 r_2 = 0$, is a quadratic equation having solutions r_1 and r_2. Given the solutions of a quadratic equation in Problems 37–40, write the equation in standard form having integral coefficients, with no common integer factors other than 1 and -1 and with leading coefficient positive.

37.   3 and 2   **38.** $-\dfrac{1}{2}$ and 3   **39.** $-\dfrac{2}{3}$ and $\dfrac{1}{2}$ **40.** $\dfrac{2}{9}$ and $-\dfrac{2}{9}$

4.3   *The Quadratic Formula*

Because the general quadratic equation

$$ax^2 + bx + c = 0 \quad (a \neq 0)$$

can be written equivalently in the form

$$x^2 + \frac{b}{a}x + \frac{c}{a} = 0,$$

the process of completing the square that we considered in Section 4.2 can be applied to obtain the **quadratic formula**,

$$x = \frac{-b \pm \sqrt{b^2 - 4ac}}{2a},$$

where the solutions, or roots, of the general quadratic equation are expressed in terms of the coefficients (see Exercise 4.3, Problem 22). The symbol $\pm$ is used to condense the two equations

$$x = \frac{-b + \sqrt{b^2 - 4ac}}{2a} \quad \text{or} \quad x = \frac{-b - \sqrt{b^2 - 4ac}}{2a}$$

into a single equation. We need only substitute the coefficients a, b, and c of a given quadratic equation in the formula to find the solution set for the equation.

Determination of number of real solutions   An examination of the quadratic formula,

$$x = \frac{-b \pm \sqrt{b^2 - 4ac}}{2a},$$

shows that, if $ax^2 + bx + c = 0, a,b,c \in R$, is to have a nonempty solution set in the set of real numbers, then $\sqrt{b^2 - 4ac}$ must be real. This, in turn, implies that only those quadratic equations for which $b^2 - 4ac \geq 0$ have real solutions. The number represented by $b^2 - 4ac$ is called the **discriminant** of the quadratic equation $ax^2 + bx + c = 0$. It yields the following information about the nature of the solution set of the equation for $a, b, c \in R$.

1. If $b^2 - 4ac = 0$, then there is precisely one real solution (multiplicity two).
2. If $b^2 - 4ac < 0$, then there are no real solutions.
3. If $b^2 - 4ac > 0$, then there are two real solutions.

Exercise 4.3

Solve for x by using the quadratic formula.

Example $\dfrac{x^2}{4} + \dfrac{x}{4} = 3$

Solution   Write an equivalent equation in standard form.

$$x^2 + x = 12$$
$$x^2 + x - 12 = 0$$

Substitute 1 for a, 1 for b, and -12 for c in the quadratic formula and simplify.

$$x = \frac{-1 \pm \sqrt{1 + 48}}{2} = \frac{-1 \pm 7}{2}$$

The solution set is $\{3, -4\}$.

1. $x^2 - 3x + 2 = 0$
2. $x^2 + 4x + 4 = 0$
3. $2x^2 = 7x - 6$
4. $3x^2 = 5x - 1$
5. $\dfrac{x^2}{3} = \dfrac{1}{2}x + \dfrac{3}{2}$
6. $\dfrac{x^2 - 3}{2} + \dfrac{x}{4} = 1$
7. $x^2 - 2\sqrt{5}x + 5 = 0$
8. $x^2 + 2\sqrt{2}x + 2 = 0$
9. $2x^2 - \sqrt{3}x - 3 = 0$
10. $2x^2 + \sqrt{7}x - 7 = 0$
11. $x^2 - kx - 2k^2 = 0$
12. $2x^2 - kx + 3 = 0$
13. $ax^2 - x + c = 0$
14. $x^2 + 2x + c + 3 = 0$

15. Determine k so that the roots of $kx^2 + 4x + 1 = 0$ will be equal. *Hint:* Use the discriminant.
16. Determine k so that the roots of $x^2 - kx + 9 = 0$ will be equal.
17. Show that if r_1 and r_2 are roots of the quadratic equation $ax^2 + bx + c = 0$, then $r_1 + r_2 = -b/a$ and $r_1 r_2 = c/a$.

Using the formulas in Problem 17, find the sums and products of the roots in 18–21.

18. $x^2 - 3x + 2 = 0$
19. $2x^2 + 3x - 6 = 0$
20. $x^2 + 2x - 3 = 0$
21. $3x^2 - 9x - 5 = 0$

22. Show that the equation $ax^2 + bx + c = 0$ $(a \neq 0)$ can be written equivalently as

$$x = \frac{-b \pm \sqrt{b^2 - 4ac}}{2a}.$$

4.4   *Equations Involving Radicals*

Equality of like powers

In order to find solution sets for equations containing radical expressions, we shall need the following result.

> **Theorem 4.3**   *If $U(x)$ and $V(x)$ are numerical expressions in x, then the solution set of $U(x) = V(x)$ is a subset of the solution set of $[U(x)]^n = [V(x)]^n$, for each natural number n.*

This theorem, which follows simply from the fact that products of equal numbers are equal numbers, permits us to raise both members of an equation to the same natural-number power with the assurance that we do not lose any solutions of the original equation in the process. On the other hand, it does not assert that the resulting equation will be equivalent to the original equation, and indeed it will not always be. The equation

$$[U(x)]^n = [V(x)]^n$$

may have additional solutions (called **extraneous solutions**) that are not solutions of $U(x) = V(x)$. Thus, if $a = b$, then $a^4 = b^4$, but the converse does not necessarily hold. That is, a^4 and b^4 may be equal, but $a \neq b$. For example, $(3)^4 = (-3)^4$, but $3 \neq -3$. The solution set of the equation $x^4 = 81$, obtained from $x = 3$ by raising each member to the fourth power, contains -3 as an extraneous real solution, since -3 does not satisfy the original equation even though it does satisfy $x^4 = 81$.

Necessity of checking solutions

Because the result of applying the foregoing process is not always an equivalent equation, each solution obtained through its use *must* be substituted for the variable in the original equation to check its validity. An application of this theorem is not an elementary transformation.

Example

Find the solution set of $\sqrt[3]{x - 1} = -1$.

Solution

If we raise each member of $\sqrt[3]{x - 1} = -1$ to the third power, we obtain

$$(\sqrt[3]{x - 1})^3 = (-1)^3,$$
$$x - 1 = -1,$$

which is equivalent to

$$x = 0.$$

Since $\sqrt[3]{0 - 1} = -1$, a solution of the original equation is 0. Moreover, 0 is the only real solution, since Theorem 4.3 guarantees that the solution set of the equation $\sqrt[3]{x - 1} = -1$ is a subset of the solution set of $x = 0$.

Example

Find the solution set of $\sqrt{x+2} + 4 = x$.

Solution

We first write the equivalent equation

$$\sqrt{x+2} = x - 4$$

and then apply Theorem 4.3. We obtain

$$(\sqrt{x+2})^2 = (x-4)^2,$$

or

$$x + 2 = x^2 - 8x + 16.$$

This last equation is equivalent to

$$x^2 - 9x + 14 = 0,$$

or

$$(x-2)(x-7) = 0,$$

which clearly has solutions 2 and 7. Upon replacing x with 2 in the original equation, however, we obtain

$$\sqrt{2+2} + 4 = 2,$$

or

$$6 = 2,$$

which is false. Hence, 2 is not a solution of the original equation; it is an extraneous root. On the other hand, 7 does satisfy the original equation, so the solution set we seek is $\{7\}$.

It is sometimes necessary to apply Theorem 4.3 more than once in solving certain equations.

Example

Find the solution set of $\sqrt{x+4} + \sqrt{9-x} = 5$.

Solution

It is helpful if this equation is first transformed so that each member of the equivalent equation contains only one of the radical expressions in one member. Thus, by adding $-\sqrt{9-x}$ to each member, we obtain

$$\sqrt{x+4} = 5 - \sqrt{9-x}.$$

An application of Theorem 4.3 leads to

$$(\sqrt{x+4})^2 = (5 - \sqrt{9-x})^2,$$
$$x + 4 = 25 - 10\sqrt{9-x} + 9 - x,$$
$$2x - 30 = -10\sqrt{9-x},$$
$$x - 15 = -5\sqrt{9-x}.$$

Applying Theorem 4.3 again, we obtain

$$(x - 15)^2 = (-5\sqrt{9 - x})^2,$$
$$x^2 - 30x + 225 = 25(9 - x),$$
$$x^2 - 5x = 0,$$
$$x(x - 5) = 0.$$

It is clear that this last equation has 0 and 5 as solutions. Since both of these satisfy the original equation, the solution set we seek is {0, 5}.

Exercise 4.4

Solve and check. If there is no solution, so state.

1. $\sqrt{x} = 8$ 　　　　　　2. $4\sqrt{x} = 9$ 　　　　　　3. $\sqrt{y + 8} = 1$

4. $\sqrt{y + 3} = 2$ 　　　　5. $x - 1 = \sqrt{2x + 1}$ 　　6. $2y + 3 = \sqrt{y + 2}$

7. $\sqrt[3]{2 - y} = 3$ 　　　8. $\sqrt[5]{7 - x} = 2$ 　　　9. $\sqrt{x}\sqrt{x + 9} = 20$

10. $\sqrt{x + 3}\sqrt{x - 9} = 8$ 　　　　　　11. $\sqrt{y^2 - y} = \sqrt{y^2 + 2y - 3}$

12. $x - 2 = \sqrt{2x^2 - 3x + 2}$ 　　　　　13. $\sqrt{y + 4} = \sqrt{y + 20} - 2$

14. $\sqrt{x} + \sqrt{2} = \sqrt{x + 2}$ 　　　　　15. $\sqrt{5 + \sqrt{x}} = \sqrt{x} - 1$

16. $\sqrt{13 + \sqrt{x}} = \sqrt{x} + 1$ 　　　　17. $(5 + x)^{1/2} + x^{1/2} = 5$

18. $(y + 7)^{1/2} + (y + 4)^{1/2} = 3$ 　　　19. $(y^2 - 3y + 5)^{1/2} - (y + 2)^{1/2} = 0$

20. $(z - 3)^{1/2} + (z + 5)^{1/2} = 4$

Solve. Leave the results in the form of an equation. Assume that denominators are not zero.

21. $r = \sqrt{\dfrac{A}{\pi}},$ for A 　　　　　　22. $t = \sqrt{\dfrac{2v}{g}},$ for g

23. $x\sqrt{xy} = 1,$ for y 　　　　　　24. $P = \pi\sqrt{\dfrac{l}{g}},$ for g

25. $x = \sqrt{a^2 - y^2},$ for y 　　　　26. $y = \dfrac{1}{\sqrt{1 - x}},$ for x

4.5 Substitution in Solving Equations

Some equations that are not polynomial equations can nevertheless be solved by means of related polynomial equations. For example, though $y + 2\sqrt{y} - 8 = 0$ is not a polynomial equation, if the variable p is substituted for the radical expression $\sqrt{y}$, we then have $p^2 + 2p - 8 = 0$, which is a polynomial equation in p.

Example   Find the solution set of $y + 2\sqrt{y} - 8 = 0$.

Solution   If we set $p = \sqrt{y}$ and substitute in the given equation, we have

$$p^2 + 2p - 8 = 0,$$
$$(p + 4)(p - 2) = 0,$$

which has -4 and 2 as solutions. Since $-4 < 0$, we must reject it as a source for solutions because $p = \sqrt{y}$, and $\sqrt{y}$ is always nonnegative. The other value, $p = 2$, leads to $\sqrt{y} = 2$ and hence $y = 4$. The solution set we seek is $\{4\}$.

The technique of substituting one variable for another—or, more generally, a variable for an expression—is not limited to cases involving radicals, but is useful in any situation in which an equation is polynomial in form. For example, in the equation

$$\left(x + \frac{1}{x}\right)^{-2} + 6\left(x + \frac{1}{x}\right)^{-1} + 8 = 0,$$

we would set

$$p = \left(x + \frac{1}{x}\right)^{-1}.$$

Similarly, in the equation

$$(y + 3)^{1/2} - 4(y + 3)^{1/4} + 4 = 0,$$

we would set

$$p = (y + 3)^{1/4}.$$

Exercise 4.5

Solve for x, y, or z.

Example $x^4 - 10x^2 + 9 = 0$

Solution   Set $x^2 = p$; then $x^4 = p^2$. Solve for p.

$$p^2 - 10p + 9 = 0$$
$$(p - 9)(p - 1) = 0$$
$$p = 9 \quad p = 1$$

Set each value of $p = x^2$ and solve for x.

$$x^2 = 9 \qquad\qquad x^2 = 1$$
$$x = 3 \quad x = -3 \qquad x = 1 \quad x = -1$$

The solution set is $\{3, -3, 1, -1\}$.

1. $x - 2\sqrt{x} - 15 = 0$ 2. $x^4 - 5x^2 + 4 = 0$
3. $2x^4 + 17x^2 - 9 = 0$ 4. $z^4 - 2z^2 - 24 = 0$

5. $(y^2 + 5y)^2 - 8y(y + 5) - 84 = 0$ 6. $y^2 - 5 - 5\sqrt{y^2 - 5} + 6 = 0$

7. $y^{2/3} - 2y^{1/3} - 8 = 0$ 8. $z^{2/3} - 2z^{1/3} - 35 = 0$

9. $y^{-2} - y^{-1} - 12 = 0$ 10. $z^{-2} + 9z^{-1} - 10 = 0$

11. $(x - 1)^{1/2} - 2(x - 1)^{1/4} - 15 = 0$ 12. $8x^{-6} + 7x^{-3} - 1 = 0$

13. $\left(x + \dfrac{4}{x}\right)^2 + \left(x + \dfrac{4}{x}\right) = 20$ 14. $\left(\dfrac{y + 2}{y - 1}\right)^2 - 5\left(\dfrac{y + 2}{y - 1}\right) = -6$

15. $\dfrac{9y^2}{(y + 2)^2} - \dfrac{9y}{y + 2} + 2 = 0$ 16. $\dfrac{y + 1}{2y^2} + \dfrac{36y^2}{y + 1} - 9 = 0$

17. $(y + 3)^{1/2} - 4(y + 3)^{1/4} = -4$ 18. $\left(x + \dfrac{1}{x}\right)^{-2} + 6\left(x + \dfrac{1}{x}\right)^{-1} = -8$

19. $(y^4 - 4y^3 + 4y^2) - 23(y^2 - 2y) + 120 = 0$

20. $(4x^4 - 12x^3 + 9x^2) - 3(2x^2 - 3x) + 2 = 0$

4.6   Solution of Linear Inequalities

Open sentences such as

$$x + 3 \geq 10 \tag{1}$$

and

$$\frac{-2y - 3}{3} < 5 \tag{2}$$

are called **inequalities**. For appropriate values of the variable, one member of an inequality represents a real number that is less than ($<$), less than or equal to ($\leq$), greater than or equal to ($\geq$), or greater than ($>$) the real number represented by the other member.

Any element of the replacement set of the variable for which an inequality is valid is called a **solution**, and the set of all solutions of an inequality is called the **solution set** of the inequality. Inequalities that are true for every element in the replacement set of the variable—such as $x^2 + 1 > 0$, $x \in R$—are called **absolute** or **unconditional inequalities**. Inequalities that are not true for every element of the replacement set are called **conditional inequalities**—for example, (1) and (2) above.

Elementary transformations

As in the case with equations, we shall solve a given inequality by generating a series of **equivalent inequalities** (inequalities having the same solution set) until we arrive at one for which the solution set is obvious. To do this, we shall need the following theorem applicable to inequalities. The proof of this theorem follows directly from the properties of order for the set of real numbers.

Theorem 4.4   *If $P(x)$, $Q(x)$, and $R(x)$ are expressions, then for all values of x for which $P(x)$, $Q(x)$, and $R(x)$ are real numbers, the sentence*

$$P(x) < Q(x)$$

(Theorem continued overleaf)

is equivalent to each of the following statements.

$$\text{I} \quad P(x) + R(x) < Q(x) + R(x),$$

$$\text{II} \quad P(x) - R(x) < Q(x) - R(x),†$$

$$\left.\begin{array}{l}\text{III} \quad P(x) \cdot R(x) < Q(x) \cdot R(x) \\[2ex] \text{IV} \quad \dfrac{P(x)}{R(x)} < \dfrac{Q(x)†}{R(x)}\end{array}\right\} for\ x \in \{x\,|\,R(x) > 0\},$$

$$\left.\begin{array}{l}\text{V} \quad P(x) \cdot R(x) > Q(x) \cdot R(x) \\[2ex] \text{VI} \quad \dfrac{P(x)}{R(x)} > \dfrac{Q(x)†}{R(x)}\end{array}\right\} for\ x \in \{x\,|\,R(x) < 0\}.$$

Similarly, the sentence

$$P(x) \le Q(x)$$

is equivalent to sentences of the form I–VI, with $<$ (or $>$) replaced by $\le$ (or $\ge$) under the same conditions as above.

The theorem can be interpreted as follows:

I and II. The addition or subtraction of the same expression to or from each member of an inequality produces an equivalent inequality in the same sense.

III and IV. If each member of an inequality is multiplied or divided by the same expression representing a positive number, the result is an equivalent inequality in the same sense.

V and VI. If each member of an inequality is multiplied or divided by the same expression representing a negative number, the result is an equivalent inequality in the opposite sense.

Note that Theorem 4.4 does not permit multiplying or dividing by zero, and variables in multipliers and divisors are restricted from values for which the expression vanishes. The result of applying any part of this theorem is an elementary transformation.

Solution of a linear inequality

Theorem 4.4 can be applied to solve inequalities in the same way that the theorems of equality are applied to solve equations.

Example

Find the solution set of $\dfrac{-x + 3}{4} > -\dfrac{2}{3}, \quad x \in R.$

Solution

Multiplying each member by 12 yields

$$3(-x + 3) > -8,$$
$$-3x + 9 > -8.$$

---

† Parts II, IV, and VI can be considered special cases of Parts I, III, and V, respectively.

Adding -9 to each member, we have

$$-3x > -17.$$

Finally, dividing each member by -3, we obtain

$$x < \frac{17}{3},$$

and the solution set is written

$$S = \left\{ x \mid x < \frac{17}{3} \right\}.$$

Graphical   The solution set in the foregoing example can be pictured on a line graph as shown
representation in Figure 4.1. The heavy line indicates points with coordinates in the solution set.

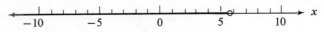

Figure 4.1

Inequalities sometimes appear in a form such as

$$-6 < 3x \le 15, \tag{3}$$

where an expression is bracketed between two inequality symbols. As observed in
Section 1.5, this means $-6 < 3x$ *and* $3x \le 15$. The solution set of such an inequality
is obtained in the same manner as the solution set of any other inequality. In (3)
above, each expression may be divided by 3 to obtain

$$-2 < x \le 5.$$

The solution set,

$$S = \{ x \mid -2 < x \le 5 \},$$

is shown on a line graph in Figure 4.2. Note here that the open dot at the left-
hand endpoint of the interval indicates that -2 *is not* a member of the solution set,
whereas the solid dot at the other end shows that 5 *is* a member of the solution set.

Figure 4.2

Exercise 4.6

Solve each inequality and represent the solution set on a line graph.

1. $x + 7 > 8$ **2.** $x - 5 \le 7$ **3.** $3x - 2 > 1 + 2x$

4. $2x + 3 \le x - 1$ **5.** $\dfrac{2x - 3}{2} \le 5$ **6.** $\dfrac{3x + 4}{3} > 12$

7. $\dfrac{x}{3} + 2 < \dfrac{x}{4}$ 8. $\dfrac{1}{2}(4 - x) > \dfrac{1}{3}x$ 9. $-6 \leq 3x < 12$

10. $3 < x + 2 \leq 8$ 11. $1 < 3x - 1 < 5$ 12. $0 < 4 - 3x < 16$

Graph each of the following sets and designate each set in simpler set notation.

Example $\{x \mid x + 2 \geq 0\} \cap \{x \mid x - 3 < 1\}$

Solution   Solve each inequality and graph. Indicate the region where the graphs overlap.

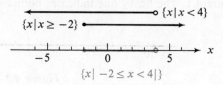

13. $\{x \mid x - 2 < 3\} \cap \{x \mid x + 4 > 2\}$ 14. $\left\{x \,\middle|\, \dfrac{1 + x}{2} \leq 3\right\} \cap \{x \mid x \leq 6\}$

15. $\{x \mid 2x - 1 > 5\} \cap \left\{x \,\middle|\, \dfrac{x - 1}{3} \geq 4\right\}$ 16. $\{x \mid 4 - x < 2\} \cap \left\{x \,\middle|\, \dfrac{2x + 5}{2} < 0\right\}$

17. $\left\{x \,\middle|\, \dfrac{3 - x}{4} < -2\right\} \cup \left\{x \,\middle|\, \dfrac{2 + 3x}{3} \leq 1\right\}$

18. $\left\{x \,\middle|\, \dfrac{3x + 2x}{3} < 0\right\} \cup \left\{x \,\middle|\, \dfrac{x + 3}{4} > 0\right\}$

19. $\{x \mid 2 < x < 5\} \cup \{x \mid x \geq 3\} \cup \{x \mid x < 0\}$

20. $\{x \mid -3 \leq x < 4\} \cup \{x \mid x = 5\} \cup \{x \mid x \geq 8\}$

Example   Determine k so that the solutions of $x^2 - 5x + k + 3 = 0$ are not real numbers.

Solution   The solutions are not real if $b^2 - 4ac < 0$. Substituting -5 for b, 1 for a, and $k + 3$ for c in $b^2 - 4ac$, we have

$$(-5)^2 - 4(1)(k + 3) < 0$$
$$25 - 4k - 12 < 0$$
$$-4k < -13$$
$$k > \dfrac{13}{4}.$$

21. Determine k so that the roots of $x^2 + 2x + k + 3 = 0$ will be real numbers.

22. Determine k so that the roots of $x^2 + 9x + k = 2$ will be real numbers.

23. Determine k so that $x^2 - 2x + 1 = k$ will have no real roots.

4.7   Solution of Quadratic Inequalities

Solving quadratic inequalities offers somewhat different problems than solving linear inequalities. For example, consider the inequality

$$x^2 + 4x < 5, \quad x \in R.$$

To determine values of x for which this condition holds, we might first rewrite the sentence equivalently as

$$x^2 + 4x - 5 < 0,$$

and then as

$$(x + 5)(x - 1) < 0.$$

It is clear here that those values and only those values of x for which the factors $x + 5$ and $x - 1$ are opposite in sign will be in the solution set. Thus the solution set can be determined analytically by finding the values of x such that

$$x + 5 < 0 \quad \text{and} \quad x - 1 > 0,$$

or else

$$x + 5 > 0 \quad \text{and} \quad x - 1 < 0.$$

Each of these two cases can be considered separately.

First, $x + 5 < 0$ and $x - 1 > 0$ imply $x < -5$ and $x > 1$, conditions which are not satisfied by any values of x. But $x + 5 > 0$ and $x - 1 < 0$ imply $x > -5$ and $x < 1$, which lead to the solution set

$$S = \{x \mid -5 < x < 1\}.$$

An alternative set notation is

$$S = \{x \mid x > -5\} \cap \{x \mid x < 1\}.$$

Use of sign graphs

One relatively easy way to visualize the solution set of a quadratic inequality is to indicate on a number line the signs associated with each factor for number replacements for the variable. Figure 4.3 shows such an arrangement, a **sign graph**, for the

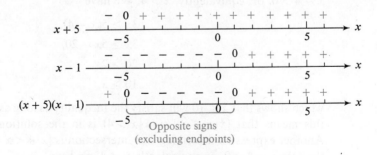

Figure 4.3

example above. This picture is constructed by first showing on the number line the places where $x + 5$ is positive ($x > -5$) and the places where it is negative ($x < -5$), and then showing on a second line those places where $x - 1$ is positive ($x > 1$) and those where it is negative ($x < 1$). The third line can then be marked by observing those parts of the first two lines where the signs are alike and those parts where the signs are opposite. It is desired that the product $(x + 5)(x - 1)$ be negative. The third line shows clearly that this occurs where $-5 < x < 1$, so that the solution set of the inequality is $\{x \mid -5 < x < 1\}$. This set can be graphed as in Figure 4.4.

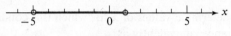

Figure 4.4

Multiplying members by a variable

Inequalities involving fractions have to be approached with care if any fraction contains a variable in the denominator. If Theorem 4.4 is invoked to multiply each member by an expression containing the variable, we have to be careful to distinguish between those values of the variable for which the expression denotes a positive and negative number, respectively. Alternatively, we can make sure that the expression by which we multiply is always positive.

Example

Find the solution set of $\dfrac{3x}{x - 4} \geq 5$.

Solution 1

First we note that $x = 4$ is not a solution. Next, applying Parts III and V of Theorem 4.4, we multiply each member of

$$\frac{3x}{x - 4} \geq 5$$

by $x - 4$, observing, as we do, that two cases arise for $x \neq 4$. First, we obtain

$$3x \geq 5(x - 4) \quad \text{for} \quad x - 4 > 0,$$

and, second,

$$3x \leq 5(x - 4) \quad \text{for} \quad x - 4 < 0.$$

We can now find the solution for each case. Let us look at the first case, in which $x - 4 > 0$, or, equivalently, $x > 4$. We have

$$3x \geq 5(x - 4),$$
$$3x \geq 5x - 20,$$
$$20 \geq 2x,$$
$$10 \geq x,$$

which shows that $x \leq 10$ will satisfy the inequality as long as $x > 4$. In terms of sets, this means that $\{x \mid x \leq 10\} \cap \{x \mid x > 4\}$ is in the solution set of the inequality. Another expression describing this intersection is $\{x \mid 4 < x \leq 10\}$. Now, examining the case $x - 4 < 0$, or equivalently $x < 4$, we have

$$3x \leq 5(x - 4),$$
$$3x \leq 5x - 20,$$
$$20 \leq 2x,$$
$$10 \leq x.$$

But x cannot be greater than or equal to 10 and at the same time less than 4. That is, $\{x\,|\,x \geq 10\} \cap \{x\,|\,x < 4\} = \emptyset$. This means that the entire solution set of the inequality is the above set, $\{x\,|\,4 < x \leq 10\}$. What we have accomplished here is to determine that the solution set is given by

$$[\{x\,|\,x \leq 10\} \cap \{x\,|\,x > 4\}] \cup [\{x\,|\,x \geq 10\} \cap \{x\,|\,x < 4\}],$$

which we express more simply as

$$\{x\,|\,4 < x \leq 10\} \cup \emptyset, \quad \text{or} \quad \{x\,|\,4 < x \leq 10\}.$$

Solution 2   Let us approach this directly by means of a sign graph. We can rewrite the given inequality equivalently as

$$\frac{3x}{x - 4} - 5 \geq 0,$$

from which we obtain

$$\frac{-2x + 20}{x - 4} \geq 0.$$

For this to be valid, $x - 4$ must not be 0, and the numerator and denominator must be of like sign. The figure shows that the quotient $(-2x + 20)/(x - 4)$ is positive or zero for x between 4 and 10, including 10 but excluding 4, a value of x for which the denominator is 0. The desired solution set is therefore $\{x\,|\,4 < x \leq 10\}$, with graph as shown on the bottom line of the sign graph.

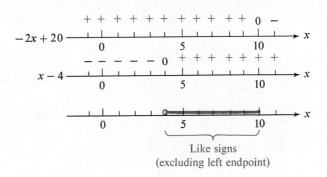

Like signs
(excluding left endpoint)

Exercise 4.7

Solve and represent each solution set on a line graph.

1. $(x + 1)(x - 2) > 0$ 2. $(x + 2)(x + 5) < 0$
3. $x(x - 2) \leq 0$ 4. $x(x + 3) \geq 0$
5. $x^2 - 3x - 4 > 0$ 6. $x^2 - 5x - 6 \geq 0$
7. $x^2 < 5$ 8. $4x^2 + 1 < 0$
9. $x^2 > -5$ 10. $x^2 + 1 > 0$

11. $\dfrac{2}{x} \leq 4$ 12. $\dfrac{3}{x - 6} > 8$

Example $\dfrac{x}{x-2} \geq 5$

Solution   We can approach this directly by means of a sign graph. We first rewrite the given inequality equivalently as

$$\frac{x}{x-2} - 5 \geq 0,$$

from which we obtain

$$\frac{-4x+10}{x-2} \geq 0.$$

For this to be valid, $x - 2$ must not be 0, and the numerator and denominator must be of like sign. The sign graph shows that the quotient $(-4x + 10)/(x - 2)$ is positive or zero for x between 2 and 5/2, including 5/2 but excluding 2, a value of x for which the denominator is 0. The desired solution set is therefore $\{x \mid 2 < x \leq 5/2\}$, with graph as shown on the last line of the sign graph.

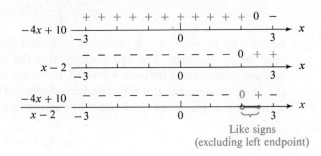

Like signs
(excluding left endpoint)

13. $\dfrac{x}{x+2} > 4$

14. $\dfrac{x+2}{x-2} \geq 6$

15. $\dfrac{2}{x-2} \geq \dfrac{4}{x}$

16. $\dfrac{3}{4x+1} > \dfrac{2}{x-5}$

17. $x(x-2)(x+3) > 0$

18. $x^3 - 4x \leq 0$

19. Show that if $a, b \in R$ and $a, b > 0$, then $(a + b)^2 > a^2 + b^2$.

20. Show that if $x, y \in R$ and $x + y = 6$, then $xy \leq 9$. *Hint:* Assume $xy > 9$, and consider $x(6 - x) > 9$.

21. Show that for $x \in R$, $x^2 + 1 \geq 2x$. *Hint:* $(x - 1)^2 \geq 0$ for all $x \in R$.

22. Show that the sum of any positive number and its reciprocal is greater than or equal to 2. *Hint:* Use Problem 21.

4.8   *Equations and Inequalities Involving Absolute Values*

Equations   In Section 1.5, we defined the absolute value of a real number x by

$$|x| = \begin{cases} x, & \text{if } x \geq 0, \\ -x, & \text{if } x < 0, \end{cases}$$

and interpreted it in terms of distance on a number line. For example, $|-5| = 5$ by definition, but 5 also denotes the distance the graph of -5 is located from the origin (Figure 4.5). More generally, then, the expression $|x - a|$ satisfies

$$|x - a| = \begin{cases} x - a, & \text{if } x - a \geq 0 \text{ or, equivalently, if } x \geq a, \\ -(x - a), & \text{if } x - a < 0 \text{ or, equivalently, if } x < a, \end{cases}$$

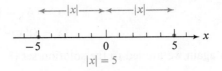

Figure 4.5

and can be interpreted on the number line as denoting the distance the graph of x is located from the graph of a (Figure 4.6).

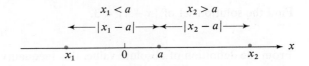

Figure 4.6

Since

$$\sqrt{x^2} = \begin{cases} x, & \text{if } x \geq 0, \\ -x, & \text{if } x < 0, \end{cases}$$

or $|x| = \sqrt{x^2}$ (see page 64), we can assert that

$$|x - a| = \sqrt{(x - a)^2}.$$

Since $(x - a)^2 = (a - x)^2$, it follows that $|x - a| = |a - x|$.

All the foregoing facts can be used effectively in finding solution sets for equations involving absolute value. Several approaches can be used.

Example   Find the solution set of $|x - 3| = 5$.

Solution 1   This can be solved by inspection. Since $x - 3$ represents the distance the graph of x is located from the graph of 3, and since by the equation this distance is 5, the two solutions of the equation are $3 + 5$, or 8, and $3 - 5$, or -2. Thus, the solution set is $\{8, -2\}$.

Solution 2   By definition, $|x - 3| = 5$ implies that

$$x - 3 = 5 \quad \text{for } x - 3 > 0,$$
$$-(x - 3) = 5 \quad \text{for } x - 3 < 0.$$

Equivalently we obtain

$$x = 8 \quad \text{and} \quad x = -2.$$

The solution set is $\{8, -2\}$.

(*Solution continued overleaf*)

Solution 3   Since $|x - 3| = \sqrt{(x - 3)^2}$, we can write the equation $|x - 3| = 5$ as

$$\sqrt{(x - 3)^2} = 5.$$

Then

$$(x - 3)^2 = 25,$$
$$x^2 - 6x + 9 = 25,$$
$$x^2 - 6x - 16 = 0,$$
$$(x - 8)(x + 2) = 0,$$

and again we are led to the solution set $\{8, -2\}$.

The choice of approach depends on the situation and is largely a matter of which is most convenient.

Inequalities   Inequalities involving absolute-value notation require some additional discussion.

Example   Find the solution set of $|x + 1| < 3$.

Solution 1   From the definition of absolute value, this inequality is equivalent to

$$x + 1 < 3 \quad \text{for} \quad x + 1 \geq 0,$$

and

$$-(x + 1) < 3 \quad \text{for} \quad x + 1 < 0,$$

so that the solution set is given by

$$S = \{x \mid x < 2 \quad \text{for} \quad x \geq -1\} \cup \{x \mid x > -4 \quad \text{for} \quad x < -1\},$$

or simply by

$$S = \{x \mid -4 < x < 2\}.$$

The graph is shown below.

Solution 2   The inequality can be written equivalently as

$$\sqrt{(x + 1)^2} < 3.$$

which, by an extension of Theorem 4.3 to inequalities involving *positive* expressions, is equivalent to

$$(x + 1)^2 < 9,$$

from which

$$x^2 + 2x + 1 < 9,$$
$$x^2 + 2x - 8 < 0,$$
$$(x + 4)(x - 2) < 0,$$

and by means of sign graphs or otherwise we have
$$S = \{x \mid -4 < x < 2\}.$$

Example   Find the solution set of $|x + 1| > 3$.

Solution   By the definition of absolute value, this inequality is equivalent to
$$x + 1 > 3 \quad \text{for} \quad x + 1 \geq 0$$

and to
$$-(x + 1) > 3 \quad \text{for} \quad x + 1 < 0,$$

so that the solution set is given by
$$S = \{x \mid x > 2 \text{ or } x < -4\}. \tag{1}$$

Alternatively, we could also write
$$S = \{x \mid x > 2\} \cup \{x \mid x < -4\},$$

where the union gives a precise meaning to the word " or " used in (1). The graph of the solution set is shown below.

Exercise 4.8

Solve.

Example $|x + 5| = 8$

Solution   Determine the solution set by inspection or write as two first-degree equations and solve each equation. Alternatively, use the method of Solution 3 shown on page 98.

$$
\begin{array}{ll}
x + 5 = 8 & -(x + 5) = 8 \\
x = 3 & -x - 5 = 8 \\
& x = -13
\end{array}
$$

The solution set is $\{3, -13\}$.

1. $|x| = 6$ 2. $|x| = 3$ 3. $|x - 1| = 4$

4. $|x - 6| = 3$ 5. $\left| x - \dfrac{2}{3} \right| = \dfrac{1}{3}$ 6. $\left| x - \dfrac{3}{4} \right| = \dfrac{1}{2}$

7. $|2x + 5| = 2$ 8. $|3x + 7| = 1$ 9. $\left| 1 - \dfrac{1}{2}x \right| = \dfrac{3}{4}$

10. $\left| 1 + \dfrac{3}{2}x \right| = \dfrac{1}{2}$ 11. $\left| \dfrac{3x + 2}{4} \right| = 6$ 12. $\left| \dfrac{2x - 1}{3} \right| = 3$

Solve and represent each solution set on a line graph.

Example $|2x - 1| \leq 7$

Solution   Rewrite without absolute-value symbol.

$$2x - 1 \leq 7 \quad \text{for } 2x - 1 \geq 0 \tag{1}$$

or

$$-(2x - 1) \leq 7 \quad \text{for } 2x - 1 < 0 \tag{2}$$

From (1), $x \leq 4$ for $x \geq 1/2$.   From (2), $x \geq -3$ for $x < 1/2$.

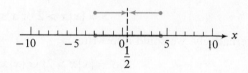

The solution set is $\{x \mid -3 \leq x \leq 4\}$.

13. $|x| < 2$	**14.** $|x - 1| > 2$	**15.** $|x + 3| \leq 4$
16. $2|x + 1| \leq 8$	**17.** $|2x - 5| < -3$	**18.** $|2x + 4| < -1$
19. $|x - 3| > 4$	**20.** $|2x + 4| > 2$	**21.** $|2x + 1| \geq 5$
22. $|3x - 5| \geq 4$	**23.** $\left|\dfrac{2x + 1}{3}\right| < 4$	**24.** $\left|\dfrac{3x - 2}{4}\right| \leq 5$

Replace each of the following sentences with a single inequality by using absolute-value notation.

Example $-3 < x < 7$

Solution $7 + (-3) = 4$. Therefore values of x are centered about $4/2$, or 2. Adding -2 to each member, we have

$$-5 < x - 2 < 5,$$
$$|x - 2| < 5.$$

25. $1 < x < 3$	**26.** $-5 \leq x \leq 9$	**27.** $-9 \leq x \leq -7$
28. $5 < x < 13$	**29.** $-7 \leq 2x \leq 12$	**30.** $-5 < 3x < 10$

In Problems 31–34, consider $a, b \in R$.

31. Show that $|-a| = |a|$. *Hint*: Consider two possible cases, a nonnegative and a negative.

32. Show that $|a - b| = |b - a|$. *Hint*: Consider two possible cases, $a - b \geq 0$ and $a - b < 0$.

33. Show that $|a^2| = |a|^2 = a^2$.

34. Show that $|ab| = |a| \cdot |b|$. *Hint*: Consider four possible cases.

4.9   *Word Problems*

Equations and inequalities can be used to express quantitative relations in word problems symbolically. The problem may be explicitly concerned with numbers, or it may be concerned with numerical measures of physical quantities. In either event, we seek the set of numbers (the solution set) for which the stated relationship holds. The following suggestions are frequently helpful in expressing the conditions of the problem symbolically:

1. Determine the quantities asked for and represent them by symbols. Since at this time we are using one variable only, all relevant quantities should be represented in terms of this variable.
2. Where applicable, draw a sketch and label all known quantities thereon; label the unknown quantities in terms of symbols.
3. Find in the problem a quantity that can be represented in two different ways and write this representation as an equation. The equation may derive from:
   a. the problem itself, which may state a relationship explicitly; for example, "What number added to 4 gives 7?" produces the equation $x + 4 = 7$;
   b. formulas or relationships that are part of your general mathematical background; for example, $A = \pi r^2$, $d = rt$.
4. Solve the resulting equation.
5. Check the results against the original problem. It is not sufficient to check the result in the equation, because the equation itself may be in error.

In some cases, the mathematical model we obtain for a physical situation is a quadratic equation that has two real solutions. It may be that one but not both of the solutions fits the physical situation. For example, if we were asked to find two consecutive *natural numbers* of which the product is 72, we would write the equation

$$x(x + 1) = 72$$

as our model. Solving this equation, we have

$$x^2 + x - 72 = 0,$$
$$(x + 9)(x - 8) = 0,$$

with solution set $\{8, -9\}$. Since -9 is not a natural number, we must reject it as a possible answer to our original question; the solution 8, however, leads to the consecutive natural numbers 8 and 9. As additional examples, observe that we would not accept -6 feet as the height of a man, or 27/4 for the number of persons in a room.

A quadratic equation used as a model for a physical situation may have two, one, or no meaningful solutions—meaningful, that is, in a physical sense. Answers to word problems should always be checked against the universal set of meaningful numbers for the original problem.

Exercise 4.9

Solve the following word problems.

In Exercises 1–12, use a mathematical model in the form of a first-degree equation in one variable.

Example   A collection of coins consisting of dimes and quarters has a value of $11.60. How many dimes and quarters are in the collection if there are 32 more dimes than quarters?

Solution   (a) Represent the unknown quantities symbolically.

Let x represent the number of quarters;
then $x + 32$ represents the number of dimes.

Write an equation relating the value of the quarters and value of the dimes to the value of the entire collection.

$$\begin{bmatrix} \text{value of} \\ \text{quarters} \\ \text{in cents} \end{bmatrix} + \begin{bmatrix} \text{value of} \\ \text{dimes} \\ \text{in cents} \end{bmatrix} = \begin{bmatrix} \text{value of} \\ \text{collection} \\ \text{in cents} \end{bmatrix}$$

$$25x \quad\quad + 10(x + 32) = \quad\quad 1160$$

(b) Solve for x.

$$25x + 10x + 320 = 1160$$
$$35x = 840$$
$$x = 24; \quad \text{therefore } x + 32 = 56$$

There are 24 quarters and 56 dimes in the collection.

Check: Do 24 quarters and 56 dimes have a value of $11.60? Yes.

1. A man has $1.15 in change consisting of two more nickels than dimes. How many dimes and how many nickels does he have?

2. The admission at a baseball game was $2.00 for adults and $1.25 for children. The receipts were $103.75 for 68 paid admissions. How many adults, and how many children, attended the game?

3. How many pounds of an alloy containing 32% silver must be melted with 25 pounds of an alloy containing 48% silver to obtain an alloy containing 42% silver?

4. How much water should be added to 9 gallons of pure acid to obtain an 18% solution?

Example   A man has an annual income of $6000 from two investments. He has $10,000 more invested at 4% than he has invested at 3%. How much does he have invested at each rate?

Solution   (a) Represent the amount invested at each rate symbolically.

Let A represent the amount in dollars invested at 3%;
then $A + 10,000$ represents the amount invested at 4%.

Write an equation relating the interest from each investment and the total interest.

$$\begin{bmatrix} \text{interest from} \\ 3\% \text{ investment} \end{bmatrix} + \begin{bmatrix} \text{interest from} \\ 4\% \text{ investment} \end{bmatrix} = [\text{total interest}]$$

$$0.03A \quad\quad + 0.04(A + 10,000) = \quad\quad 6000$$

(b) Solve for A.

$$3A + 4A + 40,000 = 600,000$$
$$7A = 560,000$$
$$A = 80,000; \text{ therefore } A + 10,000 = 90,000$$

The man had \$80,000 invested at 3% and \$90,000 invested at 4%.

Check: Does 3% of 80,000 (2400) added to 4% of 90,000 (3600) equal 6000? Yes.

5. A sum of \$2000 is invested, part at 3% and the remainder at 4%. Find the amount invested at each rate if the yearly income from the two investments is \$66.

6. A sum of \$2700 is invested, part at 4% and the remainder at 5%. Find the total yearly interest if the interest on each investment is the same.

7. A man has three times as much money invested in 3% bonds as he has in stocks paying 5%. How much does he have invested in each if his yearly income from the investments is \$1680?

8. A man has \$1000 more invested at 5% than he has invested at 4%. If his annual income from the two investments is \$698, how much does he have invested at each rate?

Example   An express train travels 150 miles in the same time that a freight train travels 100 miles. If the express goes 20 miles per hour faster than the freight, find each rate.

Solution   (a) Represent the unknown quantities symbolically.

Let r represent a rate for the freight train;

then $r + 20$ represents the rate of the express train.

The fact that the times are equal is the significant equality in the problem.

$$[t \text{ of freight}] = [t \text{ of express}]$$

Express the time of each train in terms of r (time = distance/rate).

$$\frac{100}{r} = \frac{150}{r + 20}$$

(b) Solve for r.

$$(r + 20)100 = (r)150$$
$$100r + 2000 = 150r$$
$$-50r = -2000$$
$$r = 40; \text{ therefore } r + 20 = 60$$

Freight train's rate is 40 mph; express train's rate is 60 mph.

Check: Does the time of the freight train (100/40) equal the time of the express train (150/60)? Yes.

9. An airplane travels 1260 miles in the same time that an automobile travels 420 miles. If the rate of the airplane is 120 miles per hour greater than the rate of the automobile, find the rate of each.

10. Two cars start together and travel in the same direction, one going twice as fast as the other. At the end of 3 hours they are 96 miles apart. How fast is each traveling?

11. A freight train leaves town *A* for town *B*, traveling at an average rate of 40 miles per hour. Three hours later a passenger train also leaves town *A* for town *B*, on a parallel track, traveling at an average rate of 80 miles per hour. How far from town *A* does the passenger train pass the freight train?

12. A boy walked to his friend's house at the rate of 4 miles per hour and he ran back home at the rate of 6 miles per hour. How far apart are the two houses if the round trip took 20 minutes?

In Exercises 13–23, use a mathematical model in the form of a second-degree equation in one variable.

13. Find two numbers whose sum is 15 and whose product is 56.

14. Find two numbers whose sum is − 12 and whose product is 32.

15. Find two consecutive integers whose product is 42.

16. Find two consecutive natural numbers such that the sum of their squares is 85.

17. Two airplanes with lines of flight at right angles to each other pass each other (at slightly different altitudes) at noon. One is flying at 140 miles per hour and one at 180 miles per hour. How far apart are they at 12:30 PM?

18. A box without a top is to be made from a square piece of tin by cutting a two-inch square from each corner and folding up the sides. If the box is to hold 128 cubic inches, what should be the length of each side of the original square?

19. A ball thrown vertically upward reaches a height *h* in feet given by the equation $h = 56t - 16t^2$, where *t* is the time in seconds after the throw. How long will it take the ball to reach a height of 24 feet on its way up? How long after the throw will the ball return to the height from which it was thrown?

20. The distance *s* a body falls in a vacuum is given by $s = v_0 t + \frac{1}{2}gt^2$, where *s* is measured in feet, *t* is measured in seconds, v_0 is the initial velocity in feet per second, and *g* is the constant of acceleration due to gravity (approximately 32 ft/sec/sec). How long will it take a body to fall 150 feet if v_0 is 20 feet per second? If the body starts from rest?

21. A man and his son working together can paint their house in four days. The man can do the job alone in six days less than the son can do it. How long would it take each of them to paint the house alone? *Hint*: What part of the job could each of them do in one day?

22. A theatre that is rectangular in shape seats 720 people. The number of rows needed to seat the people would be four fewer if each row held six more seats. How many seats would then be in each row?

23. Two tanks, each cylindrical in shape and 10 feet in length, are to be replaced by a single tank of the same length. If the original tanks have radii that measure six feet and eight feet, respectively, what must be the length of the radius of the single tank replacing them if it is to hold the same volume of liquid?

In Exercises 24 *and* 25, *use a mathematical model in the form of an inequality in one variable.*

Example

A student must have an average of 80 % to 90 % inclusive on five tests in a course to receive a *B*. His grades on the first four tests were 98 %, 76 %, 86 %, and 92 %. What grade on the fifth test would qualify him for a *B* in the course?

Solution

(a) Represent the unknown quantity symbolically.

Let *x* represent a grade (in percent) on the last test.

Write an inequality expressing the word sentence.

$$80 \leq \frac{98 + 76 + 86 + 92 + x}{5} \leq 90$$

(b) Solve for *x*.

$$400 \leq 352 + x \leq 450$$
$$48 \leq x \leq 98$$

Any grade equal to or greater than 48 and less than or equal to 98.

24. In the preceding example, what grade on the fifth test would qualify the student for a *B* if his grades on the first four tests were 78 %, 64 %, 88 %, and 76 %?

25. The Fahrenheit and centigrade temperatures are related by $C = \frac{5}{9}(F - 32)$. Within what range must the temperature be in Fahrenheit degrees for the temperature in centigrade degrees to lie between $-10°$ and $20°$?

Chapter Review

[4.1]   *Solve each equation.*

1. $2 + \dfrac{x}{3} = \dfrac{5}{6}$

2. $\dfrac{2x}{3} - \dfrac{2x + 5}{6} = \dfrac{1}{2}$

3. $\dfrac{x}{x + 1} + \dfrac{4}{5} = 6$

4. $1 - \dfrac{y - 2}{y - 3} = \dfrac{3}{y - 1}$

5. Solve $\dfrac{x + y}{5} = \dfrac{x - y}{3}$ for *y* in terms of *x*.

6. Solve $\dfrac{x + y}{5} = \dfrac{x - y}{3}$ for *x* in terms of *y*.

[4.2]   *Solve by factoring.*

7. $x(2x - 3) = -1$

8. $2x(x - 2) = x + 3$

Solve by extraction of roots.

9. $3x^2 = 21$

10. $(x + 4)^2 = 3$

Solve by completing the square.

11. $x^2 + 3x - 1 = 0$ **12.** $2x^2 + x - 1 = 0$

13. Write $x^2 + y^2 - 8x + 6y = 12$ in the form $(x - h)^2 + (y - k)^2 = r^2$.

14. Write $y = x^2 - bx + 2$ in the form $y = (x - a)^2 + b$.

[4.3]   *Solve for x by using the quadratic formula.*

15. $2x^2 - x - 2 = 0$ **16.** $x^2 + kx - 4 = 0$

17. Determine k so that $x^2 - 6x + k = 0$ will have exactly one real solution.

18. Determine k so that $x^2 + kx + 4 = 0$ will have exactly one real solution.

[4.4]   *Solve each equation.*

19. $x - 3\sqrt{x} + 2 = 0$ **20.** $\sqrt{x + 1} + \sqrt{x + 8} = 7$

[4.5]   *Solve each equation.*

21. $y^4 - 5y^2 + 4 = 0$ **22.** $y^{-2} - y^{-1} - 20 = 0$

[4.6]   *Solve each inequality.*

23. $\dfrac{x - 3}{4} \le 6$ **24.** $2(x - 1) > \dfrac{2}{3} x$

Graph each of the following set intersections.

25. $\{x \,|\, x + 3 < 11\} \cap \{x \,|\, 2x - 4 > 0\}$

26. $\{y \,|\, \dfrac{y + 4}{3} \le 6 + y\} \cap \{y \,|\, \dfrac{2y - 1}{3} < 1\}$

[4.7]   *Solve each inequality.*

27. $x^2 + 4x - 5 < 0$ **28.** $\dfrac{2}{x - 3} < 4$

[4.8]   *Solve each equation.*

29. $|2x + 1| = 7$ **30.** $\left| x - \dfrac{1}{3} \right| = \dfrac{2}{3}$

31. Solve $|x - 2| > 5$, and graph the solution set on a number line.

32. Write $-3 \le x \le 11$ using a single inequality involving an absolute-value symbol.

[4.9]   **33.** In a recent election, the winning candidate received 150 votes more than his opponent. How many votes did each candidate receive if there were 4376 votes cast?

34. When the length of each side of a square is increased by five inches, the area is increased by 85 square inches. Find the length of a side of the original square.

35. Clerk A can process 50 applications in four hours, and clerk B can process 50 applications in eight hours. How long will it take both clerks working together to process 100 applications?

36. A man sailed a boat across a lake and back in two and a half hours. If his rate returning was two miles per hour less than his rate going, and if the distance each way was six miles, find his rate each way.

5

Relations
and Functions I

5.1 Ordered Pairs of Real Numbers

When the order in which the numbers of a number pair are to be considered is specified, the pair is called an **ordered pair**, and the pair is denoted by a symbol such as $(3, 2)$, $(2, 3)$, $(-1, 5)$, or $(0, 3)$. Each of the two numbers in an ordered pair is called a **component** of the ordered pair, the first and second being called the **first component** and the **second component**, respectively. Having established what is meant by an ordered pair, we are ready to define a set operation involving such pairs.

Definition 5.1 *The **Cartesian product** of two sets A and B, denoted by $A \times B$, is the set of all ordered pairs (x, y) such that $x \in A$ and $y \in B$.*

$R \times R$, or R^2, and the geometric plane

The most important Cartesian product with which we shall be concerned is that formed from the set R of real numbers. The product $R \times R$, which is often denoted by R^2, is the set of all possible ordered pairs of real numbers. The fact that each member of R^2 corresponds to a point in the geometric plane, and the coordinates of each point in the geometric plane are the components of a member of R^2, is the basis for all plane graphing. As you probably recall from your earlier study of algebra, the correspondence between points in the plane and ordered pairs of real numbers is usually established through a **Cartesian** (or **rectangular**) **coordinate system**, as shown in Figure 5.1.

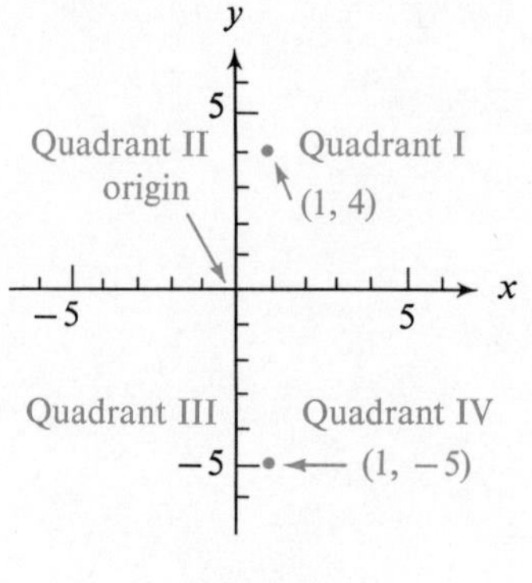

Figure 5.1

Solutions of an equation in two variables

Equations in two variables, such as

$$3x + 2y = 12, \quad x^2 y + 3x = y^5, \quad \text{and} \quad \sqrt{xy} = y^2 - 5,$$

with $x, y \in R$, have ordered pairs of numbers as solutions. For example, if the

108

components of $(2, 3)$ are substituted for the variables x and y, in that order, in the equation

$$3x + 2y = 12, \tag{1}$$

the result is

$$3(2) + 2(3) = 12,$$

which is true. Hence, $(2, 3)$ is a solution of Equation (1). In this book, the first component of an ordered pair is a value for the **abscissa** x, and the second component a value for the **ordinate** y. Since many equations (and inequalities) in two variables have an infinite number of solutions, we shall use the set-builder notation

$$\{(x, y)\,|\,\text{condition on } x \text{ and } y\}$$

to represent the set of all solutions.

Subsets of
$R \times R$

Any condition on x and y—that is, any sentence in two variables, x and y—expresses a relationship between elements in the replacement sets of the two variables, and this relationship is precisely represented by the solution set of the sentence in two variables. For x, $y \in R$, the solution set is always a subset of $R \times R$. This leads us to the following definition.

Definition 5.2   *Any subset of $R \times R$ is a **relation** in R.*

The relation is said to be in R because the components of the ordered pairs in the relation are elements of R. Alternatively, we frequently refer to the relationship as being in $R \times R$.

The set of all first components in the ordered pairs in a relation is called the **domain** of the relation, and the set of all second components is called the **range** of the relation. Thus

$$\{(2, 5,) \ (3, \ 10), \ (4, \ 15)\} \quad \substack{\text{elements in the domain} \\[6pt] \text{elements in the range}}$$

is a relation with domain $\{2, 3, 4\}$ and range $\{5, 10, 15\}$.

If a relation is defined by an equation and the domain is not specified, we shall understand that the domain is *the set of all real numbers for which a real number exists in the range* (such relations are called real-valued relations of a real variable). For example, the domain of

$$S = \left\{(x, y)\,\middle|\, y = \frac{1}{x - 2}\right\}$$

is $\{x\,|\,x \in R, \ x \neq 2\}$, because for every real number x except 2, the expression $1/(x - 2)$ represents a real number.

In a relation, each element in the domain is said to be *paired with* or *mapped onto* an element in the range. A relation, as in the above example, may be a one-to-one **mapping** (each element in the domain is mapped onto one element in the range), as illustrated in Figure 5.2. A relation may be a mapping in which one or

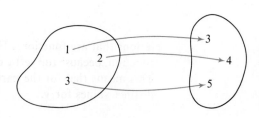

Figure 5.2

more elements in the domain are mapped onto more than one element in the range, as shown in Figure 5.3-a, or in which several elements in the domain are mapped onto the same element in the range, as shown in Figure 5.3-b.

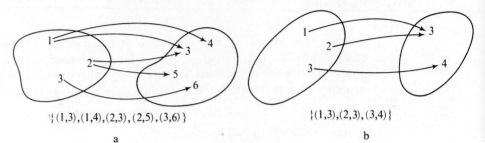

$\{(1,3),(1,4),(2,3),(2,5),(3,6)\}$ $\{(1,3),(2,3),(3,4)\}$

a   b

Figure 5.3

A special kind of relation that is very important in mathematics is called a *function*.

Definition 5.3   *A function is a relation in which no two ordered pairs have the same first component and different second components.*

Graphical characterization of a function

A function, therefore, associates each element in its domain with one and only one element in its range. In a graphical sense, this implies that no two of the ordered pairs in a function graph into points on the same vertical line.

As you should recall from your earlier study of algebra, graphs in R^2, that is, in $R \times R$, are often continuous lines and curves. Figure 5.4 shows three such graphs. Imagine a vertical line moving across each of these from left to right. If the line at any position cuts the graph of the relation in more than one point, then the relation is not a function. Thus, although Figures 5.4-a and 5.4-c show the graphs of re-

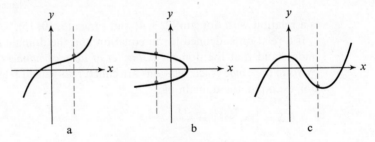

a   b   c

Figure 5.4

lations that are functions, Figure 5.4-b shows the graph of a relation that is not a function, because the vertical line shown in the figure meets the graph in two places. This means that for the particular value of x involved, the relation associates two distinct values for y.

Algebraic characterization of a function

When a relation is defined by an equation, one way to test whether or not the relation is a function is to solve the equation explicitly for the variable y representing an element in the range. This will show whether or not more than one value of y is associated with any single value of x.

Example   Is the relation $\{(x, y)\,|\,y^2 = 1 + x^2\}$ in R^2 a function?

Solution   Since $y^2 = 1 + x^2$ implies either $y = \sqrt{1 + x^2}$ or $y = -\sqrt{1 + x^2}$, the assignment of a real value to x will result in two different values for y, and hence the relation is not a function.

Function   In Section 2.1 you became acquainted with a notation that is widely used in dis-
notation   cussing functions. In general, functions are denoted by single symbols; for example, f, g, h, and F might designate functions. The symbol for a function can be used in conjunction with the variable representing an element in the domain to represent the associated element in the range. Thus $f(x)$, read "f of x" or "the value of f at x," is the element in the range of f associated with the element x in the domain.
         Suppose

$$f = \{(x, y)\,|\,y = x + 3\}.$$

The alternative notation

$$f = \{(x, f(x))\,|\,f(x) = x + 3\}$$

can be used, where $f(x)$ plays the same role as y.

Exercise 5.1

Supply the missing components so that the ordered pairs

a. $(0, \quad)$ b. $(1, \quad)$ c. $(2, \quad)$ d. $(-3, \quad)$ e. $\left(\dfrac{2}{3}, \quad \right)$

are solutions of the given equations.

1. $2x + y = 6$ **2.** $y = 9 - x^2$

3. $y = \dfrac{3x}{x^2 - 2}$ **4.** $y = 0$

5. $y = \sqrt{3x + 11}$ **6.** $y = |x - 1|$

a.   *Specify the domain and the range of each relation.*
b.   *State whether or not each relation is a function.*

Example $\{(3, 5), (4, 8), (4, 9), (5, 10)\}$

Solution   a.   Domain (the set of first components): $\{3, 4, 5\}$
              Range (the set of second components): $\{5, 8, 9, 10\}$
         b.   The relation is not a function, because two ordered pairs, $(4, 8)$ and $(4, 9)$, have the same first components.

7. $\{(2, 3), (5, 7), (7, 8)\}$ **8.** $\{(-1, 6), (0, 2), (3, 3)\}$
9. $\{(2, -1), (3, 4), (3, 6)\}$ **10.** $\{(-4, 7), (-4, 8), (3, 2)\}$
11. $\{(5, 5), (6, 6), (7, 7)\}$ **12.** $\{(0, 0), (2, 4), (4, 2)\}$

Specify the maximum domain that would yield real numbers y for elements in the range of the relation defined by each equation.

Examples   a. $y = \sqrt{16 - x^2}$ b. $y = \dfrac{1}{x(x + 2)}$

Solutions   a.   For what values of x is   b.   For what values of x is
                   $16 - x^2 \geq 0$? $x(x + 2) \neq 0$?
                   The domain is $\{x \mid -4 \leq x \leq 4\}$.   The domain is $\{x \mid x \neq 0, -2\}$.

13. $y = x + 7$ 14. $y = 2x - 3$ 15. $y = x^2$

16. $y = \dfrac{1}{x}$ 17. $y = \dfrac{1}{x - 2}$ 18. $y = \dfrac{1}{x^2 + 1}$

19. $y = \sqrt{x}$ 20. $y = \sqrt{4 - x}$ 21. $y = \sqrt{4 - x^2}$

22. $y = \sqrt{x^2 - 9}$ 23. $y = \dfrac{4}{x(x - 1)}$ 24. $y = \dfrac{x}{(x - 1)(x + 2)}$

State whether or not the given equation defines a function.

Examples   a. $x^2 y = 3$ b. $x^2 + y^2 = 36$

Solutions   Solve explicitly for y.

              a. $y = \dfrac{3}{x^2}$ b. $y = \pm\sqrt{36 - x^2}$

                   Yes. There is only one value   No. There are two values of y
                   of y associated with each   associated with values of x
                   value of x $(x \neq 0)$.   satisfying $|x| < 6$.

25. $x + y = 3$ 26. $y = -x^2$ 27. $y = \sqrt{x^2 - 5}$
28. $y = \sqrt{16 - x^2}$ 29. $x^2 + y^2 = 16$ 30. $y = \pm\sqrt{x^2}$
31. $y^2 = x^3$ 32. $y = ax^n$

If $f(x) = x + 2$, find the given element in the range.

Example $f(3)$

Solution   Substitute 3 for x.

$$f(3) = 3 + 2 = 5$$

             The element is 5.

33. $f(0)$ 34. $f(1)$ 35. $f(-3)$ 36. $f(a)$

If $g(x) = x^2 - 2x + 1$, find the given element in the range.

37. $g(-2)$ 38. $g(0)$ 39. $g(a + 1)$ 40. $g(a - 1)$

If $f(x) = x + 2$ defines a function, find the element in the domain of f associated with the given element in the range.

Example   5

Solution   Replacing $f(x)$ with 5, we have

$$5 = x + 2, \quad \text{or} \quad x = 3.$$

The element is 3.

41. 3   **42.** -2 **43.** a **44.** $a + 2$

If $g(x) = x^2 - 1$, find all elements in the domain of g associated with the given element in the range.

45. $g(x) = 0$ **46.** $g(x) = 3$ **47.** $g(x) = 8$ **48.** $g(x) = 5$

49. Suppose $f(x) = x + 2$ and $g(x) = x - 2$. Find each of the following.
    a. $f(0)$ **b.** $g(2)$ **c.** $f(g(2))$ **d.** $f(g(x))$

50. If $f(x) = x^2 - x + 1$, find each of the following.

    a. $f(x + h) - f(x)$ **b.** $\dfrac{f(x + h) - f(x)}{h}$

Any function satisfying the condition that $f(-x) = f(x)$ for all x in the domain is called an **even function**. *Any function satisfying the condition that $f(-x) = -f(x)$ for all x in the domain is an* **odd function**. *Which of the following functions are even and which are odd?*

51. **a.** $\{(x, f(x)) \mid f(x) = x^2\}$ **b.** $\{(x, f(x)) \mid f(x) = x^3\}$
52. **a.** $\{(x, f(x)) \mid f(x) = x^4 - x^2\}$ **b.** $\{(x, f(x)) \mid f(x) = x^3 - x\}$

5.2   Linear Functions

A **first-degree equation**, or **linear equation**, in x and y is an equation that can be written equivalently in the form

$$Ax + By + C = 0 \quad (A \text{ and } B \text{ not both } 0). \tag{1}$$

Graphs of first-degree equations

We shall call (1) the **standard form** for a linear equation. The graph of any such equation (technically, of its solution set) in R^2 is a straight line, although we do not prove this here.

    Since two distinct points determine a straight line, it is evident that we need find only two solutions of such an equation to determine its graph—that is, the graph of the solution set of the equation. In practice, the two solutions easiest to find are usually those with first and second components, respectively, equal to zero—that is, the solutions $(0, y_1)$ and $(x_1, 0)$.

The numbers x_1 and y_1 are called the **x-** and **y-intercepts** of the graph. As an example, consider the function

$$f = \{(x, y) \mid 3x + 4y = 12\}. \tag{2}$$

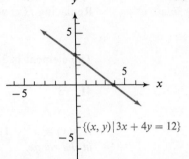

If $y = 0$, we have $x = 4$, and the x-intercept is 4. If $x = 0$, then $y = 3$, and the y-intercept is 3. Thus the graph of (2) appears as in Figure 5.5.

If the graph intersects both axes at or near the origin, then the intercepts either do not represent two separate points, or the points are too close together to be of much use in drawing the graph. It is then necessary to plot at least one other point at a distance far enough removed from the origin to establish the line with pictorial accuracy.

Figure 5.5

Equations of horizontal lines

Two special cases of linear equations are worth noting. First, an equation such as

$$y = 4$$

may be considered an equation in two variables,

$$0x + y = 4.$$

For each x, this equation assigns $y = 4$. That is, any ordered pair of the form $(x, 4)$ is a solution of the equation. For instance,

$$(1, 4), \quad (2, 4), \quad (3, 4), \quad \text{and} \quad (4, 4),$$

are all solutions of the equation. If we graph these points and connect them with a straight line, we have the graph shown in Figure 5.6-a.

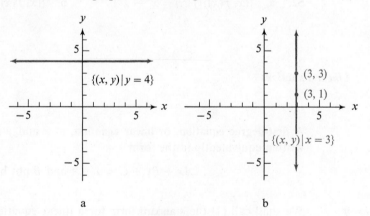

a   b

Figure 5.6

Since the equation

$$y = 4$$

assigns to each x the same value for y, the function defined by this equation is called a **constant function**.

Equations of vertical lines

The other special case of the linear equation is of the type

$$x = 3,$$

which may be looked upon as

$$x + 0y = 3.$$

Here, only one value is permissible for x, namely 3, whereas any value may be assigned to y. That is, any ordered pair of the form $(3, y)$ is a solution of this equation. If we choose two solutions, say $(3, 1)$ and $(3, 3)$, we can draw the graph shown in Figure 5.6-b. It is clear that this equation does *not* define a function with x as its first component.

If, however, $B \neq 0$ in the first-degree equation, $Ax + By + C = 0$, then this equation defines a function. Such a function is called a **linear function**. In particular, every constant function is a linear function.

Any two distinct points in a plane can be looked upon as the endpoints of a line segment. Two fundamental properties of a line segment are its **length** and its **inclination** with respect to the x-axis.

Let P_1, with coordinates (x_1, y_1), and P_2, with coordinates (x_2, y_2), be endpoints of a line segment. If we construct through P_2 a line parallel to the y-axis, and through P_1 a line parallel to the x-axis, the lines will meet at a point P_3, as shown in Figure 5.7. The x-coordinate of P_3 is evidently the same as the x-coordinate of P_2, and the

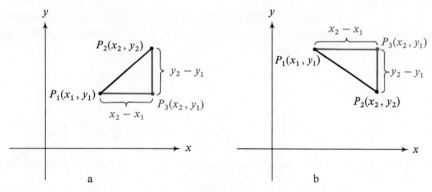

Figure 5.7

y-coordinate of P_3 is the same as that of P_1; hence the coordinates of P_3 are (x_2, y_1). By inspection, we observe that the distance from P_2 to P_3 is simply the absolute value of the difference in the y-coordinates of the two points, $|y_2 - y_1|$, and the distance between P_1 and P_3 is the absolute value of the difference of the x-coordinates of these points. $|x_2 - x_1|$.

In general, since $y_2 - y_1$ is positive or negative as $y_2 > y_1$ or $y_2 < y_1$, respectively, and $x_2 - x_1$ is positive or negative as $x_2 > x_1$ or $x_2 < x_1$, respectively, it is also convenient to designate the distances represented by $x_2 - x_1$ and $y_2 - y_1$ as positive or negative. Such distances are called **directed distances**.

Distance between two points

The Pythagorean theorem can be used to find the length, d, of the line segment from P_1 to P_2. This theorem asserts that the square on the hypotenuse of any right triangle is equal to the sum of the squares on the legs. Thus, we have

$$d^2 = (x_2 - x_1)^2 + (y_2 - y_1)^2,$$

and by considering only the positive (or nonnegative) square root of the right-hand member, we obtain

$$d = \sqrt{(x_2 - x_1)^2 + (y_2 - y_1)^2}. \tag{3}$$

Since the distances $x_2 - x_1$ and $y_2 - y_1$ are squared, it makes no difference here whether they are positive or negative—the result is the same. Equation (3) is a formula for the **distance** between any two points in the plane in terms of the coordinates of the points. The distance is always taken as positive—or 0 if the points coincide. If the points P_1 and P_2 lie on the same horizontal line, we have observed that the directed distance between them is

$$x_2 - x_1,$$

and if they lie on the same vertical line, then the directed distance is

$$y_2 - y_1.$$

If we are concerned only with distance and not direction, then these become

$$d = |x_2 - x_1| \quad \text{and} \quad d = |y_2 - y_1|.$$

The second useful property of the line segment joining two points, its inclination, can be measured by comparing the *rise* of the segment with a given *run*, as shown in Figure 5.8.

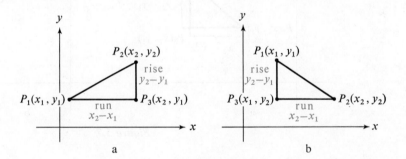

Figure 5.8

Slope of a line segment

The ratio of *rise* to *run* is called the **slope** of the line segment and is designated by the letter m. Since the rise is simply $y_2 - y_1$ and the run is $x_2 - x_1$, the slope of the line segment joining P_1 and P_2 is given by

$$m = \frac{y_2 - y_1}{x_2 - x_1}.$$

If P_2 is to the right of P_1, $x_2 - x_1$ will necessarily be positive, and the slope will be positive or negative as $y_2 - y_1$ is positive or negative. Thus positive slope indicates that a line rises to the right; negative slope indicates that it falls to the right. Since

$$\frac{y_2 - y_1}{x_2 - x_1} = \frac{-(y_1 - y_2)}{-(x_1 - x_2)} = \frac{y_1 - y_2}{x_1 - x_2},$$

the restriction that P_2 be to the right of P_1 is not necessary, and the order in which the points are considered is immaterial in determining slope.

If a line segment is parallel to the x-axis, then $y_2 - y_1 = 0$, and the line has slope 0; but if it is parallel to the y-axis, then $x_2 - x_1 = 0$, and its slope is not defined. These two special cases are shown in Figure 5.9.

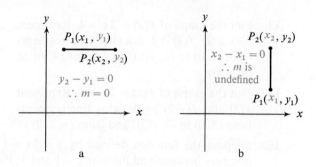

Figure 5.9

In the next section, we shall see how the slope concept is applied in discussing linear functions and their graphs.

Exercise 5.2

Graph.

Example

$3x + 4y = 24$

Solution

Determine the intercepts.

If $x = 0$, then $y = 6$;
if $y = 0$, then $x = 8$.

Sketch the line through $(0, 6)$ and $(8, 0)$, as shown.

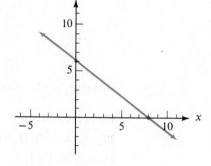

1. $y = 3x + 1$ 2. $y = x - 5$
3. $y = -2x$ 4. $2x + y = 3$
5. $3x - y = -2$ 6. $3x = 2y$ 7. $2x + 3y = 6$ 8. $3x - 2y = 8$
9. $2x + 5y = 10$ 10. $y = 5$ 11. $x = -2$ 12. $x = -3$

Example   Graph $f(x) = x - 1$. Represent $f(5)$ and $f(3)$ by drawing line segments from $(5, 0)$ to $(5, f(5))$ and from $(3, 0)$ to $(3, f(3))$.

(Solution overleaf)

Solution $f(5) = 5 - 1 = 4$

The ordinate at $x = 5$ is 4.

$f(3) = 3 - 1 = 2$

The ordinate at $x = 3$ is 2.

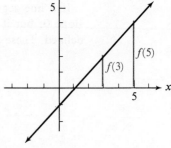

13. Plot the graph of $f(x) = 2x + 4$. Represent $f(0)$ and $f(4)$ by drawing line segments from $(0, 0)$ to $(0, f(0))$ and from $(4, 0)$ to $(4, f(4))$.

14. Plot the graph of $f(x) = 2x + 1$. Represent $f(3)$ and $f(-2)$ by drawing line segments from $(3, 0)$ to $(3, f(3))$ and from $(-2, 0)$ to $(-2, f(-2))$.

15. Suppose the function defined by $y = 2x + 1$ has as domain the set of real numbers between and including -1 and 1. Plot the graph of the function on a rectangular coordinate system. What is the range of the function?

16. Graph $x + y = 6$ and $5x - y = 0$ on the same set of axes. Estimate the co-ordinates of the point of intersection. What can you say about the coordinates of this point in relation to the two linear equations?

Find the distance between each of the given pairs of points, and find the slope of the line segment joining them.

Example $(3, -5), (2, 4)$

Solution   Consider $(3, -5)$ as P_1 and $(2, 4)$ as P_2.

$$d = \sqrt{(x_2 - x_1)^2 + (y_2 - y_1)^2} \qquad m = \frac{y_2 - y_1}{x_2 - x_1}$$

$$= \sqrt{[2 - 3]^2 + [4 - (-5)]^2} \qquad = \frac{4 - (-5)}{2 - 3}$$

$$= \sqrt{1 + 81} \qquad = \frac{9}{-1}$$

Distance, $\sqrt{82}$; slope -9

17. $(1, 1), (4, 5)$	18. $(-1, 1), (5, 9)$	19. $(-3, 2), (2, 14)$
20. $(-4, -3), (1, 9)$	21. $(2, 1), (1, 0)$	22. $(-3, 2), (0, 0)$
23. $(5, 4), (-1, 1)$	24. $(2, -3), (-2, -1)$	25. $(3, 5), (-2, 5)$
26. $(2, 0), (-2, 0)$	27. $(0, 5), (0, -5)$	28. $(-2, -5), (-2, 3)$

Find the lengths of the sides of the triangle having vertices as given.

29. $(10, 1), (3, 1), (5, 9)$ 30. $(0, 6), (9, -6), (-3, 0)$

31. $(5, 6), (11, -2), (-10, -2)$ 32. $(-1, 5), (8, -7), (4, 1)$

33. Show that the triangle described in Problem 30 is a right triangle. *Hint:* Use the converse of the Pythagorean theorem.

34. The two line segments with endpoints at $(0, -7)$, $(8, -5)$ and $(5, 7)$, $(8, -5)$ are perpendicular. Find the slope of each line segment. Compare the slopes. Do the same for the perpendicular line segments with endpoints at $(8, 0)$, $(6, 6)$ and $(-3, 3)$, $(6, 6)$. Can you make a conjecture about the slopes of perpendicular line segments?

35. The graph of a linear function contains the points $(2, -3)$ and $(6, -1)$. Find an equation that defines the function.

36. Determine algebraically whether or not the points lie on the same line.
     a. $(2, 7)$, $(-2, -5)$, $(0, 1)$ **b.** $(9, 5)$, $(-3, -1)$, $(0, 1)$

37. Show by similar triangles that the coordinates of the midpoint of the line segment joining the points $P_1(x_1, y_1)$ and $P_2(x_2, y_2)$ are given by

$$x = \frac{x_1 + x_2}{2} \quad \text{and} \quad y = \frac{y_1 + y_2}{2}.$$

38. Using the results of Problem 37, find the coordinates of the midpoint of the line segment joining:
     a. $(2, 4)$ and $(6, 8)$ **b.** $(-4, 6)$ and $(6, -10)$

5.3   Forms for Linear Equations

Point-slope form

Assuming that the slope of the line segment joining any two points on a line does not depend on the points (as can be shown by considering similar triangles), consider a line in the plane with given slope m that passes through a given point (x_1, y_1), as shown in Figure 5.10. If we choose any other point on the line and assign to it the coordinates (x, y), it is evident that the slope of the line is given by

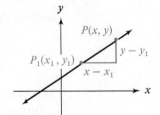

$$\frac{y - y_1}{x - x_1} = m,$$

from which

Figure 5.10

$$y - y_1 = m(x - x_1). \qquad (1)$$

Note that (1) is satisfied also by $(x, y) = (x_1, y_1)$. Since now x and y are the coordinates of *any* point on the line, (1) is an equation of the line passing through (x_1, y_1) with slope m. This is called the **point-slope form** for a linear equation.

Slope-intercept form

Now consider the equation of the line with slope m passing through a given point on the y-axis having coordinates $(0, b)$ as shown in Figure 5.11. Substituting the components of $(0, b)$ in the point-slope form of a linear equation,

$$y - y_1 = m(x - x_1),$$

we obtain

$$y - b = m(x - 0),$$

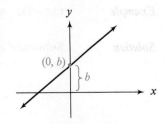

Figure 5.11

from which

$$y = mx + b. \qquad (2)$$

Equation (2) is called the **slope-intercept form** for a linear equation. Any linear equation in standard form can be written equivalently in the slope-intercept form by solving for y in terms of x if $B \neq 0$. For example,

$$2x + 3y - 6 = 0$$

can be written equivalently as

$$y = -\frac{2}{3}x + 2.$$

The slope of the line, $-2/3$, and the y-intercept, 2, can now be read directly from the last form of the equation.

Intercept form

If the x- and y-intercepts of the graph of

$$y = mx + b \qquad (3)$$

are a and b $(a, b \neq 0)$, respectively, as shown in Figure 5.12, then the slope m is clearly equal to $-b/a$. Replacing m in (3) with $-b/a$, we have

$$y = -\frac{b}{a}x + b,$$

$$ay = -bx + ab,$$

$$bx + ay = ab,$$

and multiplying each member by $\dfrac{1}{ab}$ produces

$$\frac{x}{a} + \frac{y}{b} = 1.$$

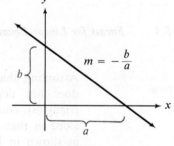

Figure 5.12

This latter form is called the **intercept form** for a linear equation.

Any of the three forms discussed in this section may be used in working with linear functions and their graphs.

Exercise 5.3

Find the equation, in standard form, of the line passing through each of the given points and having the given slope.

Example $(3, -5), \quad m = -2$

Solution   Substitute given values in the point-slope form of the linear equation.

$$y - y_1 = m(x - x_1)$$

$$y - (-5) = -2(x - 3)$$

$$y + 5 = -2x + 6$$

$$2x + y - 1 = 0$$

1. $(2, 1)$, $m = 4$ 2. $(-2, 3)$, $m = 5$ 3. $(5, 5)$, $m = -1$

4. $(-3, -2)$, $m = \dfrac{1}{2}$ 5. $(0, 0)$, $m = 3$ 6. $(-1, 0)$, $m = 1$

7. $(0, -1)$, $m = -\dfrac{1}{2}$ 8. $(2, -1)$, $m = \dfrac{3}{4}$ 9. $(-2, -2)$, $m = -\dfrac{3}{4}$

10. $(2, -3)$, $m = 0$ 11. $(-4, 2)$, $m = 0$ 12. $(-1, -2)$,   parallel to y-axis

Write each equation in slope-intercept form; specify the slope of the line and the y-intercept.

Example $2x - 3y = 5$

Solution   Solve explicitly for y.

$$-3y = 5 - 2x$$
$$3y = 2x - 5$$
$$y = \frac{2}{3}x - \frac{5}{3}$$

Compare with the general slope-intercept form $y = mx + b$.
Slope, $2/3$; y-intercept, $-5/3$

13. $x + y = 3$ 14. $2x + y = -1$ 15. $3x + 2y = 1$

16. $3x - y = 7$ 17. $x - 3y = 2$ 18. $2x - 3y = 0$

Find the equation, in standard form, of the line with the given intercepts.

Example $x = 3$; $y = -1/2$

Solution   Substitute 3 and $-1/2$ for a and b, respectively, in the intercept form $x/a + y/b = 1$.

$$\frac{x}{3} + \frac{y}{-1/2} = 1$$
$$x - 6y - 3 = 0$$

19. $x = 2$; $y = 3$ 20. $x = 4$; $y = -1$ 21. $x = -2$; $y = -5$

22. $x = -1$; $y = 7$ 23. $x = -\dfrac{1}{2}$; $y = \dfrac{3}{2}$ 24. $x = \dfrac{2}{3}$; $y = -\dfrac{3}{4}$

Write the equation, in standard form, of the line passing through the given point and parallel to the graph of the given equation. Hint: Parallel lines have the same slope.

25. $(2, 1)$; $y = 3x + 4$ 26. $(-3, 2)$; $y = -4x + 9$

27. $(-2, 5)$; $2x - 3y = 6$ 28. $(-5, -1)$; $5x - 4y = 7$

29. $(2, 3)$; $x = 6$ 30. $(5, -4)$; $y = 2$

31. Show that, for $x_2 \neq x_1$,

$$y - y_1 = \left(\frac{y_2 - y_1}{x_2 - x_1}\right)(x - x_1)$$

is an equation of the line joining the points (x_1, y_1) and (x_2, y_2). This is the **two-point form** of the linear equation.

32. Use the form given in Problem 31 to find the equation of the line through the given points.

    a. $(2, 1)$ and $(-1, 3)$ **b.** $(3, 0)$ and $(5, 0)$

    c. $(-2, 1)$ and $(3, -2)$ **d.** $(-1, -1)$ and $(1, 1)$

33. Consider the linear function

$$F = \{(x, y) \mid y = F(x)\}.$$

Assuming that $(2, 3)$ and $(-1, 4)$ are known to be in F, find $F(x)$ in terms of x.

5.4   *Inverse Relations and Functions*

Inverse relations

If the two components of each ordered pair in a relation r are interchanged, then the resulting relation is called the **inverse relation** of r and is denoted by r^{-1}. For example, each of the relations

$$\{(1,2), (3,4), (5,5)\}$$

and

$$\{(2,1), (4,3), (5,5)\}$$

is the inverse of the other,

Inverse functions

By Definition 5.3, a function is a set of ordered pairs (x, y) such that no two have the same first components and different second components. When the components of every ordered pair in a function f are interchanged, the resulting relation may or may not be a function. For example, if

$$(1, 5), \quad (2, 5), \quad \text{and} \quad (3, 6)$$

are ordered pairs in f, then

$$(5, 1), \quad (5, 2), \quad \text{and} \quad (6, 3)$$

are members of the relation formed by interchanging the components of these ordered pairs. Clearly these latter pairs cannot be members of a function since two of them, $(5, 1)$ and $(5, 2)$, have the same first components and different second components. If, however, a function f is *one-to-one*, that is, if no two different ordered pairs in f have the same second component (same value for y), then the relation obtained by interchanging the first and second components of every pair in the function will also be a function. This function is called the **inverse function** of f. The notation f^{-1} (read "f inverse") is frequently used to denote the inverse function of f.

Definition 5.4 *If the function f is such that no two of its ordered pairs with different first components have the same second component, then the **inverse function** f^{-1} is the set of ordered pairs obtained from f by interchanging the first and second components of each ordered pair in f.*

It is evident from this definition that the domain and range of f^{-1} are just the range and domain, respectively, of f. If $y = f(x)$ defines a function f, and if f is one-to-one, then $x = f(y)$ defines the inverse of f. For example, the inverse of the function defined by

$$y = 3x + 2 \tag{1}$$

is defined by

$$x = 3y + 2, \tag{2}$$

or, when y is expressed in terms of x, by

$$y = \frac{1}{3}x - \frac{2}{3}. \tag{2'}$$

Equations (2) and (2') are equivalent. In general, the equation $x = Q(y)$ is equivalent to $y = Q^{-1}(x)$.

The graphs of inverse relations are related in an interesting way. To see this, we first, in Figure 5.13, observe that the graphs of the ordered pairs (a, b) and (b, a) are always located symmetrically with respect to the graph of $y = x$. Therefore, because for every ordered pair (a, b) in Q the ordered pair (b, a) is in Q^{-1}, the graphs of $y = Q^{-1}(x)$ and $y = Q(x)$ are reflections of each other about the graph of $y = x$.

Figure 5.14 shows the graphs of the linear function

$$f = \{(x, y) \mid y = 4x - 3\} \tag{3}$$

and its inverse

$$f^{-1} = \{(x, y) \mid x = 4y - 3\} = \left\{(x, y) \mid y = \frac{1}{4}(x + 3)\right\}$$

together with the graph of $y = x$.

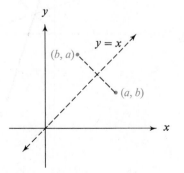

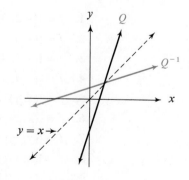

Figure 5.13   **Figure 5.14**

Since an element in the domain of the inverse Q^{-1} of a function Q is the range element in the corresponding ordered pair of Q, and vice versa, it follows that for every x in the domain of Q,

$$Q^{-1}[Q(x)] = x$$

(read " Q inverse of Q of x is equal to x"), and, for every x in the domain of Q^{-1},

$$Q[Q^{-1}(x)] = x$$

(read " Q of Q inverse of x is equal to x"). Using (3) above as an example, we note that if f is the linear function defined by

$$f(x) = 4x - 3,$$

then f is a one-to-one function. Now f^{-1} is defined by

$$f^{-1}(x) = \frac{1}{4}(x + 3),$$

and we see that

$$f^{-1}[f(x)] = \frac{1}{4}[(4x - 3) + 3] = x$$

and

$$f[f^{-1}(x)] = 4\left[\frac{1}{4}(x + 3)\right] - 3 = x.$$

Exercise 5.4

Graph each given function f and its inverse function f^{-1}, using the same set of axes.

Example $f = \{(x, y) \mid x + 3y = 6\}$

Solution   Interchange the variables
in the defining equation.

$f^{-1} = \{(x, y) \mid y + 3x = 6\}$

Graph f and f^{-1}.

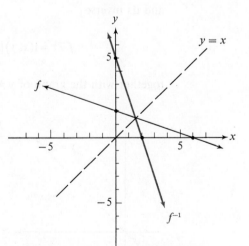

1. $f = \{(-2, 3), (4, 7), (5, 9)\}$
2. $f = \{(-3, 1), (2, -1), (3, 4)\}$
3. $f = \{(-1, -1), (2, 2), (3, 3)\}$
4. $f = \{(-4, 4), (0, 0), (4, -4)\}$
5. $f = \{(x, y) \mid y = 2x + 6\}$
6. $f = \{(x, y) \mid y = 3x - 6\}$
7. $f = \{(x, y) \mid y = 4 - 2x\}$
8. $f = \{(x, y) \mid y = 6 + 3x\}$
9. $f = \{(x, y) \mid 3x - 4y = 12\}$ 10. $f = \{(x, y) \mid x - 6y = 6\}$
11. $f = \{(x, y) \mid 4x + y = 4\}$ 12. $f = \{(x, y) \mid 2x - 3y = 12\}$

In Problems 13–18, *each equation defines a one-to-one function F in* $R \times R$. *Find an equation defining* F^{-1} *and show that* $F[F^{-1}(x)] = F^{-1}[F(x)] = x$. *Hint: Solve explicitly for y and let* $y = F(x)$.

13. $y = x$ **14.** $y = -x$ **15.** $2x + y = 4$

16. $x - 2y = 4$ **17.** $3x - 4y = 12$ **18.** $3x + 4y = 12$

5.5   *Special Functions*

Absolute-value functions

Functions involving the absolute value of one or both of the variables not only are useful in more advanced courses in mathematics; they also offer interesting properties in their own right. We recall that

$$|x| = \begin{cases} x, & \text{if } x \geq 0, \\ -x, & \text{if } x < 0. \end{cases}$$

Now consider the function defined by

$$y = |x|. \tag{1}$$

From the definition of $|x|$, for nonnegative x we have

$$y = x, \tag{2}$$

and for negative x,

$$y = -x. \tag{3}$$

If we graph (2) and (3) on the same set of axes, we have the graph of $y = |x|$ shown in Figure 5.15. In the case of any equation involving $|x|$ or $|f(x)|$, we

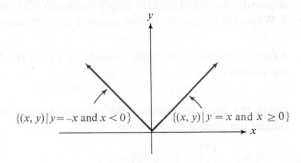

Figure 5.15

can always plot individual points to deduce the graph. For instance, if

$$y = |x| + 1, \tag{4}$$

we can find solutions by assigning values to x and computing values for y. Some solutions of (4) are

$$(-2, 3), (-1, 2), (0, 1), (1, 2), \text{ and } (2, 3),$$

which can be graphed as in Figure 5.16-a. The graph of $y = |x| + 1$, where $x \in R$, appears in Figure 5.16-b. As an alternative approach, the definition of $|x|$ implies that

$$y = |x| + 1$$

is equivalent to

$$y = x + 1 \quad \text{for } x \geq 0,$$
$$y = -x + 1 \quad \text{for } x < 0,$$

and these can be graphed separately over the specified domains.

Related graphs

It is usually advisable, where possible, to avoid graphing large numbers of points. For instance, comparing Equations (1) and (4), that is,

$$y = |x| \tag{1}$$

and

$$y = |x| + 1, \tag{4}$$

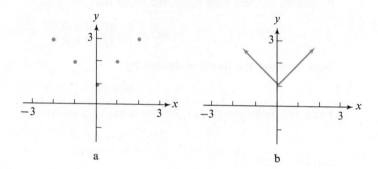

Figure 5.16

we observe that for each x the ordinate in (4) is one unit greater than that in (1); consequently, the graph of (4) is simply the graph of (1) with each ordinate increased by 1. Whenever we can, we should use such considerations to help us graph equations.

Bracket function

Another interesting function (sometimes called the **bracket function**) is defined by the equation

$$f(x) = [x], \tag{5}$$

where the brackets denote "the greatest integer contained in," or "the greatest integer not greater than." Thus

$$[2] = 2, \quad \left[\frac{7}{4}\right] = 1, \quad [-2] = -2,$$

$$\left[\frac{-3}{2}\right] = -2, \quad \text{and} \quad \left[-\frac{5}{2}\right] = -3.$$

To graph (5), we consider unit intervals along the x-axis. If $0 \leq x < 1$, $[x]$ is 0, since the greatest integer contained in any number between 0 and 1 is 0. Similarly, if $1 \leq x < 2$, $[x]$ is 1; for $-2 \leq x < -1$, $[x]$ is -2; etc. The graph of (5), therefore, is as shown in Figure 5.17. The heavy

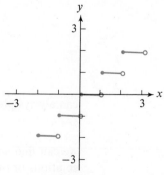

Figure 5.17

dot on the left-hand endpoint of each line segment indicates that the endpoint is a part of the graph. The function defined by (5) is sometimes called a "step function," for an obvious reason.

Exercises 5.5

Graph the function defined by the given equation over the domain $\{x \mid -5 \le x \le 5\}$.

1. $y = |x| + 2$ 2. $y = -|x| + 3$ 3. $f(x) = |x + 1|$

4. $F(x) = |x - 2|$ 5. $y = -|2x - 1|$ 6. $y = |3x + 2|$

7. $g(x) = |2x| - 3$ 8. $y = |3x| + 2$ 9. $y = |3x| - |x|$

10. $f(x) = |2x| + |x|$ 11. $y = 3|x| - x$ 12. $y = -2|x| + x$

13. $H(x) = |x^2|$ 14. $y = |x|^2$ 15. $y = |x + 1| - |x|$

16. $g(x) = |x + 1| - x$ 17. $y = [x]$ 18. $f(x) = [x] - 1$

19. $y = [x + 1]$ 20. $y = [2x]$ 21. $F(x) = [x] + x$

22. $y = \left[\dfrac{1}{2}x\right] + x$ 23. $y = [x] - x$ 24. $y = |[x]|$

25. The postage on a letter sent by first-class mail is c cents per ounce or fraction thereof. Write an equation relating the cost (C) of mailing a letter and the weight of the letter in ounces (x).

5.6   *Graphs of First-Degree Relations*

Associated equation of an inequality

An open sentence of the form

$$Ax + By + C \le 0 \quad \text{or} \quad Ax + By + C < 0,$$

A and B not both 0, is an inequality of the first degree that defines the relation

$$\{(x, y) \mid Ax + By + C \le 0\},$$

or

$$\{(x, y) \mid Ax + By + C < 0\}.$$

Such relations in R^2 can be graphed on the plane, but the graph will be a region of the plane rather than a straight line. For example, consider the relation

$$S = \{(x, y) \mid 2x + y - 3 < 0\}. \tag{1}$$

When the defining inequality is rewritten in the equivalent form

$$y < -2x + 3, \tag{2}$$

we see that solutions (x, y) are such that for each x, y is less than $-2x + 3$. The graph of the equation

$$y = -2x + 3 \tag{3}$$

is simply a straight line, as illustrated in Figure 5.18-a. To graph the relation S,

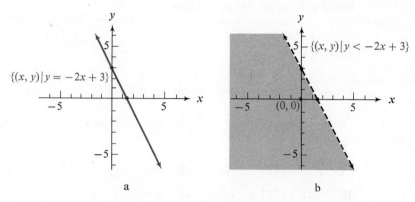

Figure 5.18

we need only observe that any point below this line has a y-coordinate that satisfies (2), and consequently the solution set of (2), which is S, corresponds to the entire region below the line. The region is indicated on the graph with shading. That the line itself is not in the graph is shown by means of a broken line, as in Figure 5.18-b. Had the inequality been

$$2x + y - 3 \leq 0,$$

the line would be a part of the graph and would be shown as a solid line. In general, the graphs of the members of

$$\{(x, y) \mid Ax + By + C < 0\} \quad \text{or} \quad \{(x, y) \mid Ax + By + C > 0\}$$

are the points in a half-plane on one side of the graph of the associated equation

$$Ax + By + C = 0,$$

depending on the constants and inequality symbols involved.

Determination of graph of an inequality

To determine which half-plane to shade in constructing graphs of first-degree inequalities, one can select any point in either half-plane and test its coordinates in the defining sentence to see whether or not the selected point lies in the graph. If the coordinates satisfy the sentence, then the half-plane containing the selected point is shaded; if not, the opposite half-plane is shaded. A very convenient point to use in this process is the origin, provided the origin is not contained in the graph of the associated equation. Thus, in the foregoing example, the replacement of x and y by 0 in

$$2x + y - 3 \leq 0$$

results in

$$2(0) + (0) - 3 \leq 0,$$

which is true, and hence the half-plane containing the origin is shaded.

Inequalities do not ordinarily define functions, according to our definition in Section 5.1, because it usually is not true that each element of the domain is associated with a unique element in the range.

Exercise 5.6

Graph each relation.

Example $\{(x, y)\,|\,2x + y \geq 4\}$

Solution   Solve the defining inequality explicitly for y.

$$y \geq 4 - 2x$$

Graph the equality $y = 4 - 2x$.

Observe that the origin is not part of the graph of the inequality, because $(0, 0)$ is not a member of the given relation; that is, $0 \not\geq 4 - 2(0)$. Hence the region above the graph of the equation $y = 4 - 2x$ is shaded.

The line is included in the graph.

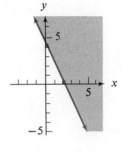

1. $\{(x, y)\,|\,y < x\}$
2. $\{(x, y)\,|\,y > x\}$
3. $\{(x, y)\,|\,y \leq x + 2\}$
4. $\{(x, y)\,|\,y \geq x - 2\}$
5. $\{(x, y)\,|\,x + y < 5\}$
6. $\{(x, y)\,|\,2x + y < 2\}$
7. $\{(x, y)\,|\,x - y < 3\}$
8. $\{(x, y)\,|\,x - 2y < 5\}$
9. $\{(x, y)\,|\,x \leq 2y - 4\}$
10. $\{(x, y)\,|\,2x \leq y + 1\}$
11. $\{(x, y)\,|\,3 \geq 2x - 2y\}$
12. $\{(x, y)\,|\,0 \geq x + y\}$

Example $\{(x, y)\,|\,x > 2\}$

Solution   Graph $\{(x, y)\,|\,x = 2\}$.

Shade region to the right of the graph of $x = 2$, that is, the set of all points such that $x > 2$.

The line is excluded from the graph.

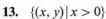

13. $\{(x, y)\,|\,x > 0\}$
14. $\{(x, y)\,|\,y < 0\}$
15. $\{(x, y)\,|\,x < 0\}$
16. $\{(x, y)\,|\,x < -2\}$
17. $\{(x, y)\,|\,-1 < x < 5\}$
18. $\{(x, y)\,|\,0 \leq y \leq 1\}$
19. $\{(x, y)\,|\,|x| < 3\}$
20. $\{(x, y)\,|\,|y| > 1\}$
21. $\{(x, y)\,|\,|x| + |y| \leq 1\}$
22. $\{(x, y)\,|\,|x| + |y| \geq 1\}$

Hint: Consider the graphs in each quadrant separately:

$$x, y \geq 0; \quad x \leq 0, y \geq 0; \quad x, y \leq 0; \quad \text{and} \quad x \geq 0, y \leq 0.$$

Chapter Review

[5.1]   *Specify the domain and the range of each relation.*

1. $\{(4, -1), (2, -4), (3, -5)\}$
2. $\{(1, 2), (1, 3), (1, 6)\}$
3. $\left\{(x, y)\,\Big|\,y = \dfrac{1}{x + 4}\right\}$
4. $\{(x, y)\,|\,y = \sqrt{x - 6}\}$

Let $f(x) = x - 3$ and $g(x) = x^2 + 4$. Find each of the following.

5. $g(3)$ **6.** $f(-2)$ **7.** $f(g(0))$ **8.** $g(x + h)$

[5.2] *Find the distance between each of the given pairs of points, and find the slope of the line segment joining them.*

9. $(2, 0)$ and $(-3, 4)$ **10.** $(-6, 1)$ and $(-8, 2)$

[5.3] *Find the equation, in standard form, of the line passing through each of the given points and having the given slope.*

11. $(2, -7)$, $m = 4$ **12.** $(-6, 3)$, $m = \dfrac{1}{2}$

Write each equation in slope-intercept form; specify the slope and the y-intercept of the line.

13. $4x + y = 6$ **14.** $3x - 2y = 16$

Find the equation, in standard form, of the line with the given intercepts.

15. $x = -3; y = 2$ **16.** $x = \dfrac{1}{3}; y = -4$

Find an equation of the line which passes through the given point and is parallel to the graph of the given equation.

17. $4x - 3y = 12$; $(-2, 1)$ **18.** $x + 6y = 0$; $(3, -4)$

[5.4] *Graph each function f and its inverse f^{-1}, using the same set of axes.*

19. $f = \{(-3, 1), (-1, 3), (2, 4)\}$ **20.** $f = \{(x, y) \mid 2x - y = 6\}$

[5.5] *Graph the function defined by the given equation over the domain $\{x \mid -5 \le x \le 5\}$.*

21. $f(x) = |x| + 4$ **22.** $g(x) = |x - 4|$
23. $g(x) = |x| + 1$ **24.** $h(x) = |x - 1|$

[5.6] *Graph the given relation.*

25. $\{(x, y) \mid 2x - y < 6\}$ **26.** $\{(x, y) \mid 2y + x > 0\}$
27. $\{(x, y) \mid -2 \le y < 3\}$ **28.** $\{(x, y) \mid 3 < x \le 5\}$

6

Relations and Functions II

6.1 *Quadratic Functions*

Graph of a quadratic function

Consider the quadratic equation in two variables,

$$y = x^2 - 4. \tag{1}$$

As with linear equations in two variables, solutions of this equation must be ordered pairs (x, y). We need replacements for both x and y in order to obtain a statement we may adjudge to be true or false. As before, such ordered pairs can be found by arbitrarily assigning values to x and computing related values for y. For instance, assigning the value -3 to x in Equation (1), we obtain

$$y = (-3)^2 - 4,$$
$$y = 5,$$

and $(-3, 5)$ is a solution. Similarly, we find that

$$(-2, 0), \quad (-1, -3), \quad (0, -4), \quad (1, -3), \quad (2, 0), \quad \text{and} \quad (3, 5)$$

are also solutions of (1). Locating the corresponding points on the plane, we have the graph in Figure 6.1-a. Clearly, these points do not lie on a straight line, and we

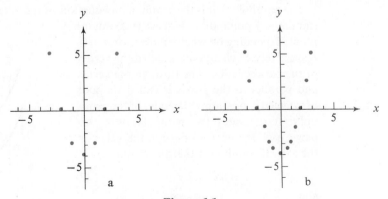

Figure 6.1

might reasonably inquire whether the graph of the solution set of (1),

$$S = \{(x, y) \mid y = x^2 - 4\},$$

forms any kind of a meaningful pattern on the plane. By graphing additional solutions of (1)—solutions with x-components between those already found—we may be able to obtain a clearer picture. Accordingly, we find the solutions

$$\left(\frac{-5}{2}, \frac{9}{4}\right), \quad \left(\frac{-3}{2}, \frac{-7}{4}\right), \quad \left(\frac{-1}{2}, \frac{-15}{4}\right), \quad \left(\frac{1}{2}, \frac{-15}{4}\right), \quad \left(\frac{3}{2}, \frac{-7}{4}\right), \quad \left(\frac{5}{2}, \frac{9}{4}\right),$$

and by graphing these points in addition to those found earlier, we have the graph in Figure 6.1-b. It now appears reasonable to connect these points in sequence, say from left to right, by a smooth curve as in Figure 6.2, and to assume that the resulting curve is a good approximation to the graph of (1). (We should realize, of course, that regardless of how many individual points are plotted, we have no absolute assurance that the smooth curve is a good approximation to the true graph; more information is needed—for example, in this case, that for $|x_2| > |x_1|$, correspondingly $y_2 > y_1$.) This curve is an example of a **parabola**.

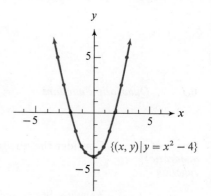

Figure 6.2

More generally, the graph of the solution set of any quadratic equation of the form

$$y = ax^2 + bx + c, \tag{2}$$

where a, b, and c are real and $a \neq 0$, is a parabola. Since for each x an equation of the form (2) will determine only one y, such an equation defines a function having as domain the entire set of real numbers and as range some subset of the reals. For example, we observe from the graph in Figure 6.2 that the range of the function defined by (1) is the set of real numbers

$$\{y \mid y \geq -4\}.$$

The parabola that is the graph of an equation of the form (2) will have a lowest (minimum) point or a highest (maximum) point, depending on whether $a > 0$ or $a < 0$, respectively. Such a point is called the **vertex** of the parabola. The line through the vertex and parallel to the y-axis is called the **axis of symmetry**, or simply the **axis**, of the parabola; it separates the parabola into two parts, each the mirror image of the other in the axis. If we observe that the graphs of

$$y = ax^2 + bx + c \tag{3}$$

and

$$y = ax^2 + bx \tag{4}$$

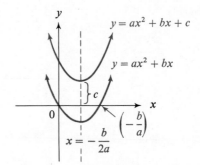

Figure 6.3

have the same axis (Figure 6.3), we can find an equation for the axis of (3) by inspecting Equation (4). Factoring the right-hand member of $y = ax^2 + bx$ yields

$$y = x(ax + b),$$

and thus we can see that 0 and $-b/a$ are the x-intercepts of the graph. Since the axis of symmetry bisects the segment with these endpoints, an equation for the axis of symmetry is

$$x = -\frac{b}{2a}.$$

Example   Find an equation for the axis of symmetry of the graph of

$$\{(x, y) \mid y = 2x^2 - 5x + 7\}.$$

Solution   By comparing the given equation to $y = ax^2 + bx + c$, we see that $a = 2$ and $b = -5$. Hence, we have

$$x = -\frac{b}{2a} = -\frac{-5}{2(2)} = \frac{5}{4},$$

and $x = 5/4$ is the desired equation for the axis.

Since the vertex of a parabola lies on its axis, obtaining an equation for this axis will give us the x-coordinate of the vertex. The y-coordinate can then easily be obtained by substitution in the equation for the parabola.

When graphing a quadratic equation in two variables, it is desirable first to select components for the ordered pairs that ensure that the more significant parts of the graph are displayed. For a parabola, these parts include the intercepts, if they exist, and the maximum or minimum point on the curve.

Example   Graph $y = x^2 - 3x - 4$.

Solution   By inspection, the y-intercept is -4. Setting $y = 0$, we have

$$0 = x^2 - 3x - 4 = (x - 4)(x + 1),$$

and the x-intercepts are 4 and -1. The x-coordinate of the minimum point is

$$x = -\frac{b}{2a} = -\frac{-3}{2(1)} = \frac{3}{2}.$$

By substituting $3/2$ for x in $y = x^2 - 3x - 4$, we obtain

$$y = \left(\frac{3}{2}\right)^2 - 3\left(\frac{3}{2}\right) - 4 = \frac{9}{4} - \frac{9}{2} - 4 = -\frac{25}{4}.$$

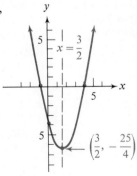

Graphing the intercepts and the coordinates of the minimum point and then sketching the curve produce the graph shown.

Solutions,
zeros,
and x-
intercepts

Consider the graph of the function

$$S = \{(x, f(x)) \mid f(x) = ax^2 + bx + c\}, \tag{5}$$

for $a \neq 0$, and the solution set of the equation

$$ax^2 + bx + c = 0. \tag{6}$$

Any value of x for which $f(x) = 0$ in (5) will be a solution of (6). Since any point on the x-axis has y-coordinate zero [that is, $f(x) = 0$], the x-intercepts of the graph of (5) are the real solutions of (6). Values of x for which $f(x) = 0$ are called the **zeros of the function**. Thus we have three different names for a single idea:

1. The *elements in R of the solution set* of the equation $ax^2 + bx + c = 0$. These are called the *solutions* or *roots* of the equation.
2. The *zeros in R of the function* defined by $f(x) = ax^2 + bx + c$.
3. The *x-intercepts* of the graph of the equation $f(x) = ax^2 + bx + c$.

We recall from Chapter 4 that a quadratic equation may have no real solution, one real solution, or two real solutions. If the equation has no real solution, we find that the graph of the related quadratic equation in two variables does not touch the x-axis: if there is one real solution, the graph is tangent to the x-axis; if there are two real solutions, the graph crosses the x-axis in two distinct points. These cases are illustrated in Figure 6.4.

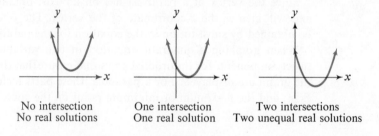

No intersection
No real solutions

One intersection
One real solution

Two intersections
Two unequal real solutions

Figure 6.4

Inverse of a
quadratic
function

In Section 5.4 we observed that the graph of the inverse of a linear function is also a straight line, and hence the inverse is also a linear function if its graph is not vertical. Because every function is a relation, every function has an inverse, but the inverse is not always a function. For example, the graphs of

$$F = \{(x, y) \mid y = x^2\}$$

and its inverse,

$$F^{-1} = \{(x, y) \mid x = y^2\} = \{(x, y) \mid y = \pm\sqrt{x}\},$$

are shown in Figure 6.5. Since F^{-1} associates two different y's with each x for all but one value in its domain, this inverse is not a function.

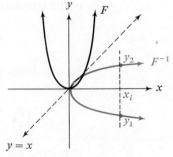

Figure 6.5

*Quadratic
inequalities*

Relations of the form

$$\{(x, y)\,|\,y < ax^2 + bx + c\} \qquad (7)$$

or

$$\{(x, y)\,|\,y > ax^2 + bx + c\} \qquad (8)$$

can be graphed in the same manner in which we graphed relations defined by *linear* inequalities in two variables in Section 5.6. We first graph the relation defined by the equation having the same members as the defining inequality and then shade an appropriate region as required. For instance, to graph

$$\{(x, y)\,|\,y < x^2 + 2\}, \qquad (7)$$

we first graph

$$\{(x, y)\,|\,y = x^2 + 2\}. \qquad (8)$$

Then, as in the graphing of linear inequalities, the coordinates of a selected point not on the graph of the associated equation can be used to determine which of the two resulting regions of the plane is the graph of the inequality. Upon substitution of 0 for x and 0 for y in $y < x^2 + 2$, we have $(0) < (0)^2 + 2$, a true statement. Hence the region containing the origin is shaded. Since the graph of (8) is not part of the graph of (7), a broken curve is used (Figure 6.6).

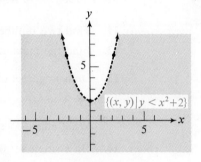

Figure 6.6

Exercise 6.1

Graph. (*Obtain analytically the intercepts and the maximum or minimum point, and then sketch the curve.*)

Example $\{(x, y)\,|\,y = x^2 - 7x + 6\}$

Solution   The y-intercept is clearly 6. Since the solutions of

$$x^2 - 7x + 6 = (x - 1)(x - 6) = 0$$

are 1 and 6, these are the x-intercepts. The axis of symmetry has equation

$$x = -\frac{b}{2a} = \frac{7}{2}.$$

Writing $7/2$ for x in $y = x^2 - 7x + 6$, we obtain

$$y = -\frac{25}{4}.$$

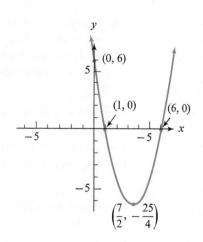

Hence the minimum point is $(7/2, -25/4)$. Using these points, you can sketch the graph as shown.

1. $\{(x, y)\,|\,y = x^2 - 5x + 4\}$ 2. $\{(x, g(x))\,|\,g(x) = x^2 - 3x + 2\}$
3. $\{(x, y)\,|\,y = x^2 - 6x - 7\}$ 4. $\{(x, y)\,|\,y = x^2 - 3x + 2\}$
5. $\{(x, f(x))\,|\,f(x) = -x^2 + 5x - 4\}$ 6. $\{(x, f(x))\,|\,f(x) = -x^2 - 8x + 9\}$

7. $\left\{(x, g(x))\,\middle|\,g(x) = \dfrac{1}{2}x^2 + 2\right\}$ 8. $\left\{(x, y)\,\middle|\,y = -\dfrac{1}{2}x^2 - 2x\right\}$

9. Graph $\{(x, f(x))\,|\,f(x) = x^2 + 1\}$. Represent $f(0)$ and $f(4)$ by drawing line segments from $(0, 0)$ to $(0, f(0))$ and from $(4, 0)$ to $(4, f(4))$.

10. Graph $\{(x, g(x))\,|\,g(x) = x^2 + 1\}$. Represent $g(-3)$ and $g(2)$ by drawing line segments from $(-3, 0)$ to $(-3, g(-3))$ and from $(2, 0)$ to $(2, g(2))$.

Solve Problems 11 *and* 12 *by completing the square.*

11. Find two numbers having sum 8 and product as great as possible.

12. Find the maximum possible area of a rectangle with perimeter 100 inches.

13. On a single set of axes, sketch the family of four curves that are the graphs of
$$y = x^2 + k \quad (k = -2, 0, 2, 4).$$
What effect does varying k have on the graph?

14. On a single set of axes, sketch the family of six curves that are the graphs of
$$y = kx^2 \quad \left(k = \frac{1}{2}, 1, 2, -\frac{1}{2}, -1, -2\right).$$
What effect does varying k have on the graph?

15. Graph the relation $\{(x, y)\,|\,x = y^2\}$.
   a. What kind of curve is the graph?
   b. Is the given relation a function? Why or why not?
   c. Does the graph of this relation have a maximum or minimum point?

16. Graph the relation $\{(x, y)\,|\,x = y^2 - 2y\}$.
   a. What kind of curve is the graph?
   b. Is the given relation a function? Why or why not?
   c. Does the graph of this relation have a maximum or minimum point?

Graph each of the following relations. The graphs are parabolas.

17. $\{(x, y)\,|\,x = y^2 - 4\}$ 18. $\{(x, y)\,|\,x = y^2 - 2y - 3\}$
19. $\{(x, y)\,|\,x = y^2 - 4y + 4\}$ 20. $\{(x, y)\,|\,x = 2y^2 + 3y - 2\}$

Graph each given function f and f⁻¹, using the same set of axes.

21. $f = \{(x, y)\,|\,y = x^2 - 4\}$
22. $f = \{(x, y)\,|\,y = x^2 + 4\}$
23. $f = \{(x, y)\,|\,y = x^2 - 4x + 4\}$
24. $f = \{(x, y)\,|\,y = x^2 - 2x - 3\}$

Graph.

Example $\{(x, y) \mid y \geq x^2 + 2x\}$

Solution   First graph $\{(x, y) \mid y = x^2 + 2x\}$.

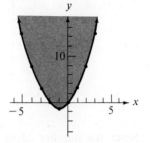

Use $(-1, 0)$ as a test point.
Is $0 \geq (-1)^2 + 2(-\ )$? Yes.
Then $(-1, 0)$ is in the graph.
Shade the portion of the plane
above the curve.

25. $\{(x, y) \mid y > x^2\}$

26. $\{(x, y) \mid y < x^2\}$

27. $\{(x, y) \mid y \geq x^2 + 3\}$

28. $\{(x, y) \mid y \leq x^2 + 3\}$

29. $\{(x, y) \mid y < 3x^2 + 2x\}$ 30. $\{(x, y) \mid y > 3x^2 + 2x\}$

31. $\{(x, y) \mid y \leq x^2 + 3x + 2\}$ 32. $\{(x, y) \mid y \geq x^2 + 3x + 2\}$

33. $\{(x, y) \mid y \geq 2x^2 - 5x + 1\}$ 34. $\{(x, y) \mid y \leq 2x^2 - 5x + 1\}$

35. Graph the set of points whose coordinates satisfy both

$$\{(x, y) \mid y \leq 4 - x^2\} \quad \text{and} \quad \{(x, y) \mid y \geq x^2 - 4\}.$$

36. Graph the set of points whose coordinates satisfy both

$$\{(x, y) \mid y \leq 1 - x^2\} \quad \text{and} \quad \{(x, y) \mid y \geq -1\}.$$

6.2   Conic Sections

In addition to equations typified by $y = ax^2 + bx + c$, there are three other types of second-degree equations in two variables with graphs that are of particular interest. We shall discuss each of them separately.

First, consider the relation

$$\{(x, y) \mid x^2 + y^2 = 25\}. \tag{1}$$

Solving the defining equation explicitly for y, we have

$$y = \pm\sqrt{25 - x^2}. \tag{2}$$

Assigning values to x, we find the following ordered pairs in the relation:

$$(-5, 0), \quad (-4, 3), \quad (-3, 4), \quad (0, 5), \quad (3, 4), \quad (4, 3), \quad (5, 0),$$
$$(-4, -3), \quad (-3, -4), \quad (0, -5), \quad (3, -4), \quad (4, -3).$$

Plotting these points on the plane, we have the graph shown in Figure 6.7-a on page 138. Connecting these points with a smooth curve, we obtain the graph in Figure 6.7-b, a **circle** with radius 5 and center at the origin.

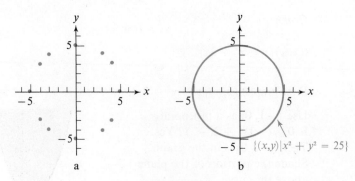

Figure 6.7

*Equation of
a circle*

Since the number 25 in the right-hand member of the defining equation of (1) clearly determines the length of the radius of the circle, we can generalize and observe that any relation defined by an equation of the form

$$x^2 + y^2 = r^2, \quad r > 0,$$

has as its graph a circle with radius r and with center at the origin. (See Problems 30 and 31, Exercise 6.2, for a more general approach.)

Note that in the preceding example it is not necessary to assign any values to x for which $|x| > 5$, because then y^2 would have to be negative. Since, except for -5 and $+5$, each permissible value for x is associated with two values for y—one positive and one negative—the relation (1) is not a function. We could represent the relationship specified by (2) by using two equations,

$$y = f(x) = \sqrt{25 - x^2} \tag{3}$$

and

$$y = g(x) = -\sqrt{25 - x^2}, \tag{3a}$$

whose graphs would appear as in Figure 6.8-a and 6.8-b, respectively. Each of

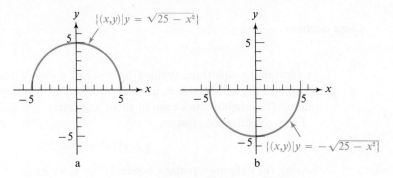

Figure 6.8

these equations does define a function. In both cases, the domain of the function is

$$\{x \mid |x| \le 5\},$$

while the ranges differ. The range of the function defined by (3) is

$$\{y \mid 0 \le y \le 5\},$$

and that of the function defined by (3a) is

$$\{y \mid -5 \le y \le 0\}.$$

The second kind of quadratic (second-degree) relation of interest, which actually has the first as a limiting case, is that typified by

$$\{(x, y) \mid 4x^2 + 9y^2 = 36\}. \tag{4}$$

We obtain members of (4) by first solving the defining equation explicitly for y,

$$y = \pm \frac{2}{3}\sqrt{9 - x^2},$$

and then assigning values to x satisfying $-3 \leq x \leq 3$. (Why these values only?) We obtain, for example,

$$(-3, 0), \quad \left(-2, \frac{2}{3}\sqrt{5}\right), \quad \left(-1, \frac{4}{3}\sqrt{2}\right), \quad (0, 2), \quad \left(1, \frac{4}{3}\sqrt{2}\right), \quad \left(2, \frac{2}{3}\sqrt{5}\right), \quad (3, 0),$$

$$\left(-2, -\frac{2}{3}\sqrt{5}\right), \quad \left(-1, -\frac{4}{3}\sqrt{2}\right), \quad (0, -2), \quad \left(1, -\frac{4}{3}\sqrt{2}\right), \quad \left(2, -\frac{2}{3}\sqrt{5}\right).$$

Equation of an ellipse

Locating the corresponding points on the plane and connecting them with a smooth curve, we have the graph shown in Figure 6.9. This curve is called an **ellipse**. In general, the graph of the relation

$$\{(x, y) \mid Ax^2 + By^2 = C;$$
$$A, B, C > 0; \quad A \neq B\}$$

is an ellipse with center at the origin.

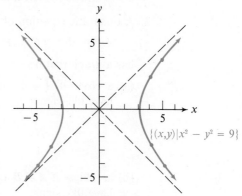

$\{(x,y) \mid 4x^2 + 9y^2 = 36\}$

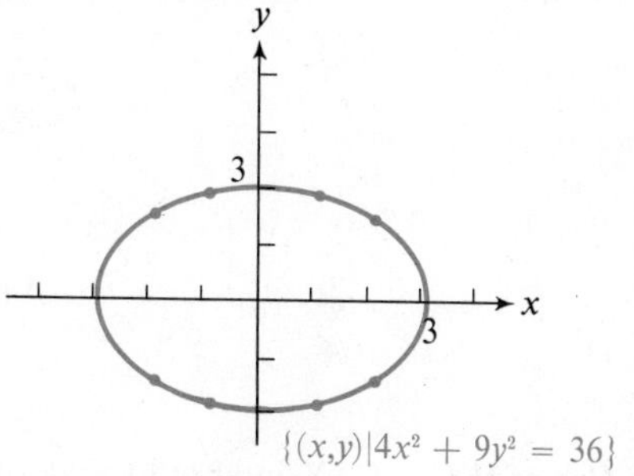

Figure 6.9

The third kind of quadratic relation with which we are presently concerned is typified by

$$\{(x, y) \mid x^2 - y^2 = 9\}.$$

Solving the defining equation for y, we obtain

$$y = \pm\sqrt{x^2 - 9},$$

which has as a part of its solution set the ordered pairs

$$(-5, 4), \quad (-4, \sqrt{7}), \quad (-3, 0), \quad (3, 0), \quad (4, \sqrt{7}), \quad (5, 4),$$
$$(-5, -4), \quad (-4, -\sqrt{7}), \quad (4, -\sqrt{7}), \quad (5, -4).$$

Equation of a hyperbola

Plotting the corresponding points and connecting them with a smooth curve, we obtain the curve shown in Figure 6.10. This curve is called a **hyperbola**. In general, the relation

$$\{(x, y) \mid Ax^2 - By^2 = C;$$
$$A, B, C > 0\}$$

graphs into a hyperbola with center at the origin, and the relation

$$\{(x, y) \mid By^2 - Ax^2 = C;$$
$$A, B, C > 0\}$$

$\{(x,y) \mid x^2 - y^2 = 9\}$

Figure 6.10

graphs into a hyperbola with center at the origin.

Asymptotes of   The dashed lines shown in Figure 6.10 are called **asymptotes** of the graph. They
a hyperbola   comprise the graph of the equation $x^2 - y^2 = 0$. While we shall not discuss the
                  notion in detail here, it is true that, in general, the graph of $Ax^2 - By^2 = C$
                  "approaches" the two straight lines in the graph of $Ax^2 - By^2 = 0$ for each
                  $A, B, C > 0$. (See Problem 35, Exercise 6.2.)

The graphs of the foregoing relations, together with the parabola of the preceding
section, are called **conic sections**, or **conics**, because such curves result from the
intersection of a plane and a right circular cone, as shown in Figure 6.11.

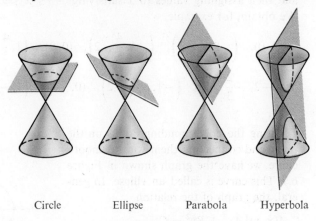

Circle   Ellipse   Parabola   Hyperbola

Figure 6.11

Use of the   You should make use of the form of the defining equation as an aid in graphing
form of an   quadratic relations. The ideas developed in this and the preceding section may be
equation in   summarized as follows:
graphing

1. A quadratic equation of the form

$$y = ax^2 + bx + c, \quad a \neq 0, \qquad (5)$$

has a graph that is a parabola, opening
upward if $a > 0$ and downward if $a < 0$.
Similarly, an equation of the form
$x = ay^2 + by + c, a \neq 0$, has a graph that
is a parabola, opening to the right if
$a > 0$ and to the left if $a < 0$.

2. A quadratic equation of the form

$$Ax^2 + By^2 = C, \quad A^2 + B^2 \neq 0, \qquad (6)$$

has a graph that is

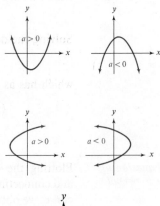

(a) a **circle** if $A = B$ and A, B, and C
have like signs;

(b) an **ellipse** if $A \neq B$ and A, B, and C
have like signs;

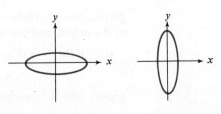

(c) a **hyperbola** if A and B are opposite in sign and $C \neq 0$;

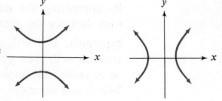

(d) **two distinct lines** through the origin if A and B are opposite in sign and $C = 0$ (see Problem 25);

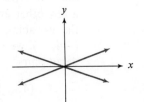

(e) **two distinct parallel lines** if one of A and $B = 0$ and the other has the same sign as C (see Problem 26);

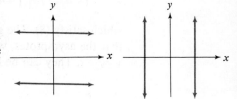

(f) **two coincident parallel lines** (one line) through the origin if one of A and $B = 0$ and also $C = 0$ (see Problem 27);

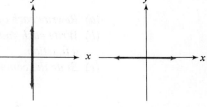

(g) a **point** if A and B are both >0 or both <0 and $C = 0$ (see Problem 28);

(h) the **null set**, $\emptyset$, if A and B are both ≥ 0 and $C < 0$, or if A and B are both ≤ 0 and $C > 0$ (see Problem 29).

Sketch of the graph of a conic section

After you recognize the general form of the curve, the graph of a few points should suffice to sketch the complete graph. The intercepts, for instance, are always easy to locate. Consider the relation

$$\{(x, y) \mid x^2 + 4y^2 = 8\}. \qquad (7)$$

By comparing the defining equation with 2(b), we note immediately that its graph is an ellipse. If $y = 0$, then $x = \pm\sqrt{8}$; and if $x = 0$, then $y = \pm\sqrt{2}$. We can accordingly sketch the graph of (7) as in Figure 6.12.

As another example, consider the relation

$$\{(x, y) \mid x^2 - y^2 = 3\}. \qquad (8)$$

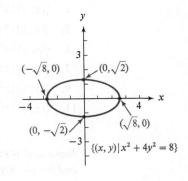

Figure 6.12

By comparing the defining equation with 2(c), we see that its graph is a hyperbola. If $y = 0$, then $x = \pm\sqrt{3}$; and if $x = 0$, then y^2 would have to be negative, an impossibility in the field of real numbers. Thus the graph does not cross the y-axis. By assigning a few other arbitrary values to one of the variables, say x, e.g., $(4, \quad)$ and $(-4, \quad)$, we can find additional ordered pairs

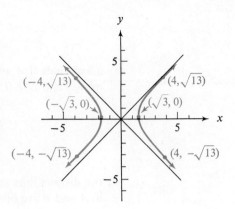

Figure 6.13

$$(4, \sqrt{13}), (4, -\sqrt{13}),$$
$$(-4, \sqrt{13}), (-4, -\sqrt{13})$$

which satisfy (8). The graph can then be sketched as shown in Figure 6.13. Observe that the asymptotes, which are the graphs of $x^2 - y^2 = 0$, are given by $y = x$ and $y = -x$. They can be helpful in sketching the hyperbola.

Exercise 6.2

(a) *Rewrite each of the defining equations equivalently with y as the left-hand member.*
(b) *Write each equation equivalently as two separate equations, each of which defines a function.*
(c) *State the common domain of each of the functions defined in (b).*

Example $\{(x, y) \mid 4x^2 + y^2 = 36\}$

Solution

(a) $y^2 = 36 - 4x^2$ (b) $y = 2\sqrt{9 - x^2}$
      $y = \pm 2\sqrt{9 - x^2}$ $y = -2\sqrt{9 - x^2}$

(c) The domain for each function is $\{x \mid -3 \le x \le 3\}$.

1. $\{(x, y) \mid x^2 + y^2 = 4\}$ 2. $\{(x, y) \mid x^2 + y^2 = 9\}$
3. $\{(x, y) \mid 9x^2 + y^2 = 36\}$ 4. $\{(x, y) \mid 4x^2 + y^2 = 4\}$
5. $\{(x, y) \mid x^2 + 4y^2 = 16\}$ 6. $\{(x, y) \mid x^2 + 9y^2 = 4\}$
7. $\{(x, y) \mid 2x^2 + 3y^2 = 24\}$ 8. $\{(x, y) \mid 4x^2 + 3y^2 = 12\}$
9. $\{(x, y) \mid x^2 - y^2 = 1\}$ 10. $\{(x, y) \mid 4x^2 - y^2 = 1\}$
11. $\{(x, y) \mid y^2 - x^2 = 9\}$ 12. $\{(x, y) \mid 4y^2 - 9x^2 = 36\}$

Name and sketch the graph of each of the following relations. Specify the intercepts, and for the hyperbolas give equations of the asymptotes.

Example $\{(x, y) \mid 4y^2 - 36 = -9y^2\}$

Solution

Rewrite the defining equation equivalently in standard form.

$$4x^2 + 9y^2 = 36$$

By inspection, the graph is an ellipse; x-intercepts are -3 and 3; y-intercepts are -2 and 2.

Sketch the graph.

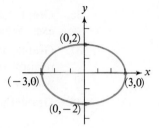

13. $\{(x, y) \mid x^2 + y^2 = 49\}$

14. $\{(x, y) \mid x^2 + y^2 = 64\}$

15. $\{(x, y) \mid 3x^2 + 25y^2 = 100\}$

16. $\{(x, y) \mid x^2 + 2y^2 = 8\}$

17. $\{(x, y) \mid 4x^2 = 4y^2\}$

18. $\{(x, y) \mid x^2 - 9y^2 = 0\}$

19. $\{(x, y) \mid x^2 = 9 + y^2\}$

20. $\{(x, y) \mid x^2 = 2y^2 + 8\}$

21. $\{(x, y) \mid 4x^2 + 4y^2 = 1\}$

22. $\{(x, y) \mid 9x^2 + 9y^2 = 2\}$

23. $\{(x, y) \mid 3x^2 - 12 = -4y^2\}$

24. $\{(x, y) \mid 12 - 3y^2 = 4x^2\}$

25. Graph $\{(x, y) \mid 4x^2 - y^2 = 0\}$. Generalize from the result and discuss the graph of any relation of the form $\{(x, y) \mid Ax^2 - By^2 = 0, \quad A, B > 0\}$.

26. Graph $\{(x, y) \mid x^2 = 4\}$. Generalize from the result and discuss the graph of any relation of the form $\{(x, y) \mid Ax^2 = C, \quad A, C > 0\}$.

27. Graph $\{(x, y) \mid x^2 = 0\}$. Generalize from the result and discuss the graph of any relation of the form $\{(x, y) \mid Ax^2 = 0, A \neq 0\}$.

28. Graph $\{(x, y) \mid 4x^2 + y^2 = 0\}$. Generalize from the result and discuss the graph of any relation of the form $\{(x, y) \mid Ax^2 + By^2 = 0, \quad A, B > 0\}$.

29. Explain why the graph of $\{(x, y) \mid x^2 + y^2 = -1\}$ is the null set. Generalize from the result and discuss the graph of any relation of the form

$$\{(x, y) \mid Ax^2 + By^2 = C, \quad A^2 + B^2 \neq 0, \quad A, B \geq 0, \quad C < 0\}.$$

30. Use the distance formula to show that the graph of the relation given by $\{(x, y) \mid x^2 + y^2 = r^2, r \geq 0\}$ is the set of all points located a distance r from the origin.

31. Show that the graph of $\{(x, y) \mid (x - 2)^2 + (y + 3)^2 = 16\}$ is a circle with center at $(2, -3)$ and radius 4. *Hint:* Use distance formula.

32. Show that the graph of $\{(x, y) \mid (x - h)^2 + (y - k)^2 = r^2, r > 0\}$ is a circle with center at (h, k) and radius r.

33. Show that the general form of an equation for an ellipse, $Ax^2 + By^2 = C$, can be written in the intercept form $\dfrac{x^2}{a^2} + \dfrac{y^2}{b^2} = 1$, where a and b are x- and y-intercepts, respectively.

34. Show that the general form of an equation for a hyperbola, $Ax^2 - By^2 = C$, can be written in the intercept form $\dfrac{x^2}{a^2} - \dfrac{y^2}{b^2} = 1$, where a is the x-intercept.

35. By solving $Ax^2 - By^2 = C$ $(A, B, C > 0)$ for y, obtain the expression

$$y = \pm \sqrt{\frac{A}{B}} |x| \left(\sqrt{1 - \frac{C}{Ax^2}} \right),$$

and argue that the graph of $Ax^2 - By^2 = C$ approaches the straight-line graphs of $y = \pm \sqrt{A/B} \, |x|$ as $|x|$ increases.

36. Graph $\{(x, y) \mid x^2 + y^2 = 25\}$ and $\{(x, y) \mid 4x^2 + y^2 = 36\}$ on the same set of axes. What is the significance of the coordinates of the points of intersection?

37. Graph the relation $\{(x, y) \mid x^2 + y^2 \leq 25\}$ by observing that the graph of the relation $\{(x, y) \mid x^2 + y^2 = a^2\}$ is a circle of radius a and then examining what $a \leq 5$ implies.

38. Graph the relation $\{(x, y) \mid 4x^2 + 9y^2 \leq 36\}$.

6.3   *Variation as a Functional Relationship*

Direct variation

There are two types of functional relationships, widely used in the sciences, to which custom has assigned special names. First, any function defined by an equation of the form

$$y = kx \quad (k \text{ a positive constant}) \tag{1}$$

furnishes an example of **direct variation**. The variable y is said to **vary directly** as the variable x. Another example of direct variation is

$$y = kx^2 \quad (k \text{ a positive constant}), \tag{1a}$$

which indicates that y varies directly as the square of x. In general,

$$y = kx^n \quad (k \text{ a positive constant and } n > 0) \tag{1b}$$

is the assertion that y varies directly as the nth power of x.

We find examples of such variation in the relationships existing between the radius of a circle and the circumference and area. Thus

$$c = 2\pi r \tag{2}$$

asserts that the circumference of a circle varies directly as the radius, while

$$A = \pi r^2 \tag{3}$$

expresses the fact that the area of a circle varies directly as the square of the radius. Since for each r, (2) and (3) associate only one value of c or A, both of these equations define functions; (2) is a linear function and (3) is a quadratic function.

Inverse variation

The second important type of variation arises from the equation

$$xy = k \quad (k \text{ a positive constant}), \tag{4}$$

in which x and y are said to **vary inversely**. When (4) is written in the form

$$y = \frac{k}{x}, \tag{5}$$

y is said to vary inversely as x. Similarly, if

$$y = \frac{k}{x^2}, \tag{5a}$$

y is said to vary inversely as the square of x, and so forth. As an example of inverse variation, consider the set of rectangles with area 24 square units. Since the area of a rectangle is given by

$$lw = A,$$

we have

$$lw = 24,$$

and the length and width can be seen to vary inversely.

Since (5) or (5a) associates only one y with each x $(x \neq 0)$, an inverse variation defines a function, with domain $\{x \,|\, x \neq 0\}$.

The names "direct" and "inverse," as applied to variation, arise from the fact that in direct variation an assignment of increasing absolute values of x results in an association with increasing absolute values of y, whereas in inverse variation an assignment of increasing absolute values of x results in an association with decreasing absolute values of y.

Constant of of variation

The constant involved in equations defining direct or inverse variation is called the **constant of variation**. If we know that one variable varies directly or inversely as another, and if we have one set of associated values for the variables, we can find the constant of variation involved. For example, suppose we know that y varies directly as x^2, and that $y = 4$ when $x = 7$. We express the fact that y varies directly as x^2 by writing

$$y = kx^2, \tag{6}$$

and then substitute 7 for x and 4 for y in (6) to obtain

$$4 = k \cdot 7^2 = k \cdot 49,$$

from which

$$k = \frac{4}{49}.$$

The equation specifically expressing the direct variation is

$$y = \frac{4}{49} x^2.$$

Joint variation

In the event that one variable varies as the product of two or more other variables, we refer to the relationship as **joint variation**. Thus, if y varies jointly as u, v, and w, we have

$$y = kuvw. \tag{7}$$

Also, direct and inverse variation may take place concurrently. That is, y may vary directly as x and inversely as z, giving rise to the equation

$$y = k\frac{x}{z}.$$

It should be pointed out that the way in which the word "variation" is used herein is a technical one, and when the ideas of direct, inverse, or joint variation are encountered, you should always think of equations of the form (1), (5), or (7). For instance, the equations

$$y = 2x + 1 \quad \text{and} \quad y = \frac{1}{2}x - 2$$

do not describe examples of variation within our meaning of the word.

Proportion-
ality and
variation

An alternative term frequently used to describe the variation relationship discussed in this section is the word "proportional." Thus, to say that "y is directly proportional to x" or "y is inversely proportional to x" is another way of describing direct and inverse variation.

The use of the word "proportion" arises from the fact that any two solutions of an equation expressing a direct variation satisfy a fractional equation of the form

$$\frac{a}{b} = \frac{c}{d},$$

which is commonly called a **proportion**. For example, consider the problem in which the volume of a gas varies directly as the absolute temperature and inversely as the pressure, and can accordingly be represented by a relation of the form

$$V = \frac{kT}{P}. \tag{8}$$

For any set of values T_1, P_1, and V_1,

$$k = \frac{V_1 P_1}{T_1}, \tag{8a}$$

and for any other set of values T_2, P_2, and V_2,

$$k = \frac{V_2 P_2}{T_2}. \tag{8b}$$

Equating the right-hand members of (8a) and (8b), we get

$$\frac{V_1 P_1}{T_1} = \frac{V_2 P_2}{T_2}, \tag{8c}$$

from which any one of the six values can be determined if the other five values are known.

Exercise 6.3

Solve. In Exercises 1–8, first find the constant of variation.

Example

If V varies directly as T and inversely as P, and $V = 40$ when $T = 300$ and $P = 30$, find V when $T = 324$ and $P = 24$.

Solution

Write an equation expressing the relationship between the variables.

$$V = \frac{kT}{P} \tag{1}$$

Substitute the initially known values for V, T, and P. Solve for k.

$$40 = \frac{k(300)}{30}$$

$$4 = k$$

Rewrite Equation (1) with k replaced by 4.

$$V = \frac{4T}{P}$$

Substitute the second set of values for T and P and solve for V.

$$V = \frac{4(324)}{24} = 54$$

1. If y varies directly as x^2, and $y = 9$ when $x = 3$, find y when $x = 4$.
2. If r varies directly as s and inversely as t, and $r = 12$ when $s = 8$ and $t = 2$, find r when $s = 3$ and $t = 6$.
3. The distance a particle falls in a certain medium is directly proportional to the square of the length of time it falls. If the particle falls 16 feet in two seconds, how far will it fall in 10 seconds?
4. In Problem 3, how far will the body fall *during* the seventh second?
5. The pressure exerted by a liquid at a given point varies directly as the depth of the point beneath the surface of the liquid. If a certain liquid exerts a pressure of 40 pounds per square foot at a depth of 10 feet, what is the pressure at 40 feet?
6. The volume (V) of a gas varies directly as its temperature (T) and inversely as its pressure (P). A gas occupies 20 cubic feet at a temperature of 300° A (absolute) and a pressure of 30 pounds per square inch. What will the volume be if the temperature is raised to 360° A and the pressure decreased to 20 pounds per square inch?
7. The maximum-safe uniformly distributed load (L) for a horizontal beam varies jointly as its breadth (b) and the square of the depth (d), and inversely as the length (l). An 8-foot beam with $b = 2$ feet and $d = 4$ feet will safely support a uniformly distributed load of up to 750 pounds. How many uniformly distributed pounds will an 8-foot beam support safely if $b = 2$ and $d = 6$?
8. The resistance (R) of a wire varies directly as its length (l) and inversely as the square of its diameter (d). Fifty feet of wire of diameter 0.012 inches has a resistance of 10 ohms. What is the resistance of 50 feet of the same type of wire if the diameter is increased to 0.015 inches?

Represent the relationship as a proportion by eliminating the constant of variation and then solving for the required variable.

Example If V varies directly as T and varies inversely as P, and $V = 40$ when $T = 300$ and $P = 30$, find V when $T = 324$ and $P = 24$.

(Solution overleaf)

Solution   Write an equation expressing the relationship between the variables.

$$V = \frac{kT}{P}$$

Solve for k.

$$k = \frac{VP}{T}$$

Write a proportion relating the variables for two different sets of conditions.

$$\frac{V_1 P_1}{T_1} = \frac{V_2 P_2}{T_2}$$

Substitute the known values of the variables.

$$\frac{(40)(30)}{300} = \frac{V_2(24)}{324}$$

Solve for V_2.

$$V_2 = \frac{(324)(40)(30)}{300(24)} = 54$$

9. Problem 1 of this set.   10. Problem 2 of this set.

11. Problem 5 of this set.   12. Problem 6 of this set.

13. Problem 7 of this set.   14. Problem 8 of this set.

15. From the formula for the circumference of a circle, $c = \pi d$, show that the ratio of the circumference of two circles equals the ratio of their respective diameters.

16. From the formula for the area of a circle, $A = \pi r^2$, show that the ratio of the areas of two circles equals the ratio of the squares of their respective radii.

17. Graph on the same set of axes the linear functions defined by $y = kx$, $x \geq 0$, when $k = 1$, 2, and 3, respectively. Note that the constant of variation and the slope of the graph of the equation are the same.

18. Graph on the same set of axes the quadratic functions defined by $y = kx^2$, $x \geq 0$, when $k = 1$, 2, and 3, respectively. What effect does a change in k have on the graph of $y = kx^2$?

19. Graph on the same set of axes the equations $y = kx$, $y = kx^2$, and $y = kx^3$, when $k = 2$ and $x \geq 0$. What effect does increasing n have on the graph of $y = kx^n$?

20. Graph on the same set of axes the functions defined by $xy = k$, for $k = 1$ and 2, and $x > 0$. What effect does a change in k have on the graph of $xy = k$?

21. Graph on the same set of axes the functions defined by $xy = k$, $x^2 y = k$, and $x^3 y = k$, for $k = 2$ and $x > 0$. What effect does increasing n have on the graph of $x^n y = k$?

22. Show that if y varies directly as x, and z varies directly as x, then $y + z$ varies directly as x.

6.4   *Polynomial Functions*

Graphs of polynomial functions

In Section 5.2, we graphed linear functions

$$\{(x, f(x))\,|\,f(x) = a_0 x + a_1\};$$

and in Section 6.1, we graphed quadratic functions

$$\{(x, f(x))\,|\,f(x) = a_0 x^2 + a_1 x + a_2\}.$$

We can graph any real polynomial function

$$\{(x, P(x))\,|\,P(x) = a_0 x^n + a_1 x^{n-1} + \cdots + a_n\}$$

by similar methods—that is, by combining the plotting of points with a consideration of certain general properties of the defining equations. In the case of the general polynomial equation, we shall lean more heavily on the use of plotted points. There is one fact about polynomial equations, however, that can be useful. This involves *turning points*, or local maximum and minimum values of y. Thus, for example, in Figure 6.14-a there is one local maximum as well as one local minimum, or a total of two turning points, and in Figure 6.14-c are one local maximum and two local minima, for a total of three turning points. In general, we have the following result.

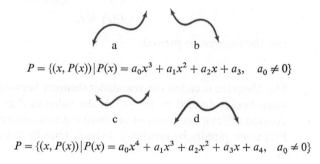

$$P = \{(x, P(x))\,|\,P(x) = a_0 x^3 + a_1 x^2 + a_2 x + a_3, \quad a_0 \neq 0\}$$

$$P = \{(x, P(x))\,|\,P(x) = a_0 x^4 + a_1 x^3 + a_2 x^2 + a_3 x + a_4, \quad a_0 \neq 0\}$$

Figure 6.14

Theorem 6.1   *If $P(x) = a_0 x^n + a_1 x^{n-1} + \cdots + a_n$ is a real polynomial of degree n, then the graph of*

$$\{(x, P(x))\,|\,P(x) = a_0 x^n + a_1 x^{n-1} + \cdots + a_n, \quad a_0 \neq 0\}$$

is a smooth curve that has at most $n - 1$ turning points.

Since the proof of this theorem involves ideas we have not discussed herein, it is omitted. As a consequence of this theorem the graphs of third- and fourth-degree polynomial functions might appear as in Figure 6.14, a, b, c, and d.

General form of a polynomial graph

To ascertain whether a graph ultimately goes up to the right, taking the general form (a) or (c) rather than (b) or (d), we can examine the leading coefficient, a_0, of the right-hand member of the defining equation; if $a_0 > 0$, then we can look for a form similar to (a) or (c), whereas if $a_0 < 0$, we can expect something similar to

(b) or (d). The graph ultimately goes up or down to the right according as $a_0 > 0$ or $a_0 < 0$. If $a_0 > 0$, then it ultimately goes down to the left, as in (a) and (d), or up to the left, as in (b) and (c), according as n is odd or even. In each case, then, the leading term $a_0 x^n$ governs the behavior of the graph of the polynomial function for large values of $|x|$.

For the actual graphing process, we can obtain ordered pairs $(x, f(x))$ for any polynomial function by direct substitution, which we used in previous sections, or by using the following.

Theorem 6.2   *If $P(x)$ is a real polynomial, then for every real number c there exists a unique real polynomial $Q(x)$ such that*

$$P(x) = (x - c)Q(x) + P(c).$$

Proof From Theorem 2.4, we know that there exists a real polynomial $Q(x)$ and a real number r such that

$$P(x) = (x - c)Q(x) + r.$$

Since this is true for all $x \in R$, then it is true for $x = c$, and we have

$$P(c) = (c - c)Q(c) + r,$$
$$P(c) = 0 \cdot Q(c) + r,$$
$$P(c) = r,$$

and the theorem is proved.

Use of the remainder theorem to find $P(c)$

This theorem is called the **remainder theorem** because it asserts that the remainder, when $P(x)$ is divided by $(x - c)$, is the value of P at c, that is, $P(c)$. Since synthetic division offers a quick means of obtaining this remainder, we can usually find values $P(c)$ more rapidly by synthetic division than by direct substitution.

Example

Given $P(x) = x^3 - x^2 + 3$, find $P(3)$ by means of the remainder theorem.

Solution

Synthetically dividing $x^3 - x^2 + 3$ by $x - 3$, we have

$$
\begin{array}{r|rrrr}
3 & 1 & -1 & 0 & 3 \\
  &   & 3 & 6 & 18 \\
\hline
  & 1 & 2 & 6 & 21
\end{array}
$$

and, by inspection, $r = P(3) = 21$.

Example

Graph $\{(x, P(x)) \mid P(x) = 2x^3 + 13x^2 + 6x\}$.

Solution

Since P is defined by a cubic polynomial with positive leading coefficient, we expect to find a graph of form similar to that in Figure 6.14-a. Now, to find points $(x, P(x))$ lying on the graph, we shall use the process of synthetic division and find $P(x)$ by the remainder theorem. Since we do not know where we should look for turning points, let us start with $x = 0$. By inspection, we have $P(0) = 0$, so that the graph includes the origin. For $x = 1$, we have

$$
\begin{array}{r|rrrr}
1 & 2 & 13 & 6 & 0 \\
  &   & 2 & 15 & 21 \\
\hline
  & 2 & 15 & 21 & 21
\end{array}
$$

and $(1, 21)$ is on the graph. For $x = 2$, we get

$$
\begin{array}{r|rrrr}
2 & 2 & 13 & 6 & 0 \\
  &   & 4 & 34 & 80 \\
\hline
  & 2 & 17 & 40 & 80
\end{array}
$$

and $(2, 80)$ is on the graph, Since the signs involved at each step in the last row of the division process here are positive, it is evident that for values $x > 2$, $P(x)$ will grow increasingly large; consequently, let us turn our attention to negative values of x. For $x = -1$, we have

$$
\begin{array}{r|rrrr}
-1 & 2 & 13 & 6 & 0 \\
   &   & -2 & -11 & 5 \\
\hline
   & 2 & 11 & -5 & 5
\end{array}
$$

and $P(-1) = 5$, so that $(-1, 5)$ is on the graph. In similar fashion, we find that the following points lie on the graph:

$$(-2, 24), \quad (-3, 45), \quad (-4, 56), \quad (-5, 45), \quad \text{and} \quad (-6, 0).$$

The graphs of the nine ordered pairs are shown in **a** of the figure. These points make the general appearance of the graph clear, and there remains only the question of whether or not the function has a zero between -1 and 0. If we let x have values $-3/4, -1/4,$ and $-1/2$, we obtain the additional pairs $(-3/4, 63/32), (-1/4, -23/32),$ $(-1/2, 0)$. Then for $-1/2 < x < 0$, we have $P(x) < 0$, and the graph can be sketched as shown in **b**.

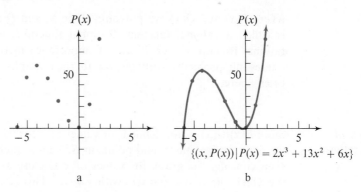

$$\{(x, P(x)) \mid P(x) = 2x^3 + 13x^2 + 6x\}$$

a   b

Exercise 6.4

Use synthetic division to find values of each polynomial for the specified values of the variable.

1. $P(x) = x^3 + 2x^2 + x - 1$; $P(1), P(2),$ and $P(3)$
2. $P(x) = x^3 - 3x^2 - x + 3$; $P(1), P(2),$ and $P(3)$

3. $P(x) = 2x^4 - 3x^3 + x + 2;$ $P(-2)$, $P(2)$, and $P(4)$
4. $P(x) = 3x^4 + 3x^2 - x + 3;$ $P(-2)$, $P(2)$, and $P(4)$
5. $P(x) = 3x^5 - x^3 + 2x^2 - 1;$ $P(-3)$, $P(2)$, and $P(3)$
6. $P(x) = 2x^6 - x^4 + 3x^3 + 1;$ $P(-3)$, $P(2)$, and $P(3)$

Graph. Use enough of the domain to include all turning points.

7. $\{(x, P(x)) \,|\, P(x) = x^3 - 4x^2 + 3x\}$
8. $\{(x, P(x)) \,|\, P(x) = x^3 - 2x^2 + 1\}$
9. $\{(x, P(x)) \,|\, P(x) = 2x^3 + 9x^2 + 7x - 6\}$
10. $\{(x, P(x)) \,|\, P(x) = 3x^3 + 2x^2 - x + 1\}$
11. $\{(x, P(x)) \,|\, P(x) = x^4\}$
12. $\{(x, P(x)) \,|\, P(x) = -x^4 + x\}$
13. $\{(x, P(x)) \,|\, P(x) = x^4 - x^3 - 2x^2 + 3x - 3\}$
14. $\{(x, P(x)) \,|\, P(x) = x^4 - 4x^3 - 4x + 12\}$

6.5   *Rational Functions*

A function defined by an equation of the form

$$y = \frac{P(x)}{Q(x)}, \tag{1}$$

where $P(x)$ and $Q(x)$ are polynomials in x, and $Q(x)$ is not the zero polynomial, is called a **rational function**. Rational functions with real coefficients, that is, rational functions over R, are of importance in the calculus and provide some interesting problems with respect to their graphs. We shall consider them only briefly here.

Vertical asymptotes

Since $P(x)/Q(x)$ is not defined for values of x for which $Q(x) = 0$, it is evident that we shall not be able to find points in R^2 having such x-coordinates. We can, however, consider the graph for values of x as close as we please to a value, say x_0, for which $Q(x_0) = 0$, but still with $x \neq x_0$. This consideration is usually described by saying that x "approaches" x_0, "grows close" to x_0, etc., and correspondingly that $Q(x)$ approaches 0. Thus, the closer $Q(x)$ approaches 0, if $P(x)$ does not approach 0 at the same time, then the larger $|y|$ becomes in (1). For example, Figure 6.15-a shows the behavior of

$$\left\{(x, y) \,|\, y = \frac{2}{x - 2}\right\}, \tag{2}$$

as x approaches 2 from the right. In situations such as this, the vertical line that the curve approaches is called a **vertical asymptote**.

Theorem 6.3   *The graph of the function over R defined by $y = P(x)/Q(x)$ has a vertical asymptote $x = a$ for each value a at which $Q(x)$ vanishes and $P(x)$ does not vanish.*

     In Figure 6.15-a we see the behavior of (2) as x approaches 2 from the right. We are also interested in its behavior as x approaches 2 from the left. Figure 6.15-b illustrates this. As long as $x > 2$, we have $x - 2 > 0$ and $2/(x - 2) > 0$; but if $x < 2$, then we have $x - 2 < 0$ and $2/(x - 2) < 0$.

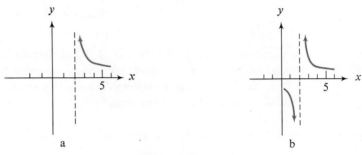

Figure 6.15

     The graphs of some rational functions have **horizontal asymptotes**, which can, in general, be identified by using the following theorem.

Theorem 6.4   *The graph of the rational function over R defined by*

$$y = \frac{a_0 x^n + a_1 x^{n-1} + \cdots + a_n}{b_0 x^m + b_1 x^{m-1} + \cdots + b_m},\tag{3}$$

where a_0, $b_0 \neq 0$, and n, m are nonnegative integers, has

     I   *a horizontal asymptote at $y = 0$ if $n < m$,*

     II   *a horizontal asymptote at $y = a_0/b_0$ if $n = m$,*

     III   *no horizontal asymptotes if $n > m$.*

     Though we shall not give a rigorous proof of this theorem here, we can certainly make the results plausible. If $n < m$, we can divide the numerator and denominator of the right-hand member of (3) by x^m to obtain, for $x \neq 0$,

$$y = \frac{\dfrac{a_0}{x^{m-n}} + \dfrac{a_1}{x^{m-n+1}} + \cdots + \dfrac{a_n}{x^m}}{b_0 + \dfrac{b_1}{x} + \cdots + \dfrac{b_m}{x^m}}.$$

Now, as $|x|$ grows larger and larger, each term containing an x in its denominator grows closer and closer to 0, and we find the expression on the right approaching $0/b_0$, so that y approaches 0. But if, as y grows close to 0, $|x|$ increases without bound, then y approaches the line $y = 0$ asymptotically, and actually must approach 0 from one of the directions shown in Figure 6.16-a (page 154). A similar argument shows that if $n = m$, then, as $|x|$ increases without bound, y approaches the line $y = a_0/b_0$ from one of the directions shown in Figure 6.16-b. If $n > m$, then as $|x|$ becomes larger and larger, so does $|y|$.

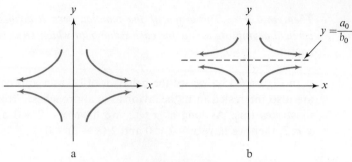

Figure 6.16

Oblique asymptotes

In particular, if $n = m + 1$, that is, if the numerator is of degree one greater than the denominator, we can argue that though the graph has no horizontal asymptote, it does have an **oblique asymptote**. We shall illustrate a special case only, but the technique involved is quite general.

Example

Find all asymptotes for the graph of the relation

$$\left\{(x, y)\,|\,y = \frac{x^2 - 4}{x - 1}\right\}.$$

Solution

We begin by observing that by Theorem 6.3, $x = 1$ is a vertical asymptote, and that by Theorem 6.4, there are no horizontal asymptotes. If, however, we rewrite $y = (x^2 - 4)/(x - 1)$ by dividing $x^2 - 4$ by $x - 1$, we obtain

$$y = x + 1 - \frac{3}{x - 1}.$$

Now, as $|x|$ grows larger and larger, $3/(x - 1)$ grows smaller and smaller, and the graph of $y = (x^2 - 4)/(x - 1)$ approaches the graph of $y = x + 1$. Hence, the graph of $y = x + 1$, which is an oblique line, is an asymptote to the curve.

Helpful items for graphing

Identifying asymptotes is one aid to the graphing of a rational function. Other helpful items are the following:

1. The zeros of the function, because these give us the x-intercepts.
2. The domain and range, because these let us know where we can expect to find parts of the graph and where we cannot.
3. Some specific points on the graph, because these give us guidelines in sketching.

Example

Graph $\left\{(x, y)\,|\,y = \frac{x - 1}{x - 2}\right\}.$

Solution

We can begin by observing that the numerator of the right-hand member will be equal to 0 when x is equal to 1. Therefore, when $x = 1$, we have $y = 0$, and 1 is an x-intercept. Also, when $x = 0$ we have $y = 1/2$, so that $1/2$ is a y-intercept. Thus we can begin our graph as shown in the figure **a** (page 155). By inspection, $y = (x - 1)/(x - 2)$ is defined for all real x except $x = 2$, so that $\{x\,|\,x \neq 2\}$ is the domain. Similarly, if we solve the defining equation for x in terms of y, we have

$$x = \frac{2y - 1}{y - 1},$$

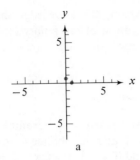

a

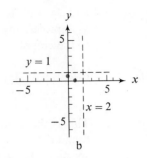

b

which is defined for all values of y except 1. Hence $\{y \mid y \neq 1\}$ is the range of the function. From the defining equation and Theorems 6.3 and 6.4, we see that there is a vertical asymptote at $x = 2$ and a horizontal asymptote at $y = 1$. We can then add this information to our graph, as indicated in **b**. Next we call our powers of observation into play. That the vertical asymptote, for example, is approached downward instead of upward from the left can be confirmed by observing that if x is just less than 2, say $2 - p$, $0 < p < 1/10$, then the denominator $x - 2$ in the expression for y is

$$2 - p - 2 = -p,$$

which is barely negative, whereas the numerator

$$x - 1 = 2 - p - 1 = 1 - p$$

is definitely positive; hence y is negative and $|y|$ large. The graph appears in **c**. To find the curve when $x > 2$, that is, to the right of the vertical asymptote, we

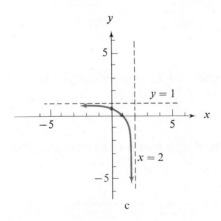

c

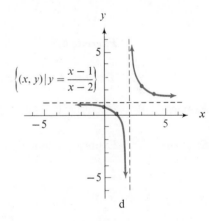

d

simply find one or two points associated with such values of x. For example, we might choose 3 and 4. If $x = 3$, then

$$y = \frac{3 - 1}{3 - 2} = 2,$$

and $(3, 2)$ is on the curve. If $x = 4$, then

$$y = \frac{4 - 1}{4 - 2} = \frac{3}{2},$$

(Solution continued)

and (4, 3/2) is on the curve. Again, the knowledge that $x = 2$ and $y = 1$ are asymptotes, together with the location of the points (3, 2) and (4, 3/2), leads us to the complete graph of the function $\{(x, y) \mid y = (x - 1)/(x - 2)\}$, as shown in **d**.

Example   Graph $\left\{ (x, y) \mid y = \dfrac{x^2 - 4}{x - 1} \right\}$.

Solution   This is the same function we investigated in the example on page 154, where we found the vertical asymptote $x = 1$ and the oblique asymptote $y = x + 1$. By inspection, if $x = 0$ then $y = 4$, and if $y = 0$ then $x = 2$ or $x = -2$, so that there is a y-intercept at 4 as well as x-intercepts at 2 and -2. We therefore have the situation shown in **a**. Without further information, we have reason to suspect that the graph will appear as shown in **b**. Plotting a few check points, say $(-1, 3/2)$, $(1/2, 15/2)$, $(3/2, -7/2)$, and $(3, 5/2)$, would tend to confirm our conjecture.

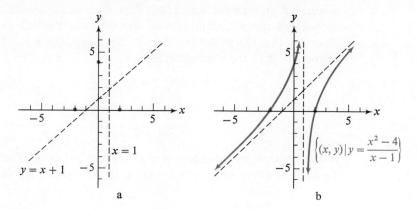

a   b

Exercise 6.5

Determine the vertical asymptotes of the graph of each function.

Example $\{(x, y) \mid x^2 y - 4y = 1\}$

Solution   Express y explicitly in terms of x by means of an equivalent equation.

$$y(x^2 - 4) = 1$$

$$y = \frac{1}{x^2 - 4} = \frac{1}{(x - 2)(x + 2)}$$

By Theorem 6.3, there are vertical asymptotes at $x = 2, -2$.

1. $\left\{ (x, y) \mid y = \dfrac{1}{x - 3} \right\}$ 2. $\left\{ (x, y) \mid y = \dfrac{1}{x + 4} \right\}$

3. $\left\{ (x, y) \mid y = \dfrac{4}{(x + 2)(x - 3)} \right\}$ 4. $\left\{ (x, y) \mid y = \dfrac{8}{(x - 1)(x + 3)} \right\}$

5. $\left\{ (x, y) \mid y = \dfrac{2x - 1}{x^2 + 5x + 4} \right\}$ 6. $\left\{ (x, y) \mid y = \dfrac{x + 3}{2x^2 - 5x - 3} \right\}$

7. $\{(x, y) \mid xy + y = 4\}$ 8. $\{(x, y) \mid x^2 y + xy = 3\}$

Graph.

9. $\left\{(x, y) \mid y = \dfrac{1}{x}\right\}$

10. $\left\{(x, y) \mid y = \dfrac{1}{x + 4}\right\}$

11. $\left\{(x, y) \mid y = \dfrac{1}{x - 3}\right\}$

12. $\left\{(x, y) \mid y = \dfrac{1}{x - 6}\right\}$

13. $\left\{(x, y) \mid y = \dfrac{4}{(x + 2)(x - 3)}\right\}$

14. $\left\{(x, y \mid) y = \dfrac{8}{(x - 1)(x + 3)}\right\}$

15. $\left\{(x, y) \mid y = \dfrac{2}{(x - 3)^2}\right\}$

16. $\left\{(x, y) \mid y = \dfrac{1}{(x + 4)^2}\right\}$

Determine any vertical, horizontal, or oblique asymptotes of the graphs of each of the following functions.

Example

$\left\{(x, y) \mid y = \dfrac{6x^2 + 1}{2x^2 + 5x - 3}\right\}$

Solution

The defining equation can be written equivalently as

$$y = \dfrac{6x^2 + 1}{(2x - 1)(x + 3)} .$$

By Theorem 6.3, there are vertical asymptotes at $x = 1/2$ and $x = -3$.

By Theorem 6.4-II, there is a horizontal asymptote at $y = 6/2 = 3$.

17. $\left\{(x, y) \mid y = \dfrac{x}{x^2 - 4}\right\}$

18. $\left\{(x, y) \mid y = \dfrac{3x + 6}{x^2 + 3x + 2}\right\}$

19. $\left\{(x, y) \mid y = \dfrac{x^2 - 9}{x - 4}\right\}$

20. $\left\{(x, y) \mid y = \dfrac{x^3 - 27}{x^2 - 1}\right\}$

21. $\left\{(x, y) \mid y = \dfrac{x^2 - 3x + 2}{x^2 - 3x - 4}\right\}$

22. $\left\{(x, y) \mid y = \dfrac{x^2}{x^2 - x - 6}\right\}$

Graph. Use information concerning the zeros of the function, and concerning vertical, horizontal, and oblique asymptotes.

23. $\left\{(x, y) \mid y = \dfrac{x}{x - 2}\right\}$

24. $\left\{(x, y) \mid y = \dfrac{x - 1}{x + 3}\right\}$

25. $\left\{(x, y) \mid y = \dfrac{2x - 4}{x^2 - 9}\right\}$

26. $\left\{(x, y) \mid y = \dfrac{3x}{x^2 - 5x + 4}\right\}$

27. $\left\{(x, y) \mid y = \dfrac{x^2 - 4}{x^3}\right\}$

28. $\left\{(x, y) \mid y = \dfrac{x - 2}{x^2}\right\}$

29. $\left\{(x, y) \mid y = \dfrac{x^2 - 4x + 4}{x - 1}\right\}$

30. $\left\{(x, y) \mid y = \dfrac{x^2 + 4}{x - 2}\right\}$

31. $\left\{(x, y) \mid y = \dfrac{x + 1}{x(x^2 - 4)}\right\}$

32. $\left\{(x, y) \mid y = \dfrac{x^2 + x - 2}{x(x^2 - 9)}\right\}$

Chapter Review

[6.1]   *Find the x-intercepts, the axis of symmetry, and the maximum or minimum point of the graph of each function by analytic methods.*

   1. $\{(x, f(x)) | f(x) = x^2 - x - 6\}$ **2.** $\{(x, f(x)) | f(x) = -x^2 + 7x - 10\}$

   3. Graph the function of Exercise 1. Draw a line segment from $(4, 0)$ to $(4, f(4))$.

   4. Graph the function of Exercise 2. Draw a line segment from $(3, 0)$ to $(3, f(3))$.

   Graph f and f^{-1} on the same set of axes.

   5. $f = \{(x, y) | y = x^2 - 9\}$ **6.** $f = \{(x, y) | y = x^2 - 4x - 5\}$

   Graph the relation defined by the given inequality.

   7. $\{(x, y) | y < x^2 - 9\}$ **8.** $\{(x, y) | y \geq x^2 + 6x + 5\}$

[6.2]   *Name and graph the relation defined by the given equation.*

   9. $x^2 - 16 = 4y^2$ **10.** $2y^2 = 8 - 2x^2$

   11. $x^2 = 36 - 4y^2$ **12.** $x^2 - y - 9 = 0$

   13. $x^2 - 9y^2 = 0$ **14.** $x^2 + 9y^2 = 0$

[6.3]   **15.** If y varies directly as x^2, and $y = 20$ when $x = 2$, find y when $x = 5$.

   16. If r varies directly as x^2 and inversely as z^3, and $r = 4$ when $x = 3$ and $z = 2$, find r when $x = 2$ and $z = 3$.

   17. The number of posts needed to string a telephone line over a given distance varies inversely as the distance between posts. If it takes 80 posts separated by 120 feet to string a wire between two points, how many posts would be required if the posts were 150 feet apart?

   18. The speed of a gear varies directly as the number of teeth it contains. If a gear with 10 teeth rotates at 240 revolutions per minute (RPM), with what speed would a gear with 18 teeth revolve under the same conditions?

[6.4]   **19.** Use synthetic division to find $P(2)$, where $P(x) = x^3 - 3x^2 + 2x + 1$.

   20. Use synthetic division to find $P(-3)$, where $P(x) = x^4 - 3x^2 - 1$.

   21. Use synthetic division to find ordered pairs in the function

$$\{(x, P(x)) | P(x) = x^3 + 4x^2 + x - 6\}$$

   for the following replacements for x: $-4, -3, -2, -1, 0, 1, 2, 3, 4$.

   22. Graph the function in Problem 21.

[6.5]   *Determine any asymptotes and graph each function.*

   23. $\left\{ (x, y) | y = \dfrac{3}{x + 2} \right\}$ **24.** $\left\{ (x, y) | y = \dfrac{2x}{(x + 3)(x - 4)} \right\}$

   25. $\left\{ (x, y) | y = \dfrac{x + 2}{x - 3} \right\}$ **26.** $\left\{ (x, y) | y = \dfrac{2x}{x^2 - 3x + 2} \right\}$

7

Exponential and Logarithmic Functions

7.1   *The Exponential Function*

Powers of the form b^x, where $b \in R$, $b > 0$, and $b \neq 1$, can be used to define functions. Notice that the base b must be restricted to positive values to ensure that b^x be real for all rational numbers x; for example, $(-1)^{1/2}$ is not a real number. If, now, we are to define a function over R using b^x, we must be able to interpret powers with irrational exponents, such as

$$b^{\pi}, \quad b^{\sqrt{2}}, \quad \text{and} \quad b^{-\sqrt{3}},$$

to be real numbers. The following theorem, which we present without proof, will be useful in doing this.

Theorem 7.1   *Let $x, y \in Q$, and let $x > y > 0$. Then*

$$b^x > b^y \text{ if } b > 1, \quad b^x = b^y \text{ if } b = 1, \quad \text{and} \quad b^x < b^y \text{ if } 0 < b < 1.$$

*Powers with
real-number
exponents*

In Section 3.5, it was observed that irrational numbers can be approximated by rational numbers to as high a degree of accuracy as desired. For example, $\sqrt{2} \approx 1.4$ or $\sqrt{2} \approx 1.414$, and so forth. Since b^x $(b > 0)$ is defined for rational x, by Theorem 7.1 we can write the sequence of inequalities

$$3^1 < 3^2,$$

$$3^{1.4} < 3^{1.5},$$

$$3^{1.41} < 3^{1.42},$$

$$3^{1.414} < 3^{1.415},$$

and so on.

This process can be continued indefinitely, yielding

$$3^1 < 3^{1.4} < 3^{1.41} < 3^{1.414} < \cdots < 3^{1.415} < 3^{1.42} < 3^{1.5} < 3^2,$$

and it seems plausible and is actually true, though we shall not prove it, that the difference between the number on the left and that on the right can be made as small as we please. This being the case, the completeness property O-3 of the real numbers guarantees that there is just one number, which we denote by $3^{\sqrt{2}}$, that lies between the number on the left and the number on the right, no matter how long this process of approximation is continued. Since we can produce the same type of argument for any irrational exponent x, we shall assume that b^x $(b > 0)$ is defined in this way for all real values of x and that the laws of exponents given in Theorem 3.1, appropriately reworded for real-number exponents, are valid for such powers.

Since for each real x there is one and only one number b^x, the equation

$$f(x) = b^x \quad (b > 0) \tag{1}$$

defines a function. Because $1^x = 1$ for all $x \in R$, (1) defines a constant function if $b = 1$. If $b \neq 1$, we say that (1) defines an **exponential function**.

Exponential functions can perhaps be visualized most clearly by considering their graphs. We illustrate two typical examples, in which $0 < b < 1$ and $b > 1$, respectively. Assigning values to x in the equations

$$f(x) = \left(\frac{1}{2}\right)^x \quad \text{and} \quad f(x) = 2^x,$$

we find some ordered pairs in each solution set and sketch the graphs of $\{(x, f(x))\mid f(x) = (\frac{1}{2})^x\}$ and $\{(x, f(x))\mid f(x) = 2^x\}$, shown in **a** and **b**, respectively, of Figure 7.1.

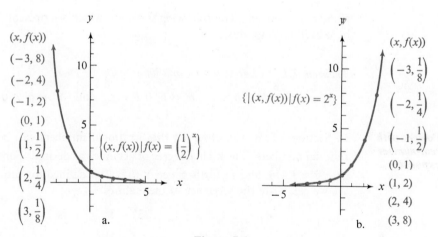

Figure 7.1

Increasing and decreasing functions

Notice that in accordance with Theorem 7.1, the graph of the function determined by $f(x) = (\frac{1}{2})^x$ goes *down* to the right, and the graph of the function determined by $f(x) = 2^x$ goes *up* to the right. For this reason, we say that the former function is a **decreasing function** and that the latter is an **increasing function**. In either case, the domain is the set of real numbers, and the range is the set of positive real numbers.

Exercise 7.1

Find the second component of each of the ordered pairs that makes the pair a solution of the corresponding equation.

1. $y = 3^x$; $(0, \ )$, $(1, \ )$, $(2, \ )$ 2. $y = -2^x$; $(-2, \ )$, $(0, \ )$, $(2, \ )$

3. $y = -5^x$; $(-2, \ )$, $(0, \ )$, $(2, \ )$ 4. $y = 4^x$; $(0, \ )$, $(1, \ )$, $(2, \ )$

5. $f(x) = \left(\dfrac{1}{2}\right)^x$; $(-3, \ )$, $(0, \ )$, $(3, \ )$ 6. $f(x) = \left(\dfrac{1}{3}\right)^x$; $(-3, \ )$, $(0, \ )$, $(3, \ )$

7. $g(x) = 10^x$; $(-2, \ )$, $(-1, \ )$, $(0, \ )$ 8. $g(x) = 10^{-x}$; $(0, \ )$, $(1, \ )$, $(2, \ )$

Graph each function.

9. $\{(x, y) \mid y = 4^x\}$ 10. $\{(x, y) \mid y = 5^x\}$

11. $\{(x, y) \mid y = 10^x\}$ 12. $\{(x, y) \mid y = 10^{-x}\}$

13. $\{(x, y) \mid y = 2^{-x}\}$ 14. $\{(x, y) \mid y = 3^{-x}\}$

15. $\left\{(x, y) \mid y = \left(\dfrac{1}{3}\right)^x\right\}$ 16. $\left\{(x, y) \mid y = \left(\dfrac{1}{4}\right)^x\right\}$

17. $\left\{(x, y) \mid y = \left(\dfrac{1}{2}\right)^{-x}\right\}$ 18. $\left\{(x, y) \mid y = \left(\dfrac{1}{3}\right)^{-x}\right\}$

19. Graph $\{(x, f(x)) \mid f(x) = 1^x\}$. Is this an exponential function? Name the function.

20. Graph $\{(x, y) \mid y = 10^x, x > 0\}$ and $\{(x, y) \mid x = 10^y, x > 0\}$ on the same set of axes.

21. Solve each equation by inspection.

     a. $10^x = \dfrac{1}{100}$ b. $\left(\dfrac{1}{2}\right)^x = 16$ c. $16^x = 8$

22. Determine an integer n such that $n < x < n + 1$.

     a. $3^x = 16.2$ b. $4^x = 87.1$ c. $10^x = 0.016$

7.2   The Logarithmic Function

Inverse of an exponential function

In the exponential function

$$\{(x, y) \mid y = b^x, \quad b > 0, b \neq 1\}, \tag{1}$$

whose graph is illustrated in Figure 7.1 for $b = 1/2$ and $b = 2$, there is only one x associated with each y. Thus, by Definition 5.4, we have the inverse function $\{(y, x)\}$, with x and y as in (1); that is, we have the function defined by

$$x = b^y \quad (b > 0, \ b \neq 1). \tag{2}$$

Observe that the relation $x > 0$ is implied by Equation (2), because there is no real number y for which b^y is not positive.

The graphs of functions of this form can be illustrated by the example

$$\{(x, y) \mid x = 10^y\}.$$

We consider x the variable denoting an element in the domain and, in the defining equation, assign arbitrary values for x, say,

$$(0.01, \ ), \quad (0.1, \ ), \quad (1, \ ), \quad (10, \ ), \quad (100, \ ),$$

to obtain the ordered pairs

$$(0.01, -2), \quad (0.1, -1), \quad (1, 0), \quad (10, 1), \quad (100, 2).$$

These can be plotted and connected with a smooth curve as in Figure 7.2.

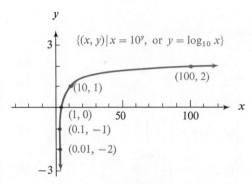

Figure 7.2

It is always useful to be able to express the variable denoting an element in the range explicitly in terms of the variable denoting an element in the domain. To do this in an equation such as that defining (2), we use the notation

$$y = \log_b x \quad (x > 0, \, b > 0, \, b \neq 1). \tag{3}$$

Functions defined by such equations are called **logarithmic functions**.

Properties of logarithmic functions

From the graph in Figure 7.2, we generalize from $\log_{10} x$ to $\log_b x$ and observe that, for $b > 1$, a logarithmic function has the following properties:

1. The domain is the set of positive real numbers, and the range is the set of all real numbers.

2. $\log_b x < 0$ for $0 < x < 1$, $\log_b x = 0$ for $x = 1$, and $\log_b x > 0$ for $x > 1$.

Logarithmic and exponential statements

It should be recognized that the equations appearing in (2) on page 161 and (3) are different equations determining the same function in the same way that $x = y + 4$ and $y = x - 4$ determine the same function, and we may use whichever equation suits our purpose. Thus, exponential statements may be written in logarithmic form, and logarithmic statements may be written in exponential form.

Examples

Write each of the following statements in logarithmic form.

a. $5^2 = 25$ b. $8^{1/3} = 2$ c. $3^{-2} = \dfrac{1}{9}$

Solutions   a. $\log_5 25 = 2$ b. $\log_8 2 = \dfrac{1}{3}$ c. $\log_3 \dfrac{1}{9} = -2$

Examples   Write each of the following statements in exponential form.

a. $\log_{10} 100 = 2$ b. $\log_3 81 = 4$ c. $\log_2 \dfrac{1}{2} = -1$

Solutions   a. $10^2 = 100$ b. $3^4 = 81$ c. $2^{-1} = \dfrac{1}{2}$

The logarithmic function associates with each number x the exponent y such that the power b^y is equal to x. In other words, we can think of $\log_b x$ as an exponent on b. Thus,

$$b^{\log_b x} = x.$$

Laws of logarithms   Since a logarithm is an exponent, the following theorem follows directly from the properties of powers with real-number exponents.

Theorem 7.2   *If x_1, x_2, and b are positive real numbers, $b \neq 1$, $m \in R$, then*

$$\text{I} \quad \log_b (x_1 x_2) = \log_b x_1 + \log_b x_2,$$

$$\text{II} \quad \log_b \frac{x_2}{x_1} = \log_b x_2 - \log_b x_1,$$

$$\text{III} \quad \log_b (x_1)^m = m \log_b x_1.$$

Proof of I   Since $x_1 = b^{\log_b x_1}$ and $x_2 = b^{\log_b x_2}$, it follows that

$$x_1 x_2 = b^{\log_b x_1} \cdot b^{\log_b x_2}$$
$$= b^{\log_b x_1 + \log_b x_2},$$

and, by the definition of a logarithm,

$$\log_b (x_1 x_2) = \log_b x_1 + \log_b x_2.$$

The validity of II and III can be established in a similar manner.

Exercise 7.2

Express in logarithmic notation.

1. $4^2 = 16$ 2. $5^3 = 125$ 3. $3^3 = 27$

4. $8^2 = 64$ 5. $\left(\dfrac{1}{2}\right)^2 = \dfrac{1}{4}$ 6. $\left(\dfrac{1}{3}\right)^2 = \dfrac{1}{9}$

7. $8^{-1/3} = \dfrac{1}{2}$ 8. $64^{-1/6} = \dfrac{1}{2}$ 9. $10^2 = 100$

10. $10^0 = 1$ 11. $10^{-1} = 0.1$ 12. $10^{-2} = 0.01$

Express in exponential notation.

13. $\log_2 64 = 6$ 14. $\log_5 25 = 2$ 15. $\log_3 9 = 2$

16. $\log_{16} 256 = 2$ 17. $\log_{1/3} 9 = -2$ 18. $\log_{1/2} 8 = -3$

19. $\log_{10} 1000 = 3$ 20. $\log_{10} 1 = 0$ 21. $\log_{10} 0.01 = -2$

Find the value of each of the following logarithms.

22. $\log_5 5$ 23. $\log_7 49$ 24. $\log_2 32$

25. $\log_4 64$ 26. $\log_5 \sqrt{5}$ 27. $\log_3 \sqrt{3}$

28. $\log_3 \dfrac{1}{3}$ 29. $\log_5 \dfrac{1}{5}$ 30. $\log_3 3$

31. $\log_2 2$ 32. $\log_{10} 10$ 33. $\log_{10} 100$

34. $\log_{10} 1$ 35. $\log_{10} 0.1$ 36. $\log_{10} 0.01$

Solve each equation for x, y, or b.

Examples   a. $\log_2 x = 3$ b. $\log_b 2 = \dfrac{1}{2}$

Solutions   Determine the solution by inspection or write in exponential form and then determine the solution.

a. $2^3 = x$ b. $b^{1/2} = 2$

$ x = 8$ $(b^{1/2})^2 = (2)^2$

$ b = 4$

37. $\log_3 9 = y$ 38. $\log_5 125 = y$ 39. $\log_b 8 = 3$

40. $\log_b 625 = 4$ 41. $\log_4 x = 3$ 42. $\log_{1/2} x = -5$

43. $\log_2 \dfrac{1}{8} = y$ 44. $\log_5 5 = y$ 45. $\log_b 10 = \dfrac{1}{2}$

46. $\log_b 0.1 = -1$ 47. $\log_2 x = 2$ 48. $\log_{10} x = -3$

49. Show that $\log_b 1 = 0$ for $b > 0$. *Hint:* Express the statement in exponential form.

50. Show that $\log_b b = 1$ for $b > 0$.

51. Show that $\log_b b^x = x$ for $b > 0$.

52. Graph $y = \log_2 x$. By examining the graph, what can you assert about $\log_2 a$ and $\log_2 b$ if $a < b$?

Express as the sum or difference of simpler logarithmic quantities.

Example $\log_b \left(\dfrac{xy}{z} \right)^{1/2}$

Solution   By Theorem 7.2-III,

$$\log_b \left(\frac{xy}{z} \right)^{1/2} = \frac{1}{2} \log_b \left(\frac{xy}{z} \right).$$

By I and II of Theorem 7.2,

$$\frac{1}{2} \log_b \left(\frac{xy}{z} \right) = \frac{1}{2} [\log_b x + \log_b y - \log_b z].$$

53. $\log_b (xy)$ **54.** $\log_b (xyz)$ **55.** $\log_b \left(\dfrac{x}{y} \right)$

56. $\log_b \left(\dfrac{xy}{z} \right)$ **57.** $\log_b x^5$ **58.** $\log_b x^{1/2}$

59. $\log_b \sqrt[3]{x}$ **60.** $\log_b \sqrt[3]{x^2}$ **61.** $\log_b \sqrt{\dfrac{x}{z}}$

62. $\log_b \sqrt{xy}$ **63.** $\log_{10} \sqrt[3]{\dfrac{xy^2}{z}}$ **64.** $\log_{10} \sqrt[5]{\dfrac{x^2 y}{z^3}}$

65. $\log_{10} \left(2\pi \sqrt{\dfrac{l}{g}} \right)$ **66.** $\log_{10} \sqrt{s(s-a)(s-b)(s-c)}$

Express as a single logarithm with coefficient 1.

Example $\dfrac{1}{2} (\log_b x - \log_b y)$

Solution   By Theorems 7.2-II and III,

$$\frac{1}{2} (\log_b x - \log_b y) = \frac{1}{2} \log_b \left(\frac{x}{y} \right) = \log_b \left(\frac{x}{y} \right)^{1/2}.$$

67. $\log_b x + \log_b y$ **68.** $\log_b x - \log_b y$

69. $2 \log_b x + 3 \log_b y$ **70.** $\dfrac{1}{4} \log_b x - \dfrac{3}{4} \log_b y$

71. $3 \log_b x + \log_b y - 2 \log_b z$ **72.** $\dfrac{1}{3} (\log_b x + \log_b y - 2 \log_b z)$

73. $\log_{10} (x - 2) + \log_{10} x - 2 \log_{10} z$

74. $\dfrac{1}{2} (\log_{10} x - 3 \log_{10} y - 5 \log_{10} z)$

75. Show that $\dfrac{1}{4} \log_{10} 8 + \dfrac{1}{4} \log_{10} 2 = \log_{10} 2$.

76. Show that $4 \log_{10} 3 - 2 \log_{10} 3 + 1 = \log_{10} 90$.

77. Show that $\log_{10} [\log_3 (\log_5 125)] = 0$.

78. Using the definition of a logarithm ($\log_b x$ is a number such that $b^{\log_b x} = x$) and the laws of exponents, prove Theorem 7.2-II.

7.3   *Logarithms to the Base 10*

There are two logarithmic functions of special interest in mathematics; one is defined by

$$y = \log_{10} x, \tag{1}$$

and the other by

$$y = \log_e x, \tag{2}$$

where e is an irrational number with decimal approximation 2.7182818, to eight digits. Because these functions possess similar properties, and because we are more familiar with the number 10, we shall, for the present, confine our attention to (1).

Determination of $\log_{10} x$

Values for $\log_{10} x$ are called **logarithms to the base 10**, or **common logarithms**. From the definition of $\log_{10} x$,

$$10^{\log_{10} x} = x \quad (x > 0); \tag{3}$$

that is, $\log_{10} x$ is the exponent that must be placed on 10 so that the resulting power is x. The problem with which we are concerned in this section is that of finding $\log_{10} x$ for each positive x. First, $\log_{10} x$ can easily be determined for all values of x that are integral powers of 10:

$$\log_{10} 10 = \log_{10} 10^1 = 1,$$
$$\log_{10} 100 = \log_{10} 10^2 = 2,$$

etc.; and, similarly,

$$\log_{10} 1 \quad = \log_{10} 10^0 \quad = 0,$$
$$\log_{10} 0.1 \quad = \log_{10} 10^{-1} = -1,$$
$$\log_{10} 0.01 \quad = \log_{10} 10^{-2} = -2,$$
$$\log_{10} 0.001 = \log_{10} 10^{-3} = -3.$$

A table of logarithms is used to find $\log_{10} x$ for $1 \le x \le 10$ (see page 346). Consider the excerpt from this table shown in Figure 7.3. Each number in the column headed by x represents the first two significant digits of x, while each number in the row opposite x contains the third significant digit of x. The digits located at the intersection of a row and a column specify the logarithm of x. For example, to find $\log_{10} 4.25$ we look at the intersection of the row opposite 4.2 under x and the column headed by 5. Thus, we see that

$$\log_{10} 4.25 = 0.6284.$$

x	0	1	2	3	4	5	6	7	8	9
3.8	.5798	.5809	.5821	.5832	.5843	.5855	.5866	.5877	.5888	.5899
3.9	.5911	.5922	.5933	.5944	.5955	.5966	.5977	.5988	.5999	.6010
4.0	.6021	.6031	.6042	.6053	.6064	.6075	.6085	.6096	.6107	.6117
4.1	.6128	.6138	.6149	.6160	.6170	.6180	.6191	.6201	.6212	.6222
4.2	.6232	.6243	.6253	.6263	.6274	.6284	.6294	.6304	.6314	.6325
4.3	.6335	.6345	.6355	.6365	.6375	.6385	.6395	.6405	.6415	.6425
4.4	.6435	.6444	.6454	.6464	.6474	.6484	.6493	.6503	.6513	.6522
4.5	.6532	.6542	.6551	.6561	.6571	.6580	.6590	.6599	.6609	.6618
4.6	.6628	.6637	.6646	.6656	.6665	.6675	.6684	.6693	.6702	.6712

Figure 7.3

The equality sign is used here in an inexact sense. More properly, we should write $\log_{10} 4.25 \approx 0.6284$; $\log_{10} 4.25$ is *irrational* and does not equal the *rational number* 0.6284. We shall follow customary usage, however, writing $=$ instead of $\approx$, and leave the intent to the context.

Characteristic and mantissa

Now suppose we wish to find $\log_{10} x$ for values of x outside the range of the table —that is, for $0 < x < 1$ or $x > 10$. This can be done quite readily by first representing the number in scientific notation—that is, as the product of a number between 1 and 10 and a power of 10—and applying Theorem 7.2-I. For example,

$$\log_{10} 42.5 = \log_{10} (4.25 \times 10^1) = \log_{10} 4.25 + \log_{10} 10^1$$
$$= 0.6284 + 1 = 1.6284,$$

$$\log_{10} 4250 = \log_{10} (4.25 \times 10^3) = \log_{10} 4.25 + \log_{10} 10^3$$
$$= 0.6284 + 3 = 3.6284.$$

Observe that the decimal portion of the logarithm is always 0.6284, and the integral portion is just the exponent on 10 when the number is written in scientific notation.

This process can be reduced to a mechanical one by considering $\log_{10} x$ to consist of two parts, an integral part (called the **characteristic**) and a nonnegative decimal fraction part (called the **mantissa**). Thus the table of values for $\log_{10} x$ for $1 < x < 10$ can be looked upon as a table of mantissas for $\log_{10} x$ for all $x > 0$.

To find $\log_{10} 43700$, we first write

$$\log_{10} 43700 = \log_{10} (4.37 \times 10^4).$$

Upon examining the table of logarithms, we find that $\log_{10} 4.37 = 0.6405$, so that

$$\log_{10} 43700 = 4.6405,$$

where we have prefixed the characteristic 4.

Now consider an example of $\log_{10} x$ for $0 < x < 1$. To find $\log_{10} 0.00402$, we write

$$\log_{10} 0.00402 = \log_{10} (4.02 \times 10^{-3}).$$

Examining the table, we find $\log_{10} 4.02$ is 0.6042. Upon adding 0.6042 to the characteristic -3, we obtain

$$\log_{10} 0.00402 = -2.3958,$$

where the decimal portion of the logarithm is no longer 0.6042 as it is in the case of all numbers $x > 1$ for which the first three significant digits of x are 402. To circumvent this situation, it is customary to write the logarithm in a form in which the fractional part is positive. In the foregoing example, we write

$$\log_{10} 0.00402 = 0.6042 - 3$$
$$= 0.6042 + (7 - 10)$$
$$= 7.6042 - 10,$$

and the fractional part is positive. The logarithms

$$6.6042 - 9,$$
$$12.6042 - 15,$$

etc., are equally valid representations, but $7.6042 - 10$ is customary in most cases.

Determination
of antilog$_{10}$ x

It is possible to reverse the process described in this section and, being given $\log_{10} x$, to find x. In this event, x is referred to as the **antilogarithm** (antilog$_{10}$) of $\log_{10} x$. For example, antilog$_{10}$ 1.6395 can be obtained by locating the mantissa, 0.6395, in the body of the $\log_{10}$ tables and observing that the associated antilog$_{10}$ is 4.36. Thus,

$$\text{antilog}_{10} \ 1.6395 = 4.36 \times 10^1 = 43.6.$$

Linear
interpolation

If we seek the common logarithm of a number that is not an entry in the table (for example, $\log_{10} 3712$), or if we seek x when $\log_{10} x$ is not an entry in the table, it is customary to use a procedure called **linear interpolation**.

A table of common logarithms is a set of ordered pairs. For each number x there is an associated number $\log_{10} x$, and we have $\{(x, \log_{10} x)\}$ displayed in convenient tabular form. Because of space limitations, only three digits for the number x and four for the number $\log_{10} x$ appear in the table. By means of linear interpolation, however, the table can be used to find approximations to logarithms for four-digit numbers.

Let us examine geometrically the concepts involved. A portion of the graph of

$$y = \log_{10} x$$

is shown in Figure 7.4a. The curvature is exaggerated to illustrate the principle

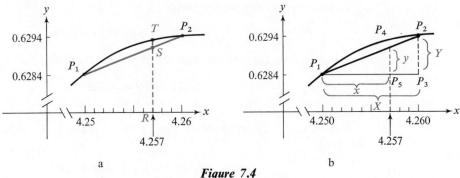

Figure 7.4

involved. We propose to use the line segment joining the points P_1 and P_2 as an approximation to the curve passing through the two points. If a large graph of

$y = \log_{10} x$ were available, the value of, say, $\log_{10} 4.257$ could be found by using the ordinate of the point T on the curve for $x = 4.257$. Since there is no way to accomplish this with a table of values only, we shall instead use the ordinate of the point S on the line segment as an approximation to the ordinate of the point T on the curve.

This can be accomplished directly from the set of numbers available in the table of logarithms. Consider Figure 7.4b, where $\overline{P_2 P_3}$ and $\overline{P_4 P_5}$ are perpendicular to $\overline{P_1 P_3}$.

From geometry, we have $\Delta P_1 P_4 P_5 \sim \Delta P_1 P_2 P_3$, where the lengths of the corresponding sides are proportional, and hence

$$\frac{x}{X} = \frac{y}{Y}. \tag{4}$$

If we know any three of these numbers, we can determine the fourth. For the purpose of interpolation, we assume all of our members have four-digit numerals; that is, we consider 4.250 instead of 4.25, and 4.260 instead of 4.26. We note that the point corresponding to 4.257 is located just $7/10$ of the distance between the points corresponding to 4.250 and 4.260, respectively, and the value Y, namely 0.0010, is just the difference between the logarithms 0.6284 and 0.6294. It follows from (4) that

$$\frac{7}{10} = \frac{y}{0.0010},$$

from which

$$y = \frac{7}{10}(0.0010) = 0.0007.$$

We now add 0.0007 to 0.6284 and thus obtain a good approximation to the required logarithm. That is, we have

$$\log_{10} 4.257 = 0.6291.$$

An example in Exercise 7.3 shows a convenient arrangement for the calculations involved in the foregoing illustrations. The antilogarithm of a number can be found by a similar procedure. With practice, however, it is possible to interpolate mentally in both procedures.

Exercise 7.3

Write the characteristic of each of the following.

Examples   a. $\log_{10} 248$ b. $\log_{10} 0.0057$

Solutions   Represent the number in scientific notation.

a. $\log_{10}(2.48 \times 10^2)$ b. $\log_{10}(5.7 \times 10^{-3})$

The exponent on the base 10 is the characteristic.

2 -3, or $7 - 10$

1. $\log_{10} 312$ 2. $\log_{10} 0.02$ 3. $\log_{10} 0.00851$
4. $\log_{10} 8.012$ 5. $\log_{10} 0.00031$ 6. $\log_{10} 0.0004$
7. $\log_{10} (15 \times 10^3)$ 8. $\log_{10} (820 \times 10^4)$

Find each logarithm.

Examples   a. $\log_{10} 16.8$ b. $\log_{10} 0.043$

Solutions   Represent the number in scientific notation.

     a. $\log_{10} (1.68 \times 10^1)$ b. $\log_{10} (4.3 \times 10^{-2})$

Determine the mantissa from the table.

     0.2253   0.6335

Add the characteristic as determined by the exponent on the base 10.

     1.2253 $8.6335 - 10$

9. $\log_{10} 6.73$ 10. $\log_{10} 891$
11. $\log_{10} 0.813$ 12. $\log_{10} 0.00214$
13. $\log_{10} 0.08$ 14. $\log_{10} 0.000413$
15. $\log_{10} (2.48 \times 10^2)$ 16. $\log_{10} (5.39 \times 10^{-3})$

Find each antilogarithm.

Example $\text{antilog}_{10} 2.7364$

Solution   Locate the mantissa in the body of the table of mantissas and determine the associated antilog_{10} (a number between 1 and 10); write the characteristic as an exponent
on the base 10.

$$5.45 \times 10^2 = 545$$

17. $\text{antilog}_{10} 0.6128$ 18. $\text{antilog}_{10} 0.2504$
19. $\text{antilog}_{10} 0.5647$ 20. $\text{antilog}_{10} 3.9258$
21. $\text{antilog}_{10} (8.8075 - 10)$ 22. $\text{antilog}_{10} (3.9722 - 5)$
23. $\text{antilog}_{10} 3.7388$ 24. $\text{antilog}_{10} 2.0086$
25. $\text{antilog}_{10} (6.8561 - 10)$ 26. $\text{antilog}_{10} (1.8156 - 4)$

Find each logarithm.

Example $\log_{10} 4.257$

Solution   Interpolate mentally or use the following procedure.

$$
\begin{array}{cc}
x & \log_{10} x \\
\end{array}
$$

$$
10 \left\{ 7 \left\{ \begin{array}{c|c} 4.250 & 0.6284 \\ 4.257 & ? \end{array} \right\} y \\ 4.260 & 0.6294 \end{array} \right\} 0.0010
$$

Set up a proportion and solve for y.

$$\frac{7}{10} = \frac{y}{0.0010}$$

$$y = 0.0007$$

Add 0.0007, the value of y, to 0.6284.

$$\log_{10} 4.257 = 0.6284 + 0.0007 = 0.6291$$

27. $\log_{10} 4.213$	**28.** $\log_{10} 8.184$	**29.** $\log_{10} 1522$
30. $\log_{10} 203.4$	**31.** $\log_{10} 37110$	**32.** $\log_{10} 72.36$
33. $\log_{10} 0.5123$	**34.** $\log_{10} 0.008351$	

Find each antilogarithm.

Example   antilog$_{10}$ 0.6446

Solution   Interpolate mentally or use the following procedure.

$$x \qquad \text{antilog}_{10}\ x$$

$$0.0010 \left\{ \begin{array}{l} 0.0002 \left\{ \begin{array}{l|l} 0.6444 & 4.410 \\ 0.6446 & ? \end{array} \right\} y \\ \\ \quad\ 0.6454 \mid 4.420 \end{array} \right\} 0.010$$

Set up a proportion and solve for y.

$$\frac{0.0002}{0.0010} = \frac{y}{0.010}$$

$$y = 0.002$$

Add 0.002, the value of y, to 4.410.

$$\text{antilog}_{10}\ 0.6446 = 4.410 + 0.002 = 4.412$$

35. antilog$_{10}$ 0.5085	**36.** antilog$_{10}$ 0.8087
37. antilog$_{10}$ 1.0220	**38.** antilog$_{10}$ 3.0759
39. antilog$_{10}$ $(8.7055 - 10)$	**40.** antilog$_{10}$ $(9.8742 - 10)$
41. antilog$_{10}$ $(2.8748 - 3)$	**42.** antilog$_{10}$ $(7.7397 - 10)$

43. If we use linear interpolation to find $\log_{10} 3.751$ and $\log_{10} 3.755$, which of the resulting approximations should we expect to be more nearly correct? Why?

44. If we use linear interpolation to find $\log_{10} 1.025$ and $\log_{10} 9.025$, which of the resulting approximations should we expect to be more nearly correct? Why?

Find each of the following by means of the table of logarithms to the base 10.

Example $10^{0.6263}$

Solution   The logarithmic function is the inverse of the exponential function. Hence, the
                   element $10^{0.6263}$ in the range of the exponential function is $\text{antilog}_{10} \, 0.6263$ in
                   the domain of the logarithmic function. From the table of logarithms on page 346,
                   we find that

$$10^{0.6263} = \text{antilog}_{10} \, 0.6263 = 4.23.$$

    45. $10^{0.9590}$ **46.** $10^{0.8241}$ **47.** $10^{3.6990}$

    48. $10^{2.3874}$ **49.** $10^{2.0531}$ **50.** $10^{1.7396}$

    51. Write the following logarithms to the base 10 in a form in which both the
integral and fractional parts are negative.

        a. $8.7321 - 10$ **b.** $6.4187 - 10$

    52. Write the following logarithms to the base 10 in a form in which the fractional
part is positive.

        a. $2 - .7113$ **b.** -4.6621

7.4   *Computations with Logarithms*

The use of the slide rule and the advent of high-speed computing devices have
almost removed the need to perform routine numerical computations with pencil
and paper by logarithms. Nevertheless, we introduce the techniques involved in
making such computations because the writing of the logarithmic equations involved sheds light on the properties of the logarithmic function and on the useful-
ness of Theorem 7.2, which we reproduce here using the base 10.
    If x_1 and x_2 are positive real numbers, $m \in R$, then

$$\text{I} \quad \log_{10}(x_1 x_2) = \log_{10} x_1 + \log_{10} x_2,$$

$$\text{II} \quad \log_{10} \frac{x_1}{x_2} = \log_{10} x_1 - \log_{10} x_2,$$

$$\text{III} \quad \log_{10} (x_1)^m = m \log_{10} x_1.$$

Before illustrating the use of these laws, let us make two observations:

L-1   If $M = N \, (M, N > 0)$, then $\log_b M = \log_b N$.
L-2   If $\log_b M = \log_b N$, then $M = N$.

Both L-1 and L-2 follow from the fact that the variables in a logarithmic function
are in one-to-one correspondence.

Example   Find the product of 3.825 and 0.00729, using logarithms.

Solution   Let $N = (3.825)(0.00729)$. By Property L-1,

$$\log_{10} N = \log_{10} [(3.825)(0.00729)].$$

Now, by Theorem 7.2-I,

$$\log_{10} N = \log_{10} 3.825 + \log_{10} 0.00729,$$

and from the table we obtain

$$\log_{10} 3.825 = 0.5826, \quad \text{and} \quad \log_{10} 0.00729 = 7.8627 - 10,$$

so that

$$\log_{10} N = (0.5826) + (7.8627 - 10)$$

$$= 8.4453 - 10.$$

The computation is completed by referring to the table for

$$N = \text{antilog}_{10} (8.4453 - 10) = 2.788 \times 10^{-2}$$

$$= 0.02788.$$

Thus we have

$$N = (3.825)(0.00729) = 0.02788.$$

Actual computation shows the product to be 0.02788425, so that the result obtained by use of logarithms is correct to four significant digits. Some error should be expected because we are using approximations to irrational numbers when we employ a table of logarithms.

Example   Compute $\dfrac{(8.21)^{1/2}(2.17)^{2/3}}{(3.14)^3}$.

Solution   Setting

$$N = \frac{(8.21)^{1/2}(2.17)^{2/3}}{(3.14)^3},$$

we obtain

$$\log_{10} N = \log_{10} \frac{(8.21)^{1/2}(2.17)^{2/3}}{(3.14)^3}$$

$$= \log_{10} (8.21)^{1/2} + \log_{10} (2.17)^{2/3} - \log_{10} (3.14)^3$$

$$= \frac{1}{2} \log_{10} 8.21 + \frac{2}{3} \log_{10} 2.17 - 3 \log_{10} 3.14.$$

The table provides values for the logarithms involved here, and the remainder of the computation is routine. In order to avoid confusion in computations of this sort, a systematic approach of some kind is desirable (see the example in the exercise set).

As observed on page 168, in computations involving numbers less than 1, it is sometimes convenient to use expressions other than such differences as 7._____ − 10 and 8._____ − 10 for negative characteristics.

Example   Compute $\sqrt[3]{0.0235}$.

Solution   Setting

$$N = \sqrt[3]{0.0235} = (0.0235)^{1/3},$$

we have

$$\log_{10} N = \log_{10} (0.0235)^{1/3} = \frac{1}{3} \log_{10} (0.0235)$$

From the table, we see that the mantissa of this logarithm is 0.3711. Because we wish to multiply by 1/3, we select $7.3711 - 9$ instead of $8.3711 - 10$ for the logarithm. Thus

$$\log_{10} N = \frac{1}{3} (7.3711 - 9) = 2.4570 - 3,$$

where we have avoided obtaining a nonintegral negative part of the logarithm. From the table,

$$N = \text{antilog}_{10} (2.4570 - 3) = 0.2864.$$

Thus,

$$\sqrt[3]{0.0235} = 0.2864.$$

Exercise 7.4

Compute each of the following using logarithms.

Example $\sqrt{\dfrac{(23.4)(0.681)}{4.13}}$

Solution   Let $P = \sqrt{\dfrac{(23.4)(0.681)}{4.13}}$.

Then $\log_{10} P = \dfrac{1}{2} (\log_{10} 23.4 + \log_{10} 0.681 - \log_{10} 4.13)$.

$$\begin{aligned}
\log_{10} 23.4 &= 1.3692 \\
\log_{10} 0.681 &= 9.8331 - 10
\end{aligned} \Big\} \text{add}$$

$$\begin{aligned}
\log_{10}(23.4)(0.681) &= 11.2023 - 10 \\
\log_{10} 4.13 &= 0.6160
\end{aligned} \Big\} \text{subtract}$$

$$\overline{\phantom{\log_{10}(23.4)(0.681) =\ } 10.5863 - 10}$$

$$\log_{10} \frac{(23.4)(0.681)}{4.13} = 10.5863 - 10 = 0.5863$$

$$\frac{1}{2} \log_{10} \frac{(23.4)(0.681)}{4.13} = \frac{1}{2} (0.5863) = 0.2931.$$

Hence, $\log_{10} P = 0.2931$, from which

$$P = \text{antilog}_{10} 0.2931 = 1.964.$$

1. $(2.32)(1.73)$
2. $(83.2)(6.12)$
3. $\dfrac{3.15}{1.37}$
4. $\dfrac{1.38}{2.52}$

5. $\dfrac{0.0149}{32.3}$
6. $\dfrac{0.00214}{3.17}$
7. $(2.3)^5$
8. $(4.62)^3$

9. $\sqrt[3]{8.12}$
10. $\sqrt[5]{75}$
11. $(0.0128)^5$
12. $(0.0021)^6$

13. $\sqrt{0.0021}$
14. $\sqrt[5]{0.0471}$
15. $\sqrt[3]{0.0214}$
16. $\sqrt[4]{0.0018}$

17. $\dfrac{(8.12)(8.74)}{7.19}$
18. $\dfrac{(0.412)^2(84.3)}{\sqrt{21.7}}$
19. $\dfrac{(6.49)^2\sqrt[3]{8.21}}{17.9}$

20. $\dfrac{(2.61)^2(4.32)}{\sqrt{7.83}}$
21. $\dfrac{(0.3498)(27.16)}{6.814}$
22. $\dfrac{(4.813)^2(20.14)}{3.612}$

23. $\sqrt{\dfrac{(4.71)(0.00481)}{(0.0432)^2}}$
24. $\sqrt{\dfrac{(2.85)^3(0.97)}{(0.035)}}$

25. Find an approximate value for $2^{\sqrt{2}}$.

26. Find an approximate value for 2^{π}.

27. The period T of a simple pendulum is given by the formula $T = 2\pi\sqrt{L/g}$, where T is in seconds, L is the length of the pendulum in feet, and $g \approx 32\ \text{ft/sec}^2$. Find the period of a pendulum 1 foot long.

28. The area A of a triangle in terms of the sides is given by the formula

$$A = \sqrt{s(s-a)(s-b)(s-c)},$$

where a, b, and c are the lengths of the sides of the triangle and s equals one-half of the perimeter. Find the area of a triangle in which the lengths of the three sides are 2.314 inches, 4.217 inches, and 5.618 inches.

7.5   *Exponential and Logarithmic Equations*

An equation in one variable in which the variable occurs in an exponent is called an **exponential equation**. Solution sets of some such equations can be found by means of logarithms.

Example   Find the solution set of $5^x = 7$.

Solution   Since $5^x > 0$ for all x, we can apply Property L-1 from the preceding section and write

$$\log_{10} 5^x = \log_{10} 7,$$

and from Theorem 7.2-III,

$$x \log_{10} 5 = \log_{10} 7.$$

(Solution continued)

Dividing each member by $\log_{10} 5$, we obtain

$$x = \frac{\log_{10} 7}{\log_{10} 5} = \frac{0.8451}{0.6990} \approx 1.209.$$

The solution set is

$$\left\{ \frac{\log_{10} 7}{\log_{10} 5} \right\},$$

and a decimal approximation for the single solution is 1.209. Note that for this approximation, the logarithms are *divided*, not subtracted.

Computation of interest

Both exponential and logarithmic functions are of very great importance in applied mathematics. One such application of logarithms is in the computation of compound interest. The interest I on a given amount P of money for a definite number n of periods at a specified rate r per period is called **simple interest** and can be computed by the familiar formula $I = Prn$. If, however, the interest accruing to an amount of money is periodically added to the amount, and over the next period this new total is earning interest, we say that the principal is earning **compound interest**. For example, if the sum of one dollar is earning interest at a rate r per year compounded annually, the amount present, A, is given

after one year by $A = 1 + r$,

after two years by $A = (1 + r) + r(1 + r) = (1 + r)^2$,

after three years by $A = (1 + r)^2 + r(1 + r)^2 = (1 + r)^3$,

and

after n years by $A = (1 + r)^n$.

For each dollar invested under such an arrangement, we have $(1 + r)^n$ dollars after n years, so that P dollars invested under the same arrangement, after n years, would amount to

$$A = P(1 + r)^n. \tag{1}$$

If the interest is compounded t times yearly, then the rate per period is r/t instead of r, where r is the stated rate per year, and the number of periods is increased to tn, so that (1) becomes

$$A = P\left(1 + \frac{r}{t}\right)^{tn}. \tag{2}$$

Specifically, the amount of P dollars compounded *semiannually* for n years at a yearly rate of interest r will be

$$A = P\left(1 + \frac{r}{2}\right)^{2n}, \tag{2a}$$

and when compounded *quarterly*,

$$A = P\left(1 + \frac{r}{4}\right)^{4n}. \tag{2b}$$

Although from a practical standpoint problems similar to the following example are handled by means of tables, for illustrative purposes we shall use logarithms both in the example and in the exercise set.

Example   What rate is necessary in order that \$2500 compounded quarterly will amount to \$4800 in twelve years?

Solution   From Equation (2b), we have

$$4800 = 2500\left(1 + \frac{r}{4}\right)^{48}.$$

Hence,

$$\left(1 + \frac{r}{4}\right)^{48} = 1.92,$$

$$\left(1 + \frac{r}{4}\right) = 1.92^{1/48}.$$

Solving for r by means of logarithms, we obtain

$$\log_{10}\left(1 + \frac{r}{4}\right) = \log_{10} 1.92^{1/48}$$

$$= \frac{1}{48}\log_{10} 1.92$$

$$= \frac{1}{48}(0.2833),$$

$$\log_{10}\left(1 + \frac{r}{4}\right) = 0.0059$$

$$\text{antilog}_{10}\, 0.0059 = 1 + \frac{r}{4} = 1.014.$$

Thus we have

$$r = 4(1.014 - 1),$$

$$r = 0.056.$$

The required rate of interest is about 5.6%.

Computation   An interesting application of logarithms occurs in the field of chemistry. The
of pH   chemist defines the pH (hydrogen potential) of a solution by

$$p\text{H} = \log_{10}\frac{1}{[\text{H}^+]}, \tag{3}$$

where the symbol $[\text{H}^+]$ denotes a numerical value for the concentration of hydrogen ions in aqueous solution in moles per liter. From (3) we obtain

$$p\text{H} = \log_{10}[\text{H}^+]^{-1},$$

and therefore

$$p\text{H} = -\log_{10}[\text{H}^+]. \tag{3a}$$

Values of pH are usually given to the nearest tenth of a unit.

Example   Find the pH of a solution for which $[\text{H}^+] = 4.0 \times 10^{-5}$.

Solution   From Equation (3), we have

$$p\text{H} = \log_{10} \frac{1}{4 \times 10^{-5}} = \log_{10}(2.5 \times 10^4),$$
$$p\text{H} = 4.4.$$

Exercise 7.5

Solve. Leave each solution in logarithmic form using the base 10.

Example $3^{x-2} = 16$.

Solution   By Property L-1,

$$\log_{10} 3^{x-2} = \log_{10} 16.$$

By Theorem 7.2-III,

$$(x - 2)\log_{10} 3 = \log_{10} 16,$$
$$x - 2 = \frac{\log_{10} 16}{\log_{10} 3},$$
$$x = \frac{\log_{10} 16}{\log_{10} 3} + 2.$$

The solution set is $\left\{ \dfrac{\log_{10} 16}{\log_{10} 3} + 2 \right\}$.

1. $2^x = 7$ 2. $3^x = 4$ 3. $3^{x+1} = 8$
4. $2^{x-1} = 9$ 5. $7^{2x-1} = 3$ 6. $3^{x+2} = 10$
7. $4^{x^2} = 15$ 8. $8^{x^2} = 21$ 9. $3^{-x} = 10$
10. $2.13^{-x} = 8.1$ 11. $3^{1-x} = 15$ 12. $4^{2-x} = 10$

Solve. Leave each result in the form of an equation equivalent to the given equation.

13. $y = x^n$,   for n 14. $y = Cx^{-n}$,   for n
15. $y = e^{kt}$,   for t 16. $y = Ce^{-kt}$,   for t

Solve.

Example $\log_{10}(x + 9) + \log_{10} x = 1$.

Solution   By Theorem 7.2-I, the given equation is equivalent to $\log_{10}(x+9)(x) = 1$ for $x+9$ and x greater than 0. Write this equation in exponential form, and solve for x.

$$x^2 + 9x = 10^1$$
$$x^2 + 9x - 10 = 0$$
$$(x+10)(x-1) = 0$$
$$x = -10, \quad x = 1$$

Since neither $\log_{10}(-10+9)$ nor $\log_{10}(-10)$ is defined, -10 does not satisfy the original equation; the solution set, therefore, is simply $\{1\}$.

17. $\log_{10} x + \log_{10} 2 = 3$ 18. $\log_{10}(x-1) - \log_{10} 4 = 2$
19. $\log_{10} x + \log_{10}(x+21) = 2$ 20. $\log_{10}(x+3) + \log_{10} x = 1$
21. $\log_{10}(x+2) + \log_{10}(x-1) = 1$ 22. $\log_{10}(x-3) - \log_{10}(x+1) = 1$

Solve for the variable n (nearest year), r (nearest $\frac{1}{2}\%$), or A (accuracy obtainable on four-place table of mantissas).

Example $(1+r)^{12} = 1.127$.

Solution   Equate $\log_{10}$ of each member and apply Theorem 7.2-III.

$$12 \log_{10}(1+r) = \log_{10} 1.127 = 0.0519$$

Multiply each member by $\frac{1}{12}$.

$$\log_{10}(1+r) = \frac{1}{12}(0.0519) = 0.0043$$

Determine $\text{antilog}_{10} 0.0043$ and solve for r.

$$\text{antilog}_{10} 0.0043 = 1.01$$
$$1 + r = 1.01$$
$$r = 0.01 \text{ or } 1\%$$

23. $(1+0.03)^{10} = A$ 24. $(1+0.04)^8 = A$
25. $(1+r)^6 = 1.34$ 26. $(1+r)^{10} = 1.48$
27. $100\left(1 + \dfrac{r}{2}\right)^{10} = 113$ 28. $40\left(1 + \dfrac{r}{4}\right)^{12} = 50.9$

Example $40(1 + 0.02)^n = 51.74$

Solution   Divide each member by 40; equate $\log_{10}$ of each member, and apply Theorem 7.2, Parts II and III.

$$(1.02)^n = \frac{51.74}{40}$$

$$\log_{10}(1.02)^n = \log_{10} \frac{51.74}{40}$$

(Solution continued)

$$n \log_{10} (1.02) = \log_{10} 51.74 - \log_{10} 40$$

$$n(0.0086) = 1.7138 - 1.6021$$

$$n = \frac{0.1117}{0.0086}$$

$$n = 13 \text{ years}$$

29. $(1 + 0.04)^n = 2.19$ 30. $(1 + 0.04)^n = 1.60$
31. $150(1 + 0.01)^{4n} = 240$ 32. $60(1 + 0.02)^{2n} = 116$

33. Find the amount of $5000 invested at 4% for 10 years when compounded annually. When compounded semi-annually.

34. Two men, A and B, each invested $10,000 at 4% for 20 years with a bank that computed interest quarterly. A withdrew his interest at the end of each three-month period, but B let his investment be compounded. How much more did B earn than A from his investment over the period of 20 years?

Calculate the pH of a solution with given hydrogen-ion concentration.

35. $[H^+] = 10^{-7}$ 36. $[H^+] = 4.0 \times 10^{-5}$ 37. $[H^+] = 2.0 \times 10^{-8}$
38. $[H^+] = 8.5 \times 10^{-3}$ 39. $[H^+] = 6.3 \times 10^{-7}$ 40. $[H^+] = 5.7 \times 10^{-7}$

Calculate the hydrogen-ion concentration $[H^+]$ of a solution with given pH.

Example $pH = 7.4$

Solution   Substitute 7.4 for pH in the relationship $pH = \log_{10} \dfrac{1}{[H^+]}$.

$$\log_{10} \frac{1}{[H^+]} = 7.4$$

Equate antilog$_{10}$ of each member and solve for $[H^+]$.

$$\frac{1}{[H^+]} = \text{antilog}_{10}\ 7.4 = 2.5 \times 10^7$$

$$[H^+] = \frac{1}{2.5 \times 10^7} = 0.4 \times 10^{-7} = 4 \times 10^{-8}$$

41. $pH = 3.0$ 42. $pH = 4.2$ 43. $pH = 5.6$
44. $pH = 8.3$ 45. $pH = 7.2$ 46. $pH = 6.9$

7.6 Logarithms to the Base e

The number e ($e \approx 2.7182818$) mentioned in Section 7.3 is of great mathematical interest and importance, and logarithms to the base e are frequently encountered in practical situations. Logarithms to the base e are called **natural logarithms**. It is possible to determine $\log_e x$ provided we have a table of $\log_{10} x$. Indeed, given a table of $\log_b x$, we can always find $\log_a x$ for any $a > 0$, $a \neq 1$. For example, if we wanted to express $\log_8 9$ in terms of $\log_{10}$, we could proceed as follows. First we let

$$\log_8 9 = N.$$

Then from our definition of a logarithm, we have

$$8^N = 9.$$

Using Theorem 7.2-III with logarithms to the base 10, we obtain

$$N \log_{10} 8 = \log_{10} 9,$$

from which

$$N = \frac{\log_{10} 9}{\log_{10} 8}.$$

Hence,

$$\log_8 9 = \frac{\log_{10} 9}{\log_{10} 8}.$$

Conversion to another base

We can do this for the general case as follows. Since $x > 0$ and $a \neq 1$, we have

$$x = a^{\log_a x}. \tag{1}$$

We can equate the logarithms to the base b, $b > 0$ and $b \neq 1$, of each member of (1) to obtain

$$\log_b x = \log_b a^{\log_a x}.$$

By Theorem 7.2-III, this yields

$$\log_b x = \log_a x \cdot \log_b a,$$

from which we obtain

$$\log_a x = \frac{\log_b x}{\log_b a}. \tag{2}$$

Conversion to base e

Equation (2) gives us a means of finding $\log_a x$ when we have a table of logarithms to the base b. In particular, if $a = e$ and $b = 10$, we have

$$\log_e x = \frac{\log_{10} x}{\log_{10} e}. \tag{3}$$

Since $\log_{10} e \approx 0.4343$, Equation (3) can be written

$$\log_e x = \frac{\log_{10} x}{0.4343}, \tag{4}$$

or

$$\log_e x = 2.3026 \log_{10} x, \tag{5}$$

which gives us a direct means of approximating $\log_e x$ when we have a table of logarithms to the base 10.

A table of $\log_e x$ will be found on page 349. We cannot, however, do as we did with a table of $\log_{10} x$—that is, simply list mantissas for logarithms over a certain interval and then manipulate characteristics to take care of all other intervals. Our numeration system is based on 10 and not on e. With the help of logarithms, though, we can use Formula (5), above, and the table of $\log_{10}$ on page 346 to find $\log_e x$.

Example   Find $\log_e 278$.

Solution   We shall do this in two ways.

1. Since

$$\log_e x = 2.3026 \log_{10} x,$$

using the $\log_{10}$ table, we have

$$\begin{aligned} \log_e 278 &= 2.3026 \log_{10} 278 \\ &= 2.3026(2.4440) \\ &= 5.63. \end{aligned}$$

2. Alternatively, we observe that $278 = 2.78 \times 10^2$. Hence, using the $\log_e$ table, we have

$$\begin{aligned} \log_e 278 = \log_e(2.78 \times 10^2) &= \log_e 2.78 + 2 \log_e 10 \\ &= 1.0223 + 2(2.3026) \\ &= 1.0223 + 4.6052 \\ &= 5.6275 \approx 5.63. \end{aligned}$$

Example   Find $\log_e 0.278$.

Solution   This time, let us simply observe that $0.278 = 2.78 \times 10^{-1}$. Hence

$$\begin{aligned} \log_e 0.278 = \log_e(2.78 \times 10^{-1}) &= \log_e 2.78 + \log_e 10^{-1} \\ &= \log_e 2.78 - \log_e 10 \\ &= 1.0224 - 2.3026 \\ &= -1.2802. \end{aligned}$$

As one would expect, the logarithm is negative.

The process used in the preceding examples can be reversed to find $\text{antilog}_e x$. If, however, a table for the exponential function $\{(x, y) \mid y = e^x\}$ is available (see Table II on page 348), the $\text{antilog}_e x$ can be read directly from this table, since $\text{antilog}_e x = e^x$.

Examples   a. Find $\text{antilog}_e 3.2$.   b. Find $\text{antilog}_e (-3.2)$.

Solutions   The values can be read directly from Table II.

a. $\text{antilog}_e 3.2 = e^{3.2}$ b. $\text{antilog}_e(-3.2) = e^{-3.2}$

$\qquad\qquad = 24.533$ $= 0.0408$.

Exercise 7.6

Find each logarithm to the nearest hundredth.

Example $\log_3 7$

Solution   Use logarithms to base 10 to evaluate.

$$\log_3 7 = \frac{\log_{10} 7}{\log_{10} 3} = \frac{0.8451}{0.4771} = 1.77$$

1. $\log_2 10$	**2.** $\log_2 5$	**3.** $\log_5 240$	**4.** $\log_3 18$
5. $\log_7 8.1$	**6.** $\log_5 60$	**7.** $\log_{100} 38$	**8.** $\log_{100} 240$

Find each logarithm directly from Table III on page 349.

9. $\log_e 3$	**10.** $\log_e 8$	**11.** $\log_e 17$	**12.** $\log_e 98$
13. $\log_e 327$	**14.** $\log_e 107$	**15.** $\log_e 450$	**16.** $\log_e 605$

Find each antilogarithm directly from Table II on page 348.

17. antilog$_e$ 0.50	**18.** antilog$_e$ 1.5	**19.** antilog$_e$ 3.4
20. antilog$_e$ 4.5	**21.** antilog$_e$ 0.231	**22.** antilog$_e$ 1.43
23. antilog$_e$ $-(0.15)$	**24.** antilog (-0.95)	**25.** antilog$_e$ (-2.5)
26. antilog$_e$ (-4.2)	**27.** antilog$_e$ (-0.255)	**28.** antilog$_e$ (-3.65)

Solve without using the tables of logarithms.

29. If $\log_2 8 = 3$, find $\log_8 2$.   **30.** If $\log_4 16 = 2$, find $\log_{16} 4$.

31. If $\log_{10} 3 = 0.4771$, find $\log_3 10$.   **32.** If $\log_{10} e = 0.4343$, find $\log_e 10$.

33. If $\log_{10} 5 = 0.6990$, find $\log_5 100$.   **34.** If $\log_{10} 3 = 0.4771$, find $\log_3 100$.

35. Show that $\log_9 7 = (1/2) \log_3 7$. By a similar argument, show that for $a, b > 0$, $a \ne 1$, $\log_{a^2} b = (1/2) \log_a b$. *Hint:* In the first case, $\log_9 7 = 1/\log_7 9$ and $9 = 3^2$.

36. Explain why $y = \log_x x$, $x > 1$, defines a function. What is its domain? What is its range? Graph the function.

37. Show that $(\log_{10} 4 - \log_{10} 2) \log_2 10 = 1$.

38. Show that $(2 \log_2 3)(\log_9 2 + \log_9 4) = 3$.

39. The amount of a certain radioactive element remaining at any time t is given by $y = y_0 e^{-0.4t}$, where t is measured in seconds and y_0 is the amount present initially. How much of the element would remain after three seconds if 40 grams were present initially?

40. The number of bacteria present in a culture is related to time by $N = N_0 e^{0.04t}$, where N_0 represents the amount of bacteria present at time $t = 0$, and t is time in hours. If 10,000 bacteria are present 10 hours after the beginning of the experiment, how many were present when $t = 0$?

41. The intensity I (in lumens) of a light beam, after passing through a thickness t (in centimeters) of a medium having an absorption coefficient of 0.1, is given by $I = 1000e^{-0.1t}$. How many centimeters of the material would reduce the illumination to 800 lumens?

42. Using the information given in Problem 41, find the thickness of a medium which would reduce the illumination to 600 lumens.

Chapter Review

[7.1]   **1.** Graph the function $\{(x, y) \,|\, y = 3^x\}$.

2. Solve by inspection: $10^x = \dfrac{1}{1000}$.

[7.2]   *Express in logarithmic notation.*

3. $3^4 = 81$ **4.** $3^{-2} = \dfrac{1}{9}$

Express in exponential notation.

5. $\log_2 8 = 3$ **6.** $\log_{10} 0.0001 = -4$

Solve.

7. $\log_2 16 = y$ **8.** $\log_{10} x = 3$

Express as the sum or difference of simpler logarithmic quantities.

9. $\log_{10} \sqrt[3]{xy^2}$ **10.** $\log_{10} \dfrac{2R^3}{\sqrt{PQ}}$

Express as a single logarithm with coefficient 1.

11. $2 \log_b x - \dfrac{1}{3} \log_b y$ **12.** $\dfrac{1}{3}(2 \log_{10} x + \log_{10} y) - 3 \log_{10} z$

[7.3]   *Find each logarithm. Interpolate as required.*

13. $\log_{10} 42$ **14.** $\log_{10} 0.00314$
15. $\log_{10} 682.4$ **16.** $\log_{10} 0.04142$

Find each antilogarithm. Interpolate as required.

17. $\text{antilog}_{10} \, 1.8287$ **18.** $\text{antilog}_{10} \, (8.6684 - 10)$
19. $\text{antilog}_{10} \, 0.4240$ **20.** $\text{antilog}_{10} \, (9.8224 - 10)$

[7.4] *Compute by means of logarithms.*

21. $(8.73)(41.6)$ 22. $\dfrac{85.9}{1.73}$ 23. $\dfrac{(4.78)(62.3)^2}{12.8}$ 24. $\sqrt[3]{4.3}\,\sqrt[5]{0.024}$

[7.5] *Solve. Leave solutions in logarithmic form, using the base* 10.

25. $5^x = 7$ 26. $3^{x+1} = 80$

27. Solve $\log_{10}(x-3) + \log_{10}(x) = 1$.
28. Solve $(1 + .05)^n = 1.215$ for n to the nearest integer.

[7.6] *Find each value using the table of logarithms to the base* 10.

29. $\log_3 11$ 30. $\log_2 120$

31. Find $\log_e 130$, using Table III.
32. Find $\text{antilog}_e\, 4.8$, using Table II.
33. Find $\text{antilog}_e\, (-0.24)$, using Table II.
34. Given $y = y_0\, e^{-0.4t}$, find y if $y_0 = 30$ and $t = 5$.

8 Systems of Equations and Inequalities

8.1   Systems of Linear Equations in Two Variables

In Chapter 5, we observed that the solution set of an open sentence in two variables, such as

$$ax + by + c = 0$$

or

$$ax + by + c \leq 0,$$

might contain infinitely many ordered pairs of numbers. It is often necessary to consider pairs of such sentences and to inquire whether or not the solution sets of the sentences contain ordered pairs in common. More specifically, we are interested in determining the members of the intersection of their solution sets.

We shall begin by considering the system

$$a_1 x + b_1 y + c_1 = 0 \quad (a_1, b_1 \text{ not both } 0)$$
$$a_2 x + b_2 y + c_2 = 0 \quad (a_2, b_2 \text{ not both } 0)$$

and studying $A \cap B$, where

$$A = \{(x, y) \mid a_1 x + b_1 y + c_1 = 0\}$$

and

$$B = \{(x, y) \mid a_2 x + b_2 y + c_2 = 0\}.$$

In a geometric sense, because the graphs of both A and B are straight lines, we are confronted with three possibilities, as illustrated in Figure 8.1 on page 187:

a.   The graphs are the same line.
b.   The graphs are parallel but distinct lines.
c.   The graphs intersect in one and only one point.

These possibilities lead, correspondingly, to the conclusion that one and only one of the following is true for any given system of two such linear equations in x and y:

a.   The intersection of the two solution sets contains all those ordered pairs found in either one of the given solution sets; that is, the solution sets are equal.
b.   The intersection of the two solution sets is the null set.
c.   The intersection of the two solution sets contains exactly one ordered pair.

186

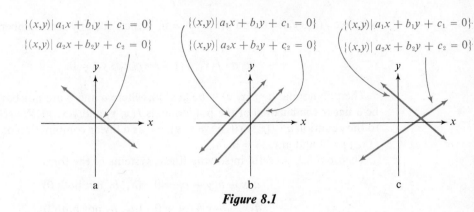

Figure 8.1

Dependency and consistency   In case (a), the left-hand members of the two equations in standard form are said to be **linearly dependent**, and the equations are **consistent**; in case (b), the left-hand members are **linearly independent**, and the equations are **inconsistent**; and in case (c), the left-hand members are linearly independent and the equations are consistent. If the two left-hand members are linearly dependent, then one of them can be obtained from the other through multiplication by a constant. For example,

$$2x + 4y - 8 \quad \text{and} \quad 6x + 12y - 24$$

are linearly dependent, but

$$2x + 4y - 8 \quad \text{and} \quad 6x + 12y - 23$$

are not.

You will recall that equivalent open sentences are defined to be open sentences that have the same solution set. Systems of open sentences may be said to be equivalent in a similar sense.

Definition 8.1   *If the the solution set of one system of open sentences is equal to (the same as) the solution set of another system of open sentences, then the systems are **equivalent**.*

Generation of equivalent systems   In seeking the solution set of a system of equations, our procedure will be to generate equivalent systems until we arrive at a system of which the solution set is obvious. The following theorem establishes one way to do this.

Theorem 8.1   *Any ordered pair that satisfies both the equations*

$$f(x, y) = 0$$
$$g(x, y) = 0$$

also satisfies the equation

$$a \cdot f(x, y) + b \cdot g(x, y) = 0,$$

for all real numbers a and b.

Proof   For any ordered pair (x_1, y_1) in the solution set of the system

$$f(x, y) = 0$$
$$g(x, y) = 0,$$

we have $f(x_1, y_1) = 0$ and $g(x_1, y_1) = 0$. Hence, for any real numbers a and b, we have

$$a \cdot f(x_1, y_1) + b \cdot g(x_1 \cdot y_1) = 0.$$

The polynomial $a \cdot f(x, y) + b \cdot g(x, y)$, where a and b are not both 0, is said to be a **linear combination** of the polynomials $f(x, y)$ and $g(x, y)$. We sometimes refer to the equation $a \cdot f(x, y) + b \cdot g(x, y) = 0$ as a linear combination of the equations $f(x, y) = 0$ and $g(x, y) = 0$.

Theorem 8.1 is useful in solving linear systems of the form

$$a_1 x + b_1 y + c_1 = 0 \quad (a_1, b_1 \text{ not both } 0) \tag{1}$$

$$a_2 x + b_2 y + c_2 = 0 \quad (a_2, b_2 \text{ not both } 0). \tag{2}$$

By appropriate choice of multipliers a and b, the linear combination

$$a(a_1 x + b_1 y + c_1) + b(a_2 x + b_2 y + c_2) = 0 \tag{3}$$

will be free of one variable; that is, the coefficient of one variable will be 0. We can then form an equivalent system by substituting Equation (3) for either Equation (1) or Equation (2).

Example   Solve

$$\begin{aligned} x - 3y + 5 &= 0 \\ 2x + y - 4 &= 0. \end{aligned} \tag{4}$$

Solution   We first form the linear combination

$$1(x - 3y + 5) + 3(2x + y - 4) = 0,$$

or

$$7x - 7 = 0,$$

from which

$$x - 1 = 0.$$

Replacing $x - 3y + 5 = 0$ with $x - 1 = 0$, we then have the equivalent system

$$\begin{aligned} x - 1 &= 0 \\ 2x + y - 4 &= 0. \end{aligned} \tag{5}$$

We can next replace the second equation in (5) with the linear combination

$$-2(x - 1) + 1(2x + y - 4) = 0,$$

or

$$y - 2 = 0,$$

to obtain the equivalent system

$$\begin{aligned} x - 1 &= 0 \\ y - 2 &= 0, \end{aligned} \tag{6}$$

from which, by inspection, we can obtain the unique solution $(1, 2)$. Hence the solution set of (4) is $\{(1, 2)\}$. Graphs of the systems (4), (5), and (6) appear in Figure 8.2. The point of intersection in each case has coordinates $(1, 2)$.

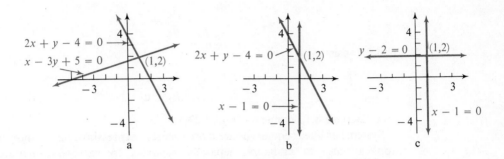

Figure 8.2

Solution of a system by substitution

Note that we could have proceeded somewhat differently from (5), as follows. Since any solution of the system (5) must be of the form $(1, y)$ (because any such ordered pair is a solution of $x - 1 = 0$), and when x is replaced by 1 in $2x + y - 4 = 0$, we obtain $y = 2$, the only ordered pair that satisfies both of the equations in (5) is $(1, 2)$. Hence, again the solution set of (4) is $\{(1, 2)\}$. In this latter way of proceeding from (5) to the solution of (4), we have used what is known as the **method of substitution**.

Choice of multipliers for linear combinations

Observe that the multipliers used in the foregoing example were first 3 and 1, and later -2 and 1. These were chosen because they produced coefficients that were additive inverses, first for the terms in y and later for the terms in x. In general, the linear combination

$$b_2(a_1x + b_1y + c_1) - b_1(a_2x + b_2y + c_2) = 0$$

will always be free of y, and

$$a_2(a_1x + b_1y + c_1) - a_1(a_2x + b_2y + c_2) = 0$$

will be free of x. For example, to form a useful linear combination of the equations in the system

$$2x + 6y + 7 = 0$$
$$5x + 4y + 3 = 0,$$

we might use 5 and -2, or -5 and 2, if we wish to obtain an equation that is free of x; and we might use 2 and -3, or -2 and 3, if we wish the result to obtain an equation that is free of y.

Criteria for dependent or inconsistent equations

If the coefficients of the variables in one equation in a system are proportional to the corresponding coefficients in the other equation, then the equations might be either dependent or inconsistent. The equations in the system

$$a_1x + b_1y + c_1 = 0$$
$$a_2x + b_2y + c_2 = 0 \quad (a_2, b_2, c_2 \neq 0)$$

are dependent if

$$\frac{a_1}{a_2} = \frac{b_1}{b_2} = \frac{c_1}{c_2},$$

and inconsistent if

$$\frac{a_1}{a_2} = \frac{b_1}{b_2} \neq \frac{c_1}{c_2}.$$

(See Exercise 8.1, Problems 26 and 27.)

Systems of linear equations are quite useful in expressing relationships in practical applications. By assigning separate variables to represent separate physical quantities, we can usually decrease the difficulty in symbolically representing these relationships.

Exercise 8.1

Solve each system.

Example

$$\frac{2}{3}x - y = 2 \tag{1}$$

$$x + \frac{1}{2}y = 7 \tag{2}$$

Solution   Multiply each member of Equation (1) by 3 and each member of Equation (2) by 2.

$$2x - 3y = 6 \tag{1'}$$

$$2x + y = 14 \tag{2'}$$

Form a linear combination by adding -1 times (1') to 1 times (2') and solve for y.

$$4y = 8$$

$$y = 2 \tag{3}$$

Equations (2) and (3) constitute an equivalent system for (1) and (2).

Substitute 2 for y in any of (1), (2), (1'), or (2'), and solve for x. In this example, (2) is used.

$$x + \frac{1}{2}(2) = 7$$

$$x = 6$$

The solution set is $\{(6, 2)\}$.

1. $x - y = 1$
   $x + y = 5$

2. $2x - 3y = 6$
   $x + 3y = 3$

3. $3x + y = 7$
   $2x - 5y = -1$

4. $2x - y = 7$
   $3x + 2y = 14$

5. $5x - y = -29$
   $2x + 3y = 2$

6. $6x + 4y = 12$
   $3x + 2y = 12$

7. $5x + 2y = 3$
$x = 0$

8. $2x - y = 0$
$x = -3$

9. $3x - 2y = 4$
$y = -1$

10. $x + 2y = 6$
$x = 2$

11. $\dfrac{1}{4}x - \dfrac{1}{3}y = -\dfrac{5}{12}$
$\dfrac{1}{10}x + \dfrac{1}{5}y = \dfrac{1}{2}$

12. $\dfrac{2}{3}x - y = 4$
$x - \dfrac{3}{4}y = 6$

13. $\dfrac{1}{7}x - \dfrac{3}{7}y = 1$
$2x - y = -4$

14. $\dfrac{1}{3}x - \dfrac{2}{3}y = 2$
$x - 2y = 6$

15. Find a and b so that the graph of $ax + by + 3 = 0$ passes through the points $(-1, 2)$ and $(-3, 0)$.

16. Find a and b so that the solution set of the system
$$ax + by = 4$$
$$bx - ay = -3$$
is $\{(1, 2)\}$.

17. Recall that the slope-intercept form of the equation of a straight line is given by $y = mx + b$. Find an equation of the line that passes through the points $(0, 2)$ and $(3, -8)$.

18. Find an equation of the line that passes through the points $(-6, 2)$ and $(4, 1)$.

19. Find a linear relationship between centigrade temperature, C and Fahrenheit temperature, F, given that $F = 32°$ when $C = 0°$, and $F = 212°$ when $C = 100°$.

20. A man has \$1.80 in nickels and dimes, with three more dimes than nickels. How many dimes and nickels does he have?

21. How many pounds of an alloy containing 45% silver must be melted with an alloy containing 60% silver to obtain 40 lbs of a 48% silver alloy?

22. A man has three times as much money invested in 3% bonds as he has in stocks paying 5%. How much does he have invested in each if his yearly income from the investments is \$1680?

23. A man has \$1000 more invested at 5% than he has invested at 4%. If his annual income from the two investments together is \$698, how much does he have invested at each rate?

24. An airplane travels 1260 miles in the same time that an automobile travels 420 miles. If the rate of the airplane is 120 miles per hour greater than the rate of the automobile, find the rate of each.

25. Two cars start together and travel in the same direction, one going twice as fast as the other. At the end of 3 hours, they are 96 miles apart. How fast is each traveling?

26. Show that the linear polynomials
$$a_1x + b_1y + c_1$$
$$a_2x + b_2y + c_2 \quad (a_2, b_2, c_2 \neq 0)$$
are dependent if and only if
$$\frac{a_1}{a_2} = \frac{b_1}{b_2} = \frac{c_1}{c_2}.$$
Hint: Write the equations in slope-intercept form.

27. Show that the equations in the system

$$a_1 x + b_1 y + c_1 = 0$$
$$a_2 x + b_2 y + c_2 = 0 \quad (a_2, b_2, c_2 \neq 0)$$

are inconsistent if and only if

$$\frac{a_1}{a_2} = \frac{b_1}{b_2} \neq \frac{c_1}{c_2}.$$

Hint: Write the equations in slope-intercept form.

8.2   *Systems of Linear Equations in Three Variables*

Solutions of an equation in three variables

A solution of an equation in three variables, such as

$$x + 2y - 3z + 4 = 0, \tag{1}$$

is an ordered triple of numbers (x, y, z), because all three of the variables must be replaced before we can decide whether or not the sentence is true. Thus, $(0, -2, 0)$ and $(-1, 0, 1)$ are solutions of (1), whereas $(1, 1, 1)$ is not. There are, of course, infinitely many members in the solution set of such an equation.

Solution of a system

The solution set of a system of linear (first-degree) equations in three variables is the intersection of the solution sets of the separate equations in the system. We are primarily interested in systems involving three equations, such as

$$x + 2y - 3z + 4 = 0 \tag{1}$$
$$2x - y + z - 3 = 0 \tag{2}$$
$$3x + 2y + z - 10 = 0, \tag{3}$$

where the solution set is

$$\{(x, y, z) \mid x + 2y - 3z + 4 = 0\} \cap \{(x, y, z) \mid 2x - y + z - 3 = 0\}$$
$$\cap \{(x, y, z) \mid 3x + 2y + z - 10 = 0\}.$$

We can find the members of this set by methods analogous to those used in the preceding section.

Theorem 8.2   *If any equation in the system*

$$f(x, y, z) = 0$$
$$g(x, y, z) = 0$$
$$h(x, y, z) = 0$$

is replaced by a linear combination, with nonzero coefficients, of itself and any one of the other equations in the system, then the result is an equivalent system.

The proof of this theorem, which follows from the definition of a linear combination, and of equivalent systems, is omitted here. Now, to see how this theorem applies, let us examine the foregoing system (1), (2), and (3). If we begin by replacing Equation (2) with the linear combination formed by multiplying Equation (1) by -2 and Equation (2) by 1, that is, with

$$-2(x + 2y - 3z + 4) + 1(2x - y + z - 3) = 0,$$

or

$$-5y + 7z - 11 = 0,$$

we obtain the equivalent system

$$x + 2y - 3z + 4 = 0 \tag{1}$$

$$-5y + 7z - 11 = 0 \tag{2'}$$

$$3x + 2y + z - 10 = 0, \tag{3}$$

where (2') is free of x. Next, if we replace (3) by the sum of -3 times (1) and 1 times (3), we have

$$x + 2y - 3z + 4 = 0 \tag{1}$$

$$-5y + 7z - 11 = 0 \tag{2'}$$

$$-4y + 10z - 22 = 0, \tag{3'}$$

where both (2') and (3') are free of x. If now (3') is replaced by the sum of 4 times (2') and -5 times (3'), we have

$$x + 2y - 3z + 4 = 0 \tag{1}$$

$$-5y + 7z - 11 = 0 \tag{2'}$$

$$-22z + 66 = 0, \tag{3''}$$

which is equivalent to the original (1), (2), and (3). But, from (3''), we see that for any solution of (1), (2'), and (3''), $z = 3$; that is, the solution will be of the form $(x, y, 3)$. If 3 is substituted for z in (2'), we have

$$-5y + 7(3) - 11 = 0, \quad \text{or} \quad y = 2,$$

and any solution of the system must be of the form $(x, 2, 3)$. Substituting 2 for y and 3 for z in (1) gives

$$x + 2(2) - 3(3) + 4 = 0, \quad \text{or} \quad x = 1,$$

so that the single member of the solution set of (1), (2'), and (3'') is (1, 2, 3). Therefore, the solution set of the original system is $\{(1, 2, 3)\}$.

The foregoing process of solving a system of linear equations can be reduced to a series of mechanical procedures as follows.

Example   Solve.

$$x + 2y - z + 1 = 0 \tag{1}$$

$$x - 3y + z - 2 = 0 \tag{2}$$

$$2x + y + 2z - 6 = 0 \tag{3}$$

(Solution overleaf)

Solution

Multiply (1) by -1 and add the result to 1 times (2); multiply (1) by -2 and add the result to 1 times (3).

$$-5y + 2z - 3 = 0 \tag{4}$$

$$-3y + 4z - 8 = 0 \tag{5}$$

Multiply (4) by -3 and add the result to 5 times (5).

$$14z - 31 = 0 \tag{6}$$

Now (1), (4), and (6) constitute a set of equations equivalent to (1), (2), and (3).

$$x + 2y - \quad z + \ 1 = 0 \tag{1}$$

$$-5y + \ 2z - \ 3 = 0 \tag{4}$$

$$14z - 31 = 0 \tag{6}$$

Solve for z in (6).

$$z = \frac{31}{14}$$

Substitute $31/14$ for z in (4).

$$-5y + 2\left(\frac{31}{14}\right) - 3 = 0$$

$$y = \frac{2}{7}$$

Substitute $2/7$ for y and $31/14$ for z in either (1), (2), or (3); say (1).

$$x + 2\left(\frac{2}{7}\right) - \frac{31}{14} + 1 = 0$$

$$x = \frac{9}{14}$$

The solution set is $\left\{\left(\frac{9}{14}, \frac{2}{7}, \frac{31}{14}\right)\right\}$.

Linear dependence and inconsistent equations

If at any step in the procedure used in the examples above, the resulting linear combination vanishes or yields a contradiction, the system contains linearly dependent left-hand members or else inconsistent equations, or both, and it either has an infinite number of members or else has no member in its solution set.

Graphs in three dimensions

By establishing a three-dimensional Cartesian coordinate system as shown in Figure 8.3, a one-to-one correspondence can be established between the points in a three-dimensional space and ordered triples of real numbers. If this is done, it can be shown that the graph of a linear equation in three variables is a plane. For example, the equation

$$x + y + z = 1$$

represents the plane through the points $(1, 0, 0)$, $(0, 1, 0)$, and $(0, 0, 1)$, as illustrated in Figure 8.3. Hence the solution set of a system of three linear

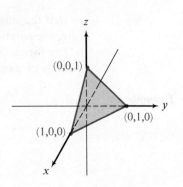

Figure 8.3

equations in three variables consists of the coordinates of the common intersection of three planes. Figure 8.4 shows the possibilities for the relative positions of the plane graphs of three linear equations in three variables. In case (a), the common intersection consists of a single point, and hence the solution set of the corresponding system of three equations contains a single member. In cases (b), (c), and (d), the intersection is a line or a plane, and the solution of the corresponding system has infinitely many members. In cases (e), (f), (g) and (h), the three planes have no common intersection, and the solution set of the corresponding system is the null set. Linear equations corresponding to cases (a), (b), (c), and (d) are consistent, and the others inconsistent.

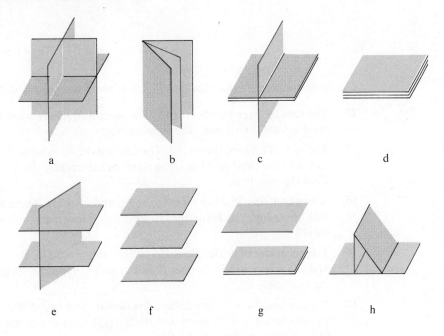

a   b   c   d

e   f   g   h

Figure 8.4

The use of linear combinations to solve systems of linear equations can be extended to cover cases of n equations in n variables. Clearly, as n grows larger, the time and effort necessary to find the solution set of a system increase correspondingly. Fortunately, modern high-speed computers can handle such computations in stride for fairly large values of n. There are analytic methods other than those exhibited here which become preferable for very large systems.

Exercise 8.2

Solve.

1. $x + y + z = 2$
$2x - y + z = -1$
$x - y - z = 0$

2. $x + y + z = 1$
$2x - y + 3z = 2$
$2x - y - z = 2$

3. $x + y + 2z = 0$
$2x - 2y + z = 8$
$3x + 2y + z = 2$

4. $2x - 3y + z = 3$
     $x - y - 3z = -1$
   $-x + 2y - 3z = -4$

5. $x - 2y + z = -1$
     $2x + y - 3z = 3$
     $3x + 3y - 2z = 10$

6. $x - 2y + 4z = -3$
     $3x + y - 2z = 12$
     $2x + y - 3z = 11$

7. $4x - 2y + 3z = 4$
     $2x - y + z = 1$
     $3x - 3y + 4z = 5$

8. $x + 5y - z = 2$
     $3x - 9y + 3z = 6$
     $x - 3y + z = 4$

9. $x + z = 5$
     $y - z = -4$
     $x + y = 1$

10. $5y - 8z = -19$
      $5x - 8z = 6$
      $3x - 2y = 12$

11. $x - \dfrac{1}{2}y - \dfrac{1}{2}z = 4$

      $x - \dfrac{3}{2}y - 2z = 3$

      $\dfrac{1}{4}x + \dfrac{1}{4}y - \dfrac{1}{4}z = 0$

12. $x + 2y + \dfrac{1}{2}z = 0$

      $x + \dfrac{3}{5}y - \dfrac{2}{5}z = \dfrac{1}{5}$

      $4x - 7y - 7z = 6$

Use three variables in the solution of each of the following problems.

13. The sum of three numbers is 15. The second equals two times the first and the third equals the second. Find the numbers.

14. The sum of three numbers is 2. The first number is equal to the sum of the other two, and the third number is the result of subtracting the first from the second. Find the numbers.

15. A box contains \$6.25 in nickels, dimes, and quarters. There are 85 coins in all, with three times as many nickels as dimes. How many coins of each kind are there?

16. The perimeter of a triangle is 155 in. The side x is 20 in. shorter than the side y, and the side y is 5 in. longer than the side z. Find the lengths of the sides of the triangle.

17. A man had \$446 in ten-dollar, five-dollar, and one-dollar bills. There were 94 bills in all and 10 more five-dollar bills than ten-dollar bills. How many bills of each kind did he have?

18. Find values for a, b, and c so that the graph of $x^2 + y^2 + ax + by + c = 0$ will contain the points $(0, 0)$, $(6, 0)$, and $(0, 8)$.

19. The equation for a circle can be written $x^2 + y^2 + ax + by + c = 0$. Find the equation of the circle with graph containing the points $(2, 3)$, $(3, 2)$, and $(-4, -5)$.

20. Find values for a, b, and c so that the graph of $y = ax^2 + bx + c$ contains the points $(-1, 2)$, $(1, 6)$, and $(2, 11)$.

21. Three solutions of the equation $ax + by + cz = 1$ are $(0, 2, 1)$, $(6, -1, 2)$, and $(0, 2, 0)$. Find the coefficients a, b, and c.

22. Three solutions of the equation $ax + by + cz = 1$ are $(2, 1, 0)$, $(-1, 3, 2)$, and $(3, 0, 0)$. Find the coefficients a, b, and c.

23. Show that the system

$$x + y + 2z = 2$$
$$2x - y - z = 3$$

has an infinite number of members in its solution set. List two ordered triples that are solutions. *Hint:* Express x in terms of z alone, and express y in terms of z alone.

24. Give a geometric argument to show that any system of two consistent equations in three variables has an infinite number of members in its solution set.

25. Show that the system

$$x + y + z = 3 \tag{a}$$
$$x - 2y - z = -2 \tag{b}$$
$$x + y + 2z = 4 \tag{c}$$
$$2x - y + 2z = 4 \tag{d}$$

has $\emptyset$ as its solution set. *Hint:* Does the solution set of the system (a), (b), and (c) have any member that satisfies (d)?

8.3   *Systems of Nonlinear Equations*

Substitution   The substitution method is sometimes a convenient means of finding solution sets of systems in which nonlinear equations are present.

Example   Solve.

$$x^2 + y^2 = 25 \tag{1}$$
$$x + y = 7 \tag{2}$$

Solution   Equation (2) can be written equivalently in the form

$$y = 7 - x, \tag{3}$$

and we can replace y in (1) by $(7 - x)$ from (3). This produces

$$x^2 + (7 - x)^2 = 25, \tag{4}$$

which has as a solution set those values of x for which the ordered pair (x, y) is a common solution of (1) and (2). We can now find the solution set of (4):

$$x^2 + 49 - 14x + x^2 = 25,$$
$$x^2 - 7x + 12 = 0,$$
$$(x - 3)(x - 4) = 0,$$

which is satisfied if x is either 3 or 4. Now, by replacing x in the *first-degree equation* (3) by each of these numbers in turn, we obtain

$$y = 7 - 3 = 4$$

and

$$y = 7 - 4 = 3,$$

respectively, so that the solution set of the system (1) and (2) is $\{(3, 4), (4, 3)\}$.

If, in the foregoing example, we had substituted the values 3 and 4 that we obtained for x in the *second-degree equation* (1), we would have obtained some extraneous solutions that do not satisfy Equation (2). (See Exercise 8.3, Problem 31.)

Linear combination

If both of the equations in a system of two equations in two variables are of the second degree in both variables, the use of linear combinations of the equations often provides a simpler means of solution than does substitution.

Example

Solve.

$$3x^2 - 7y^2 + 15 = 0 \tag{5}$$

$$3x^2 - 4y^2 - 12 = 0 \tag{6}$$

Solution

By forming the linear combination of -1 times (5) and 1 times (6), we obtain

$$3y^2 - 27 = 0$$
$$y^2 = 9,$$

from which

$$y = 3 \quad \text{or} \quad y = -3,$$

and we have the y-components of the members of the solution set of the system (5) and (6). Substituting 3 for y in either (5) or (6), say (5), we have

$$3x^2 - 7(3)^2 + 15 = 0,$$
$$x^2 = 16,$$

from which

$$x = 4 \quad \text{or} \quad x = -4.$$

Thus, the ordered pairs $(4, 3)$ and $(-4, 3)$ are solutions of the system. Substituting -3 for y in (5) or (6) [this time we shall use (6)] gives us

$$3x^2 - 4(-3)^2 - 12 = 0,$$
$$x^2 = 16,$$

so that

$$x = 4 \quad \text{or} \quad x = -4.$$

Thus the ordered pairs $(4, -3)$ and $(-4, -3)$ are solutions of the system, and accordingly the complete solution set is $\{(4, 3), (4, -3), (-4, 3), (-4, -3)\}$.

Linear combination and substitution

The solution of some systems requires the application of both linear combinations *and* substitution.

Example

Solve.

$$x^2 + y^2 = 5 \tag{7}$$

$$x^2 - 2xy + y^2 = 1 \tag{8}$$

Solution

By forming the linear combination of 1 times (7) and -1 times (8), we obtain

$$2xy = 4,$$
$$xy = 2. \tag{9}$$

The system (7) and (8) is equivalent to the system (7) and (9). This latter system can be solved by substitution. From (9), we have

$$y = \frac{2}{x}.$$

Replacing y in (7) by $2/x$, we obtain

$$x^2 + \left(\frac{2}{x}\right)^2 = 5,$$

$$x^2 + \frac{4}{x^2} = 5, \tag{10}$$

$$x^4 + 4 = 5x^2,$$

$$x^4 - 5x^2 + 4 = 0, \tag{11}$$

which is a quadratic in x^2. The left-hand member of (11), when factored, yields

$$(x^2 - 1)(x^2 - 4) = 0,$$

from which we obtain

$$x^2 - 1 = 0 \quad \text{or} \quad x^2 - 4 = 0,$$

so that

$$x = 1, \quad x = -1, \quad x = 2, \quad x = -2.$$

Since the step from (10) to (11) was not an elementary transformation, we are careful to note that these values of x all satisfy (10). Substituting $1, -1, 2$, and -2 in turn for x in (9), we obtain the corresponding values for y, and thus the solution set of the system (7) and (9) or, equivalently, the solution set of the system (7) and (8), is $\{(1, 2), (-1, -2), (2, 1), (-2, -1)\}$.

There are other techniques involving substitution in conjunction with linear combinations that are useful in handling systems of higher-degree equations, but they all bear similarity to those illustrated.

Use of graphs as a check   In solving a set of equations, it is frequently helpful to sketch the graphs of the equations as a rough check on the algebraic solution. Any attempt to solve a set of two second-degree equations graphically on the real plane may produce only approximations to such real solutions as exist. We can expect ordinarily to find at most four and as few as no points of intersection, depending on the types of equations involved and on the coefficients and constants in the given equations.

Approxima-tions to solutions   When the degree of any equation in a set of equations is greater than two, or when the left-hand member of one of the equations of the form $f(x, y) = 0$ is not a polynomial, it may be very difficult or impossible to find common solutions analytically. In this case, we can at least obtain approximations to any real solutions by graphical methods. Consider the system of equations

$$y = 2^{-x}$$
$$y = \sqrt{0.01x}, \tag{12}$$

and the graphs of these equations in Figure 8.5 on page 200.

Examining Figure 8.5-b, an enlargement of part of Figure 8.5-a, we observe that the curves intersect at approximately (2.6, 0.16), and from geometric considerations we conclude that this ordered pair approximates the only member in the solution set of (12).

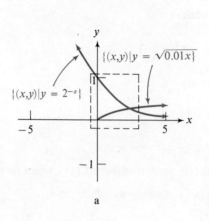 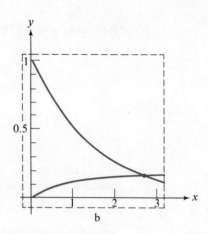

a   b

Figure 8.5

This result shows, in particular, that the equation

$$2^{-x} - \sqrt{0.01x} = 0$$

has just one root, approximately 2.6, and suggests a method of approximating solutions of similar equations.

Exercise 8.3

Solve by the method of substitution. Check the solutions by sketching the graphs of the equations and estimating the coordinates of any points of intersection.

Example

$y = x^2 + 2x + 1$ $\qquad\qquad$ (1)

$y - x = 3$ $\qquad\qquad\qquad$ (2)

Solution

Solve Equation (2) explicitly for y.

$$y = x + 3 \qquad\qquad (2')$$

Substitute $x + 3$ for y in (1).

$$x + 3 = x^2 + 2x + 1 \qquad (3)$$

Solve for x.

$$x^2 + x - 2 = 0$$

$$(x + 2)(x - 1) = 0$$

$$x = -2, \quad x = 1$$

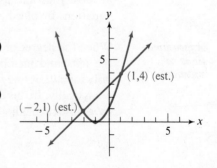

Substitute each of these values in turn in (2′) to determine values for y.

$$\text{If } x = -2, \text{ then } y = 1.$$
$$\text{If } x = 1, \quad \text{then } y = 4.$$

The solution set is $\{(-2, 1), (1, 4)\}$.

1. $y = x^2 - 5$ $y = 4x$	**2.** $y = x^2 - 2x + 1$ $y + x = 3$	**3.** $x^2 + y^2 = 13$ $x + y = 5$
4. $x^2 + 2y^2 = 12$ $2x - y = 2$	**5.** $x + y = 1$ $xy = -12$	**6.** $2x - y = 9$ $xy = -4$

Solve using linear combinations.

7. $x^2 + y^2 = 10$ $9x^2 + y^2 = 18$	**8.** $x^2 + 4y^2 = 52$ $x^2 + y^2 = 25$
9. $x^2 - y^2 = 7$ $2x^2 + 3y^2 = 24$	**10.** $x^2 + 4y^2 = 25$ $4x^2 + y^2 = 25$
11. $4x^2 - 9y^2 + 132 = 0$ $x^2 + 4y^2 - 67 = 0$	**12.** $16y^2 + 5x^2 - 26 = 0$ $25y^2 - 4x^2 - 17 = 0$
13. $x^2 - xy + y^2 = 7$ $x^2 + y^2 = 5$	**14.** $3x^2 - 2xy + 3y^2 = 34$ $x^2 + y^2 = 17$
15. $3x^2 + 3xy - y^2 = 35$ $x^2 - xy - 6y^2 = 0$	**16.** $x^2 - xy + y^2 = 21$ $x^2 + 2xy - 8y^2 = 0$

Solve by graphing. Approximate components of solutions to the nearest half unit.

17. $y = 10^x$ $x + y = 2$	**18.** $y = 2^x$ $y - x = 2$
19. $y = \log_{10} x$ $y = x^2 - 2x + 1$	**20.** $y = 10^x$ $y = x^2$
21. $y = 10^{-x}$ $y = \log_{10} x$	**22.** $y = 2^{x-3}$ $y = \log_2 x$
23. $10^{-x} - \log_{10} x = 0$	**24.** $2^{x-3} - \log_2 x = 0$

25. How many real solutions are possible for simultaneous systems of linearly independent equations that consist of:
   a. two linear equations in two variables?
   b. one linear equation and one quadratic equation in two variables?
   c. two quadratic equations in two variables?
   Support each of your answers with sketches.

26. The sum of the squares of two positive numbers is 13. If twice the first number is added to the second, the sum is 7. Find the numbers.

27. The sum of two numbers is 6 and their product is 35/4. Find the numbers.

28. The annual income from an investment is \$32. If the amount invested were \$200 more and the rate 1/2% less, the annual income would be \$35. What are the amount and rate of the investment?

29. At a constant temperature, the pressure P and volume V of a gas are related by the equation $PV = K$. The product of the pressure (in pounds per square inch) and the volume (in cubic inches) of a certain gas is 30 inch-pounds. If the temperature remains constant as the pressure is increased 4 lbs per sq. in., the volume is decreased by 2 cu in. Find the original pressure and volume of the gas.

30. What relationships must exist between the numbers a and b so that the solution set of the system

$$x^2 + y^2 = 25$$
$$y = ax + b$$

has two ordered pairs of real numbers? One ordered pair of real numbers? No ordered pairs of real numbers? *Hint:* Use substitution and consider the nature of the roots of the resulting quadratic equation.

31. Consider the system

$$x^2 + y^2 = 8 \qquad\qquad (1)$$
$$xy = 4. \qquad\qquad (2)$$

We can solve this system by substituting $4/x$ for y in (1) to obtain

$$x^2 + \frac{16}{x^2} = 8,$$

from which we obtain $x = 2$ or $x = -2$. Now if we obtain the y-components of the solution from (2), we find that for $x = 2$ we have $y = 2$, and that for $x = -2$ we have $y = -2$. But if we seek y-components from (1), for $x = 2$ we have $y = \pm 2$, and for $x = -2$, $y = \pm 2$. Discuss the fact that we seem to obtain two more solutions from (1) than from (2). What is the solution set of the system?

8.4   *Systems of Inequalities*

Graphs of systems of inequalities

In Chapters 5 and 6, we observed that the graph of the solution set of an inequality in two variables might be a region in the plane. The graph of the solution set of a system of inequalities in two variables, which consists of the intersection of the graphs of the inequalities in the system, might also be a plane region.

Example

Graph the solution set of the system

$$x + 2y \le 6$$
$$2x - 3y \ge 12.$$

Solution

Graphing each inequality by the method discussed in Section 5.6, we obtain the figure shown, where the doubly shaded region is the graph of the solution set of the system. We have graphed

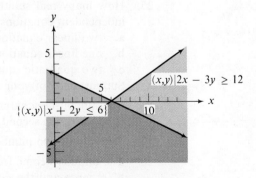

$$\{(x, y) \mid x + 2y \le 6\} \cap \{(x, y) \mid 2x - 3y \ge 12\}.$$

Example

Graph the solution set of the system

$$y \geq x + 2$$
$$y \geq x^2.$$

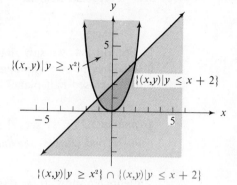

Solution

The graph of each inequality is shaded as shown in the figure. The doubly shaded region constitutes the graph of the solution set of the system. We have graphed

$$\{(x, y) \mid y \geq x^2\} \cap \{(x, y) \mid y \leq x + 2\}.$$

Example

Graph the solution set of the system

$$y > 2$$
$$x > -2$$
$$x + y > 1.$$

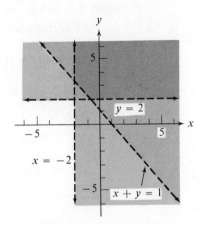

Solution

The triply shaded region in the figure constitutes the graph of the solution set of the system. We have graphed

$$\{(x, y) \mid y > 2\} \cap \{(x, y) \mid x > -2\}$$
$$\cap \{(x, y) \mid x + y > 1\}.$$

Exercise 8.4

By double or triple shading, indicate the region in the plane representing the solution set of each system.

1. $y \geq x + 1$
$y \geq 5 - x$

2. $y \geq 4$
$x \geq 2$

3. $y - x \geq 0$
$y + x \geq 0$

4. $2y - x \geq 1$
$x < -3$

5. $y + 3x < 6$
$y > 2$

6. $x > 3$
$x + y \geq 5$

7. $x = 3$
$x + y < 4$

8. $y = 2$
$2y + x < 3$

9. $x - 3y < 6$
$y + x = 1$

10. $2x + y \geq 4$
$x - y = -2$

11. $y > x^2 + 1$
$x + y > 4$

12. $y < x^2 + 4$
$x - y \leq 4$

13. $x^2 + y^2 < 25$
$y > 3$

14. $x^2 + y^2 \geq 25$
$y > x^2$

15. $9x^2 + 4y^2 \geq 36$
$y < x^2$

16. $y \geq 2$
$x \geq 2$
$y \geq x$

17. $y \leq 3$
$x \leq 2$
$y < x$

18. $x^2 + y^2 \leq 36$
$x \geq 3$
$y \geq 3$

19. $x^2 + y^2 \leq 25$
$y \geq x^2 - 4$
$y \leq -x^2 + 4$

20. $y \leq \log_{10} x$
$y \geq x - 1$

21. $y \leq 10^x$
$y \leq 2 - x^2$

22. $9 \leq x^2 + y^2 \leq 16$
$-1 \leq x - y \leq 1$

23. $9 \leq x^2 + y^2 \leq 16$
$x^2 + 1 \leq y \leq x^2 + 3$

24. $x^2 - 2 \leq y \leq 2 - x^2$
$|x| \geq 1$

8.5   Convex Sets—Polygonal Regions

In Section 8.4, we saw that the graph of the solution set of a system of linear inequalities in two variables is simply the common intersection of a number of open or closed half-planes. Any such intersection is an example of a *convex set*.

Definition 8.2   *A set $\mathscr{S}$ of points is a* **convex set** *if and only if, for each two points P and Q in $\mathscr{S}$, the line segment $\overline{PQ}$ lies entirely in $\mathscr{S}$.*

Figure 8.6 shows three sets of points in a plane. In **a** and **b** are pictured examples of convex sets, while the set in **c** is not convex because part of the line segment $\overline{PQ}$ does not lie in the set. If the boundary of a convex set is a (closed) polygon, then the boundary is called a **convex polygon**, and the region enclosed by the polygon (including the boundary) is called a **closed convex polygonal set**.

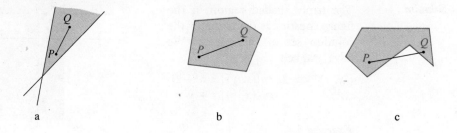

a   b   c

Figure 8.6

We have two immediate results from Definition 8.2 which are intuitively true and whose proofs are omitted.

Theorem 8.3   *Any half-plane is a convex set.*

Theorem 8.4   *The intersection of two convex sets is a convex set.*

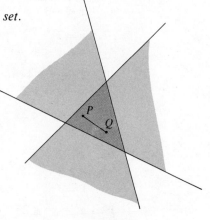

As suggested by Figure 8.7, this latter result extends to any number of convex sets. If the intersection of the graphs of a system of linear inequalities constitutes a (closed) polygonal set $\mathscr{P}$, then we can locate the set $\mathscr{P}$ in the plane by graphing the equations associated with the given inequalities and identifying the vertices $\mathscr{P}$.

Figure 8.7

Example

Locate and identify by shading the polygonal set specified by the system

$$x + y \le 5$$
$$x - y \le 2$$
$$x - y \ge -2$$
$$x \ge 0$$
$$y \ge 0.$$

Solution

Graph the associated equations

$$x + y = 5$$
$$x - y = 2$$
$$x - y = -2$$
$$x = 0$$

and

$$y = 0,$$

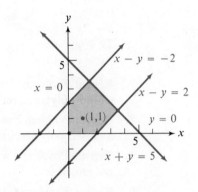

as shown. Note which half-plane
is determined by each of the five
inequalities, and shade the intersection of the five half-planes, The intersection is
the pentagonal region shown. To check the result, note that (1, 1) is in the region
and its coordinates satisfy each given inequality.

We should not conclude from the foregoing example that, simply because the
five equations associated with the set of inequalities determine a convex pentagon,
the pentagon is necessarily the graph of the system of inequalities. Thus, if the
inequality $y \ge 0$ in the example is replaced with $y \le 0$, then the graph of the
equations remains unchanged, but the graph of the system of inequalities is
the triangular region shown in Figure 8.8**a**. On the other hand, if the inequality $x + y \le 5$ is replaced with $x + y \ge 5$, then the intersection of the graphs
becomes the (infinite) rectangular region shown in **b** of the figure. Finally
we observe that if both of the foregoing changes are made, that is, if
$x + y \ge 5$ replaces $x + y \le 5$ and $y \le 0$ replaces $y \ge 0$, then the intersection of
the graphs of the inequalities in the system is ∅.

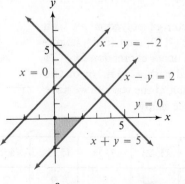

a

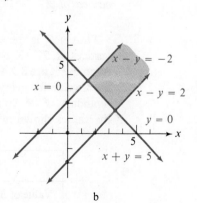

b

Figure 8.8

Exercise 8.5

Graph the convex polygonal set defined by the given system of inequalities.

1. $x - y \leq 2$
   $x - y \geq -2$
   $x + y \geq 2$
   $x + y \leq 6$

2. $0 \leq x \leq 5$
   $0 \leq y \leq 5$
   $x + y \leq 6$

3. $0 \leq x \leq 5$
   $0 \leq y \leq 4$
   $x + 2y \leq 10$
   $2x + y \leq 10$

4. $0 \leq x$
   $0 \leq y \leq 4$
   $x - 2y \leq 6$
   $2x + y \leq 12$
   $2x - y \leq 10$

5. $0 \leq x \leq 5$
   $y \geq x - 3$
   $x + y \leq 9$
   $3y \leq 2x + 12$

6. $0 \leq x$
   $0 \leq y \leq 6$
   $x + 2y \leq 13$
   $2x + y \leq 11$
   $3x + y \leq 15$

7. Repeat Problem 1 with $x - y \leq 2$ replaced with $x - y \geq 2$. Is the graph a closed polygonal set?

8. Repeat Problem 3 with $x + 2y \leq 10$ replaced with $x + 2y \geq 10$. Is the graph a closed polygonal set?

9. Repeat Problem 4 with $0 \leq y \leq 4$ replaced with $y \leq 4$. Is the graph a closed polygonal set?

10. Repeat Problem 6 without the condition $x \geq 0$. Is the graph a closed polygonal set?

8.6   Linear Programming

For each ordered pair $(x, y) \in R^2$, the linear expression $ax + by$ has a value. In particular, if $\mathcal{P} \subset R^2$, and $\mathcal{P}$ is a closed polygonal subset of R^2, then $ax + by$ has a value for each $(x, y) \in \mathcal{P}$. It can be shown that the following result holds for the values of $ax + by$ over $\mathcal{P}$.

Theorem 8.5   *If $\mathcal{P} \subset R^2$ is a closed convex polygonal set, then the expression $ax + by$ takes on its maximum and minimum values over $\mathcal{P}$ at one of the vertices of $\mathcal{P}$.*

Thus if $\mathcal{P}$ is the closed convex polygonal set with vertices $(1, 2)$, $(3, 1)$, $(5, 3)$, $(4, 5)$, and $(3, 6)$, as pictured in Figure 8.9, then the linear expression $3x - 2y$ has a value for every ordered pair (x, y) in, or on the boundary of, $\mathcal{P}$. In particular, at the vertices we have the values shown in the table below.

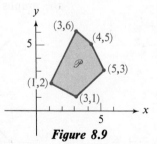

Figure 8.9

Vertex	(1, 2)	(3, 1)	(5, 3)	(4, 5)	(3, 6)
Value of $3x - 2y$	-1	7	9	2	-3

By inspection, the maximum of these values is 9 and the minimum is -3. By Theorem 8.5, then, 9 is the greatest value $3x - 2y$ has over the entire set $\mathscr{P}$, and its least value over $\mathscr{P}$ is -3.

Theorem 8.5 has many important applications.

Example

A company manufactures two kinds of electric shavers, one using a cord and the other a cordless model. The company can make up to 500 cord models and 400 cordless models per day, but it can make a total of only 600 shavers per day. It takes 2 man-hours to manufacture the cord model and 3 man-hours to manufacture the cordless model, and the company has available at most 1400 man-hours per day. If there is a profit of \$2.50 on each cord shaver and \$3.50 on each cordless shaver, how many of each kind should the company make each day in order to realize the greatest profit?

Solution

Let $x =$ number of cord shavers made in one day;

       $y =$ number of cordless shavers made in one day.

We have the following restraints on x and y:

$$0 \le x \le 500, \qquad 0 \le y \le 400,$$
$$x + y \le 600, \qquad 2x + 3y \le 1400.$$

Solving the associated equations for these inequalities in pairs yields the vertices of the polygonal region shown in the figure. We wish to maximize

$$2.5x + 3.5y.$$

We find the values for this expression at the vertices to be:

Vertex	Profit
(0, 0)	$ 0
(0, 400)	$1400
(100, 400)	$1650
(400, 200)	$1700
(500, 100)	$1601
(500, 0)	$1250

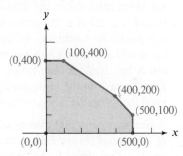

Clearly, the most profitable numbers of shavers are 400 cord models and 200 cordless models.

Exercise 8.6

Find the maximum and minimum values of the given expression over the given closed convex polygonal set for the specified set of Exercise 8.5, page 206.

1. $x + 3y$;   the set in Problem 1.

2. $3x + 5y$;   the set in Problem 2.

3. $3x + 5y$;   the set in Problem 3.

4. $x + 2y$;   the set in Problem 4.

5. $5x - 4y$;   the set in Problem 5.

6. $10x - 2y$;   the set in Problem 6.

For each Problem 7–10, use Tables A and B below, which show the number of units of labor, machinery, materials, and overhead necessary for a manufacturer to produce one unit of each of two products x and y. The tables also show the total available units of each of these items.

Table A

	Units needed for 1 unit of product x	Units needed for 1 unit of product y	Total units available
Labor	1	4	40
Machinery	1	3	31
Material	5	2	60
Overhead	1	1	15

Table B

	Units needed for 1 unit of product x	Units needed for 1 unit of product y	Total units available
Labor	2	7	63
Machinery	5	1	50
Material	1	2	21
Overhead	1	1	14

7. Using Table A, find the maximum profit if each unit of product x earns a profit of $50 and each unit of product y earns a profit of $60.

8. Using Table A find the maximum profit if each unit of product x earns a profit of $60 and each unit of product y earns a profit of $50.

9. Using Table B, find the maximum profit if each unit of product x earns a profit of $50 and each unit of product y earns a profit of $60.

10. Using Table B, find the maximum profit if each unit of product x earns a profit of $60 and each unit of product y earns a profit of $50.

11. An electronics company makes two kinds of electronic ranges, a standard model that earns a $50 profit and a deluxe model that earns a $60 profit. The company has machinery capable of producing any number of deluxe ranges up to 400 per month and any number of standard ranges up to 500 per month, but it has enough man-hours available to produce only 600 ranges of both kinds in a month. How many of each kind of range should the company produce to realize a maximum profit?

12. A pharmacy has 300 ounces of a drug it can use to make two different kinds of medicines. From each ounce of the drug, 15 bottles of medicine A or 25 bottles of medicine B can be produced. The pharmacy must keep all the bottles in its own storage room, which has a capacity of 6000 bottles in all. At least 600 bottles of medicine A and 1250 bottles of medicine B must be retained in inventory and the rest will be used. If medicine A sells for $2.75 per bottle and medicine B sells for $2.00 per bottle, how many ounces of the drug should be devoted to each medicine to maximize the total return?

Chapter Review

[8.1]   *Solve each system.*

1. $x + 5y = 18$
   $x - y = -3$

2. $x + 5y = 11$
   $2x + 3y = 8$

3. $2x - 3y = 8$
   $3x + 2y = 7$

4. Find values for a and b so that the graph of $ax + by = 19$ passes through the points $(2, 3)$ and $(-3, 5)$.

[8.2]   *Solve each system.*

5. $x + 3y - z = 3$
     $2x - y + 3z = 1$
     $3x + 2y + z = 5$

6. $x + y + z = 2$
     $3x - y + z = 4$
     $2x + y + 2z = 3$

7. $2x + 3y - z = -2$
     $x - y + z = 6$
     $3x - y + z = 10$

8. Find value for a, b, and c so that the graph of $y = ax^2 + bx + c$ contains the points $(-1, 9)$, $(0, 4)$, and $(1, 3)$.

[8.3]   *Solve each system.*

9. $x^2 + y = 3$
     $5x + y = 7$

10. $x^2 + 3xy + x = -12$
      $2x - y = 7$

11. $2x^2 + 5y^2 - 53 = 0$
      $4x^2 + 3y^2 - 43 = 0$

12. Approximate the solution of the system

$$y = 3^x$$
$$y = 3 - x$$

by graphical methods.

[8.4]   *By double shading, indicate the region representing the solution set of each system.*

13. $y > x^2 - 4$
      $y < 2 - x$

14. $y + x^2 < 0$
      $y + 3 > 0$
      $y < x - 3$

[8.5]   *Graph the polygonal set defined by each system of inequalities.*

15. $0 \le x \le 3$
      $0 \le y \le 4$
      $x + y \le 5$

16. $0 \le x \le 3$
      $0 \le y$
      $y \le 2 + x$
      $x + y \le 4$

[8.6]   17. Find the maximum and minimum values of $2x + y$ over the set in Problem 15.

18. Find the maximum and minimum values of $3x + 2y$ over the set in Problem 16.

19. Using Table A on page 208, find the maximum profit if each unit of product x earns a profit of $80 and each unit of product y earns a profit of $60.

20. Using Table B on page 208, find the maximum profit if each unit of product x earns a profit of $10 and each unit of product y earns a profit of $20.

9

Matrices and Determinants

9.1 *Definitions; Matrix Addition*

Matrices are today much used in mathematics and engineering, and also in the physical, social, and life sciences.

A **matrix** is a rectangular array of numbers (or other suitable entities), which are called the **entries** or **elements** of the matrix. In this book, we shall consider only real numbers as entries. A matrix is customarily displayed in a pair of brackets or parentheses (we shall use brackets). Thus

$$\begin{bmatrix} 1 & 2 & 3 \\ 4 & 5 & 6 \end{bmatrix} \quad \text{and} \quad \begin{bmatrix} 2 \\ 1 \end{bmatrix}$$

are matrices. The **order**, or **dimension**, of a matrix is the ordered pair having as first component the number of (horizontal) **rows** and as second component the number of (vertical) **columns** in the matrix. Thus,

$$\begin{bmatrix} 1 & 2 & 3 \\ 4 & 5 & 6 \end{bmatrix}, \quad \begin{bmatrix} 1 \\ 2 \\ 3 \end{bmatrix}, \quad \text{and} \quad \begin{bmatrix} a_1 & a_2 & a_3 & a_4 \\ b_1 & b_2 & b_3 & b_4 \\ c_1 & c_2 & c_3 & c_4 \\ d_1 & d_2 & d_3 & d_4 \end{bmatrix}$$

are 2×3 (read "two by three"), 3×1 (read "three by one"), and 4×4 (read "four by four") matrices, respectively. Note that the number of *rows* is given first, and then the number of *columns*. A matrix consisting of a single row is called a **row matrix** or a **row vector**, whereas a matrix consisting of a single column is called a **column matrix** or a **column vector**.

Matrices are frequently represented by capital letters. Thus, we might want to talk about the matrices A and B, where

$$A = \begin{bmatrix} a_1 & a_2 \\ b_1 & b_2 \end{bmatrix} \quad \text{and} \quad B = [b_1 \quad b_2].$$

To show that A is a 2×2 matrix, we can write $A_{2 \times 2}$. Similarly, $B_{1 \times 2}$ is a matrix with one row and two columns.

210

To represent the entries of a matrix, either single or double subscript notation is employed. Consider any 3×3 matrix, A. We can represent A by

$$A = \begin{bmatrix} a_1 & a_2 & a_3 \\ b_1 & b_2 & b_3 \\ c_1 & c_2 & c_3 \end{bmatrix},$$

where a different letter is used for each row and a single subscript denotes the column in which each particular entry is located. Alternatively, we can use a different letter for each column, and let the subscript denote the row. Thus, we might write

$$A = \begin{bmatrix} a_1 & b_1 & c_1 \\ a_2 & b_2 & c_2 \\ a_3 & b_3 & c_3 \end{bmatrix}.$$

In either event, problems would clearly arise if we wanted to talk about a matrix containing a large number of rows or columns, because we would run out of letters.

A much more useful convention involves double subscripts, where a single letter, say a, is used to denote an entry in a matrix, and then *two* subscripts are appended, the first subscript telling in which *row* the entry occurs, and the second telling in which *column*. Thus, we write

$$A = \begin{bmatrix} a_{11} & a_{12} & a_{13} \\ a_{21} & a_{22} & a_{23} \\ a_{31} & a_{32} & a_{33} \end{bmatrix},$$

where a_{21} is the element in the second *row* and first *column*, a_{33} is the element in the third *row* and third *column*, and, if we wish to generalize, a_{ij} is the element in the *i*th *row* and *j*th *column*.

Definition 9.1   *Two matrices, A and B, are **equal** if and only if both matrices are of the same order and $a_{ij} = b_{ij}$ for each i, j.*

Thus,

$$\begin{bmatrix} 2 & 1 \\ 3 & 0 \end{bmatrix} = \begin{bmatrix} \dfrac{4}{2} & 2-1 \\ \sqrt{9} & 0 \end{bmatrix}, \quad \text{but} \quad \begin{bmatrix} 2 & 1 \\ 3 & 0 \end{bmatrix} \neq \begin{bmatrix} 2 & 3 \\ 1 & 0 \end{bmatrix}.$$

Definition 9.2   *The **transpose** of a matrix A, denoted by A^t, is the matrix in which the rows are the columns of A and the columns are the rows of A.*

Thus,

$$\begin{bmatrix} 2 & 1 \\ 3 & 0 \end{bmatrix}^t = \begin{bmatrix} 2 & 3 \\ 1 & 0 \end{bmatrix} \quad \text{and} \quad \begin{bmatrix} 1 & 2 & 3 \\ 4 & 5 & 6 \end{bmatrix}^t = \begin{bmatrix} 1 & 4 \\ 2 & 5 \\ 3 & 6 \end{bmatrix}.$$

Definition 9.3   *The **sum** of two matrices of the same order, $A_{m \times n}$ and $B_{m \times n}$, is the matrix $(A + B)_{m \times n}$, in which the entry in the ith row and jth column is $a_{ij} + b_{ij}$, for $i = 1, 2, 3, \ldots, m$ and $j = 1, 2, 3, \ldots, n$.*

For example,

$$\begin{bmatrix} 3 & 1 & 2 \\ 2 & 1 & 4 \end{bmatrix} + \begin{bmatrix} 1 & 0 & 2 \\ -1 & 3 & 0 \end{bmatrix} = \begin{bmatrix} 3+1 & 1+0 & 2+2 \\ 2+(-1) & 1+3 & 4+0 \end{bmatrix}$$

$$= \begin{bmatrix} 4 & 1 & 4 \\ 1 & 4 & 4 \end{bmatrix}.$$

The sum of two matrices of different order is not defined.

Definition 9.4   *A matrix with each entry equal to* 0 *is a **zero matrix**.*

Zero matrices are generally denoted by the symbol **0**. This distinguishes the zero matrix from the real number 0. For example,

$$\mathbf{0}_{2 \times 4} = \begin{bmatrix} 0 & 0 & 0 & 0 \\ 0 & 0 & 0 & 0 \end{bmatrix}$$

is the 2×4 zero matrix.

Definition 9.5   *The **negative of a matrix** $A_{m \times n}$, denoted by $-A_{m \times n}$, is formed by replacing each entry in the matrix $A_{m \times n}$ with its additive inverse.*

For example, if

$$A_{3 \times 2} = \begin{bmatrix} 3 & -1 \\ 2 & -2 \\ -4 & 5 \end{bmatrix}, \quad \text{then} \quad -A_{3 \times 2} = \begin{bmatrix} -3 & 1 \\ -2 & 2 \\ 4 & -5 \end{bmatrix}.$$

The sum $B_{m \times n} + (-A_{m \times n})$ is called the **difference** of $B_{m \times n}$ and $A_{m \times n}$ and is denoted $B_{m \times n} - A_{m \times n}$.

Properties
of sums

At this point, we are able to establish the following facts concerning sums of matrices with real-number entries.

Theorem 9.1   *If A, B, and C are $m \times n$ matrices with real-number entries, then:*

I $(A + B)_{m \times n}$ *is a matrix with real-number*   *Closure law for addition.*
    entries.

II $(A + B) + C = A + (B + C)$.   *Associative law for addition.*

III   *The matrix $\mathbf{0}_{m \times n}$ has the property that for*   *Additive-identity law.*
     every matrix $A_{m \times n}$,

$$A + \mathbf{0} = A \quad \text{and} \quad \mathbf{0} + A = A.$$

IV   *For every matrix $A_{m \times n}$, the matrix $-A_{m \times n}$*   *Additive-inverse law.*
     has the property that

$$A + (-A) = \mathbf{0} \quad \text{and} \quad (-A) + A = \mathbf{0}.$$

V $A + B = B + A$.   *Commutative law for addition.*

Proof of 9.1-III Since each entry of the zero matrix is 0, it follows that the entries of $A_{m \times n} + \mathbf{0}_{m \times n}$ are $a_{ij} + 0 = a_{ij}$ and the entries of $\mathbf{0}_{m \times n} + A_{m \times n}$ are $0 + a_{ij} = a_{ij}$, and part III is proved.

For example,

$$\begin{bmatrix} a_{11} & a_{12} \\ a_{21} & a_{22} \end{bmatrix} + \begin{bmatrix} 0 & 0 \\ 0 & 0 \end{bmatrix} = \begin{bmatrix} a_{11} & a_{12} \\ a_{21} & a_{22} \end{bmatrix}.$$

Proof of 9.1-IV Let the entries of $A_{m \times n}$ and $-A_{m \times n}$ be a_{ij} and $-a_{ij}$, respectively. Since each entry of $A + (-A)$ is $a_{ij} - a_{ij}$, or 0, we have $A + (-A) = \mathbf{0}$. Similarly, $(-A) + A = \mathbf{0}$, and part IV is proved.

For example, if

$$A = \begin{bmatrix} 1 & -1 & 2 \\ 3 & -1 & 1 \end{bmatrix},$$

then

$$A + (-A) = \begin{bmatrix} 1 & -1 & 2 \\ 3 & -1 & 1 \end{bmatrix} + \begin{bmatrix} -1 & 1 & -2 \\ -3 & 1 & -1 \end{bmatrix} = \begin{bmatrix} 0 & 0 & 0 \\ 0 & 0 & 0 \end{bmatrix} = \mathbf{0}.$$

The proof of Parts I, II, and V of Theorem 9.1 are left as exercises.

Exercise 9.1

State the order and find the transpose of each matrix.

Example

$$\begin{bmatrix} 2 & 4 \\ 1 & -3 \\ 6 & 0 \end{bmatrix}$$

Solution

3×2 matrix; $\begin{bmatrix} 2 & 4 \\ 1 & -3 \\ 6 & 0 \end{bmatrix}^t = \begin{bmatrix} 2 & 1 & 6 \\ 4 & -3 & 0 \end{bmatrix}$

1. $\begin{bmatrix} 6 & -1 \\ 2 & 3 \end{bmatrix}$

2. $\begin{bmatrix} 4 & 1 \\ 0 & -2 \end{bmatrix}$

3. $\begin{bmatrix} 2 & -7 & 3 \\ 1 & 4 & 0 \end{bmatrix}$

4. $\begin{bmatrix} -3 & 1 \\ 6 & 0 \\ 0 & 2 \end{bmatrix}$

5. $\begin{bmatrix} 2 & 3 & -1 \\ 4 & 0 & 1 \\ -2 & 3 & 1 \end{bmatrix}$

6. $\begin{bmatrix} 4 & -1 & -2 \\ 3 & 0 & 0 \\ 2 & 1 & 1 \end{bmatrix}$

7. $\begin{bmatrix} 4 & -3 & -1 & 0 \\ 2 & 1 & 1 & 6 \end{bmatrix}$

8. $\begin{bmatrix} -2 & 1 & 3 & 2 \\ 4 & 0 & 0 & -2 \\ -1 & 3 & 2 & 4 \end{bmatrix}$

Write each sum or difference as a single matrix.

Example

$$\begin{bmatrix} 2 & 1 & 4 \\ 3 & -1 & 0 \end{bmatrix} + \begin{bmatrix} 6 & 3 & 0 \\ -2 & 1 & 0 \end{bmatrix}$$

(Solution overleaf)

Solution $\begin{bmatrix} 2 & 1 & 4 \\ 3 & -1 & 0 \end{bmatrix} + \begin{bmatrix} 6 & 3 & 0 \\ -2 & 1 & 0 \end{bmatrix} = \begin{bmatrix} 2+6 & 1+3 & 4+0 \\ 3-2 & -1+1 & 0+0 \end{bmatrix} = \begin{bmatrix} 8 & 4 & 4 \\ 1 & 0 & 0 \end{bmatrix}$

9. $\begin{bmatrix} 2 & 3 \\ 1 & 6 \end{bmatrix} + \begin{bmatrix} 1 & -2 \\ 2 & 3 \end{bmatrix}$ 10. $\begin{bmatrix} 4 & -1 & 3 \\ 2 & 1 & 0 \end{bmatrix} + \begin{bmatrix} 3 & -1 & 0 \\ 4 & 0 & -2 \end{bmatrix}$

11. $\begin{bmatrix} 3 & 0 & -1 \\ 2 & 1 & 2 \end{bmatrix} + \begin{bmatrix} 6 & -1 & 0 \\ 0 & 2 & 4 \end{bmatrix}$ 12. $[1 \quad 3 \quad 5 \quad 7] + [0 \quad -2 \quad 1 \quad 3]$

13. $\begin{bmatrix} 4 & -3 \\ 2 & 1 \end{bmatrix} - \begin{bmatrix} 6 & 0 \\ -2 & 1 \end{bmatrix}$ 14. $\begin{bmatrix} 4 & -1 & 2 \\ 3 & 1 & -4 \end{bmatrix} - \begin{bmatrix} -1 & -1 & 2 \\ 3 & 1 & 4 \end{bmatrix}$

15. $\begin{bmatrix} 10 & 3 & 2 \\ 5 & 1 & 7 \\ 6 & 1 & 9 \end{bmatrix} - \begin{bmatrix} 8 & 12 & 15 \\ -2 & 5 & 6 \\ -3 & 1 & 9 \end{bmatrix}$ 16. $\begin{bmatrix} 3 & -1 & 2 \\ 4 & -2 & 1 \\ 6 & 3 & 2 \end{bmatrix} - \begin{bmatrix} 2 & -1 & 2 \\ 4 & -1 & 1 \\ 6 & 3 & 1 \end{bmatrix}$

17. $\begin{bmatrix} 4 \\ 3 \\ -1 \end{bmatrix} + \begin{bmatrix} 6 \\ 0 \\ -2 \end{bmatrix}$ 18. $\begin{bmatrix} 2 & 3 \\ 1 & 0 \\ -1 & 2 \end{bmatrix} + \begin{bmatrix} -2 & 0 \\ -3 & 0 \\ 4 & -1 \end{bmatrix}$

19. $\begin{bmatrix} 2 & 3 & 4 \\ -1 & 6 & 2 \\ 1 & 0 & 3 \end{bmatrix} + \begin{bmatrix} 0 & 0 & 0 \\ 0 & 0 & 0 \\ 0 & 0 & 0 \end{bmatrix}$ 20. $\begin{bmatrix} 2 & -3 \\ 4 & -1 \\ -2 & 1 \end{bmatrix} + \begin{bmatrix} -2 & 3 \\ -4 & 1 \\ 2 & -1 \end{bmatrix}$

21. Use Theorem 9.1 to argue that $X + A = B$ and $X = B - A$ are equivalent matrix equations in the system of 2×2 matrices.

Solve each of the following matrix equations.

Example $X + \begin{bmatrix} 2 & 3 \\ 1 & 7 \end{bmatrix} = \begin{bmatrix} 9 & -4 \\ 2 & 0 \end{bmatrix}$

Solution $X = \begin{bmatrix} 9 & -4 \\ 2 & 0 \end{bmatrix} - \begin{bmatrix} 2 & 3 \\ 1 & 7 \end{bmatrix} = \begin{bmatrix} 7 & -7 \\ 1 & -7 \end{bmatrix}$

22. $X + \begin{bmatrix} 3 & -1 \\ 2 & 1 \end{bmatrix} = \begin{bmatrix} 5 & 1 \\ -3 & 5 \end{bmatrix}$ 23. $X - \begin{bmatrix} -1 & 0 \\ 0 & 0 \end{bmatrix} = \begin{bmatrix} 3 & -1 \\ 2 & 1 \end{bmatrix}^t$

24. $X + \begin{bmatrix} 3 & 2 \\ -1 & 4 \end{bmatrix} = \begin{bmatrix} 2 & 6 \\ 1 & 5 \end{bmatrix} + \begin{bmatrix} -4 & -8 \\ -2 & 0 \end{bmatrix}$

25. $\begin{bmatrix} 1 & 3 \\ -1 & 0 \end{bmatrix}^t - \begin{bmatrix} 0 & 1 \\ 1 & 0 \end{bmatrix} = \begin{bmatrix} 2 & -2 \\ -1 & 3 \end{bmatrix}^t - X$

26. Show that $[A_{2 \times 2} + B_{2 \times 2}]^t = A_{2 \times 2}^t + B_{2 \times 2}^t$. Does an analogous result seem valid for $n \times n$ matrices?

27. Prove Theorem 9.1-I.   28. Prove Theorem 9.1-II.

29. Prove Theorem 9.1-V.

9.2   *Matrix Multiplication*

We shall be interested in two kinds of products involving matrices: (1) the product of a matrix and a real number, and (2) the product of two matrices.

Definition 9.6   The **product** of a real number c and an $m \times n$ matrix A with entries a_{ij} is the matrix cA with corresponding entries ca_{ij}, where $i = 1, 2, 3, \ldots, m$ and $j = 1, 2, 3, \ldots, n$.

For example,

$$3\begin{bmatrix} 2 & 1 \\ 0 & 5 \end{bmatrix} = \begin{bmatrix} 3 \times 2 & 3 \times 1 \\ 3 \times 0 & 3 \times 5 \end{bmatrix} = \begin{bmatrix} 6 & 3 \\ 0 & 15 \end{bmatrix}.$$

Properties of products of matrices and real numbers

The following theorem states some simple algebraic laws for the multiplication of matrices by real numbers.

Theorem 9.2   If A and B are $m \times n$ matrices, and $c, d \in R$, then

I cA is an $m \times n$ matrix,   V $1A = A$,

II $c(dA) = (cd)A$,   VI $(-1)A = -A$,

III $(c + d)A = cA + dA$,   VII $0A = \mathbf{0}$,

IV $c(A + B) = cA + cB$,   VIII $c\mathbf{0} = \mathbf{0}$.

Proof We shall prove only Part IV, leaving the remaining parts as exercises. Since the elements of $A + B$ are of the form $a_{ij} + b_{ij}$, it follows, by definition, that the elements of $c(A + B)$ are of the form $c(a_{ij} + b_{ij})$. But, since a_{ij}, b_{ij}, and c denote real numbers, $c(a_{ij} + b_{ij}) = ca_{ij} + cb_{ij}$. Now, the elements of cA are of the form ca_{ij}, and those of cB are of the form cb_{ij}, so that the elements of $cA + cB$ are of the form $ca_{ij} + cb_{ij}$ and part IV is proved.

Products of matrices

Turning now to the *product of two matrices*, we have the following definition.

Definition 9.7   The **product** of the matrices $A_{m \times p}$ and $B_{p \times n}$ is the matrix $(AB)_{m \times n}$ with entries determined as follows: The entry c_{ij} in the ith row and jth column of $(AB)_{m \times n}$ is found by multiplying the first element in the ith row of A by the first element of the jth column of B, to this product adding the product of the second element in the ith row of A with the second element in the jth column of B, to this sum adding the product of the third element in the ith row of A with the third element in the jth column of B, and so on.

For example, the product of the matrices

$$A_{2 \times 2} = \begin{bmatrix} a_{11} & a_{12} \\ a_{21} & a_{22} \end{bmatrix} \quad \text{and} \quad B_{2 \times 2} = \begin{bmatrix} b_{11} & b_{12} \\ b_{21} & b_{22} \end{bmatrix}$$

is the matrix

$$(AB)_{2 \times 2} = \begin{bmatrix} a_{11}b_{11} + a_{12}b_{21} & a_{11}b_{12} + a_{12}b_{22} \\ a_{21}b_{11} + a_{22}b_{21} & a_{21}b_{12} + a_{22}b_{22} \end{bmatrix}.$$

The following schematic shows how to find the entry in the first row and first column of AB.

$$\begin{bmatrix} a_{11} & a_{12} \\ a_{21} & a_{22} \end{bmatrix}\begin{bmatrix} b_{11} & b_{12} \\ b_{21} & b_{22} \end{bmatrix} = \begin{bmatrix} a_{11}b_{11} + a_{12}b_{21} & \\ & \end{bmatrix}$$

Example   If $A = \begin{bmatrix} 1 & 2 \\ -1 & 3 \end{bmatrix}$ and $B = \begin{bmatrix} 2 & 1 \\ 1 & 1 \end{bmatrix}$, find AB and BA.

Solution $AB = \begin{bmatrix} 1 & 2 \\ -1 & 3 \end{bmatrix}\begin{bmatrix} 2 & 1 \\ 1 & 1 \end{bmatrix} = \begin{bmatrix} 2+2 & 1+2 \\ -2+3 & -1+3 \end{bmatrix} = \begin{bmatrix} 4 & 3 \\ 1 & 2 \end{bmatrix}$

$BA = \begin{bmatrix} 2 & 1 \\ 1 & 1 \end{bmatrix}\begin{bmatrix} 1 & 2 \\ -1 & 3 \end{bmatrix} = \begin{bmatrix} 2-1 & 4+3 \\ 1-1 & 2+3 \end{bmatrix} = \begin{bmatrix} 1 & 7 \\ 0 & 5 \end{bmatrix}$

Properties of products   The foregoing example shows very clearly that the multiplication of matrices, in general, is *not commutative*. Thus, when discussing products of matrices, we must specify the *order* in which the matrices are to be considered as factors. For the product AB, we say that A is *right-multiplied* by B, and that B is *left-multiplied* by A.

Note that the definition of the product of two matrices, A and B, requires that the matrix A have the same number of *columns* as B has *rows*; the result, AB, then has the same number of rows as A and the same number of columns as B. Such matrices A and B are said to be **comformable** for multiplication. The fact that two matrices are conformable in the order AB, however, does not mean that they necessarily are conformable in the order BA.

Example   If $A = \begin{bmatrix} 3 & 1 \\ 1 & 0 \\ 2 & 1 \end{bmatrix}$ and $B = \begin{bmatrix} 1 & -1 \\ 2 & 1 \end{bmatrix}$, find AB.

Solution   Since A is a 3×2 matrix, and B is a 2×2 matrix, they are conformable for multiplication in the order AB. We have

$$AB = \begin{bmatrix} 3 & 1 \\ 1 & 0 \\ 2 & 1 \end{bmatrix}\begin{bmatrix} 1 & -1 \\ 2 & 1 \end{bmatrix} = \begin{bmatrix} 3+2 & -3+1 \\ 1+0 & -1+0 \\ 2+2 & -2+1 \end{bmatrix} = \begin{bmatrix} 5 & -2 \\ 1 & -1 \\ 4 & -1 \end{bmatrix}.$$

Note that the matrices A and B in the example above are not conformable in the order BA.

In much of the matrix work in this book, we shall focus our attention on matrices having the same number of rows as columns. For brevity, a matrix of order $n \times n$ is often called a **square matrix** of order n. Although many of the ideas we shall discuss are applicable to conformable matrices of any order, we shall apply the notions only to square matrices.

Theorem 9.3   *If A, B, and C are $n \times n$ square matrices, then*

$$(AB)C = A(BC).$$

If A is a square matrix, then A^2, A^3, etc. denote AA, $(AA)A$, etc.

Theorem 9.4   *If A, B, and C are $n \times n$ square matrices, then*

$$A(B + C) = AB + AC$$

and

$$(B + C)A = BA + CA.$$

The proofs of these theorems involve some complicated symbolism and are omitted here, but you will be asked to show their validity for the case of 2×2 matrixes in the exercises. Observe that, because matrix multiplication is not, in general, commutative, we must establish both the left-hand and the right-hand distributive property.

Definition 9.8   *The **principal diagonal** of a square matrix is the ordered set of entries a_{jj}, extending from the upper left-hand corner to the lower right hand corner of the matrix. Thus, the principal diagonal contains a_{11}, a_{22}, a_{33}, etc.*

For example, the principle diagonal of

$$\begin{bmatrix} 1 & 3 & -1 \\ 5 & 2 & 3 \\ 6 & 4 & 0 \end{bmatrix}$$

consists of 1, 2, and 0, in that order.

Definition 9.9   *A **diagonal matrix** is a square matrix in which all entries not in the principal diagonal are 0.*

Thus,

$$\begin{bmatrix} 4 & 0 \\ 0 & 2 \end{bmatrix} \quad \text{and} \quad \begin{bmatrix} 1 & 0 & 0 \\ 0 & 1 & 0 \\ 0 & 0 & 0 \end{bmatrix}$$

are diagonal matrices.

Definition 9.10   *$I_{n \times n}$ denotes the diagonal matrix having 1's for entries on the principal diagonal.*

For example,

$$I_{2 \times 2} = \begin{bmatrix} 1 & 0 \\ 0 & 1 \end{bmatrix} \quad \text{and} \quad I_{4 \times 4} = \begin{bmatrix} 1 & 0 & 0 & 0 \\ 0 & 1 & 0 & 0 \\ 0 & 0 & 1 & 0 \\ 0 & 0 & 0 & 1 \end{bmatrix}.$$

The following properties are consequences of the definitions we have adopted.

Theorem 9.5   *For each matrix $A_{n \times n}$,*

$$A_{n \times n} I_{n \times n} = I_{n \times n} A_{n \times n} = A_{n \times n}.$$

Furthermore, $I_{n \times n}$ is the unique matrix having this property for all matrices $A_{n \times n}$.

Accordingly, $I_{n \times n}$ is the **identity element for multiplication** in the set of $n \times n$ square matrices. The proof of this theorem, for the illustrative case $n = 2$, is left as an exercise.

***Order of
multiplication***   The following result relates the order in which matrices can be multiplied by real numbers and by other matrices.

Theorem 9.6   *If A and B are n × n square matrices, and a is a real number, then*

$$a(AB) = (aA)B = A(aB).$$

The proof for the illustrative case $n = 2$ is left as an exercise.

Exercise 9.2

Write each product as a single matrix.

Examples   a. $3\begin{bmatrix} 2 & 1 \\ -1 & 3 \\ 2 & 0 \end{bmatrix}$ b. $\begin{bmatrix} 3 & 1 & -1 \\ 0 & -1 & 2 \end{bmatrix} \cdot \begin{bmatrix} 1 & -1 \\ 0 & 2 \\ 1 & 0 \end{bmatrix}$

Solutions   a. $\begin{bmatrix} 6 & 3 \\ -3 & 9 \\ 6 & 0 \end{bmatrix}$ b. $\begin{bmatrix} 3+0-1 & -3+2+0 \\ 0+0+2 & 0-2+0 \end{bmatrix} = \begin{bmatrix} 2 & -1 \\ 2 & -2 \end{bmatrix}$

1. $-5\begin{bmatrix} 0 & 1 & -1 \\ 3 & -1 & 2 \end{bmatrix}$

2. $2\begin{bmatrix} 2 & 1 & 3 & -2 \\ 4 & 2 & 0 & -1 \\ 0 & 0 & -1 & 2 \end{bmatrix}$

3. $[1 \ \ -2] \cdot \begin{bmatrix} 3 \\ 2 \end{bmatrix}$

4. $[3 \ \ -2 \ \ 2] \cdot \begin{bmatrix} 1 \\ 0 \\ -2 \end{bmatrix}$

5. $\begin{bmatrix} 3 & -1 \\ 2 & 1 \end{bmatrix} \cdot \begin{bmatrix} 1 & -4 \\ 2 & 1 \end{bmatrix}$

6. $\begin{bmatrix} 1 & -5 \\ 0 & 2 \end{bmatrix} \cdot \begin{bmatrix} 3 & 1 \\ -1 & 2 \end{bmatrix}$

7. $\begin{bmatrix} 4 & -5 \\ 7 & 3 \end{bmatrix} \cdot \begin{bmatrix} 5 & -1 \\ -2 & 7 \end{bmatrix}$

8. $\begin{bmatrix} 1 & -2 \\ -3 & 1 \end{bmatrix} \cdot \begin{bmatrix} 5 & 1 \\ 0 & 2 \end{bmatrix}$

9. $\begin{bmatrix} -3 & 1 & 0 \\ 2 & 1 & 1 \end{bmatrix} \cdot \begin{bmatrix} 2 & 0 \\ 1 & -1 \\ 3 & 0 \end{bmatrix}$

10. $\begin{bmatrix} 1 & -1 & 0 \\ 2 & 1 & 3 \end{bmatrix} \cdot \begin{bmatrix} 4 & -1 \\ 2 & 0 \\ 1 & 1 \end{bmatrix}$

11. $\begin{bmatrix} -1 & 0 & 1 \\ 2 & 1 & 0 \\ 1 & 0 & 0 \end{bmatrix} \cdot \begin{bmatrix} 0 & 1 & 3 \\ 1 & 0 & 2 \\ -1 & 1 & 1 \end{bmatrix}$

12. $\begin{bmatrix} 2 & -3 & 1 \\ 0 & 1 & -1 \\ 2 & 0 & 0 \end{bmatrix} \cdot \begin{bmatrix} 1 & 0 & 0 \\ 0 & 1 & 0 \\ 0 & 0 & 1 \end{bmatrix}$

13. $\begin{bmatrix} 2 & -2 & -1 \\ 1 & 1 & -2 \\ 1 & 0 & -1 \end{bmatrix} \cdot \begin{bmatrix} -1 & -2 & 5 \\ -1 & -1 & 3 \\ -1 & -2 & 4 \end{bmatrix}$

14. $\begin{bmatrix} -1 & -2 & 5 \\ -1 & -1 & 3 \\ -1 & -2 & 4 \end{bmatrix} \cdot \begin{bmatrix} 2 & -2 & -1 \\ 1 & 1 & -2 \\ 1 & 0 & -1 \end{bmatrix}$

Let $A = \begin{bmatrix} 1 & -2 \\ 1 & 0 \end{bmatrix}$ and $B = \begin{bmatrix} -1 & 2 \\ -1 & 1 \end{bmatrix}$. *Compute each of the following products.*

15. AB 16. BA 17. $(AB)A$

18. $(BA)B$ 19. $A^t B$ 20. AB^t

Find a matrix X satisfying each matrix equation.

21. $3X + \begin{bmatrix} 1 & 0 \\ 2 & 1 \end{bmatrix} = \begin{bmatrix} -2 & 3 \\ -1 & -2 \end{bmatrix}$ **22.** $2X + 3\begin{bmatrix} 1 & 1 \\ 0 & 1 \end{bmatrix} = \begin{bmatrix} 7 & -1 \\ 3 & -5 \end{bmatrix}$

23. $X + 2I = \begin{bmatrix} 3 & -1 \\ 1 & 2 \end{bmatrix}$ **24.** $3X - 2I = \begin{bmatrix} 7 & 3 \\ 6 & 4 \end{bmatrix}$

25. Show that if $A = \begin{bmatrix} -1 & 2 \\ 0 & 1 \end{bmatrix}$ and $B = \begin{bmatrix} 1 & 0 \\ -1 & 2 \end{bmatrix}$, then

    a. $(A + B)(A + B) \neq A^2 + 2AB + B^2$,

    b. $(A + B)(A - B) \neq A^2 - B^2$.

26. **a.** Show that for each matrix $A_{2 \times 2}$,

$$A_{2 \times 2} \cdot I_{2 \times 2} = I_{2 \times 2} \cdot A_{2 \times 2} = A_{2 \times 2}.$$

    b. Show that if, for a given matrix $B_{2 \times 2}$ and for *all* $A_{2 \times 2}$,

$$A_{2 \times 2} \cdot B_{2 \times 2} = B_{2 \times 2} \cdot A_{2 \times 2} = A_{2 \times 2},$$

    then $B_{2 \times 2} = I_{2 \times 2}$.

27. Show that

$$\left(\begin{bmatrix} a_{11} & a_{12} \\ a_{21} & a_{22} \end{bmatrix} \cdot \begin{bmatrix} b_{11} & b_{12} \\ b_{21} & b_{22} \end{bmatrix} \right) \cdot \begin{bmatrix} c_{11} & c_{12} \\ c_{21} & c_{22} \end{bmatrix} = \begin{bmatrix} a_{11} & a_{12} \\ a_{21} & a_{22} \end{bmatrix} \cdot \left(\begin{bmatrix} b_{11} & b_{12} \\ b_{21} & b_{22} \end{bmatrix} \cdot \begin{bmatrix} c_{11} & c_{12} \\ c_{21} & c_{22} \end{bmatrix} \right).$$

28. Show that

$$\begin{bmatrix} a_{11} & a_{12} \\ a_{21} & a_{22} \end{bmatrix} \cdot \left(\begin{bmatrix} b_{11} & b_{12} \\ b_{21} & b_{22} \end{bmatrix} + \begin{bmatrix} c_{11} & c_{12} \\ c_{21} & c_{22} \end{bmatrix} \right)$$

$$= \begin{bmatrix} a_{11} & a_{12} \\ a_{21} & a_{22} \end{bmatrix} \cdot \begin{bmatrix} b_{11} & b_{12} \\ b_{21} & b_{22} \end{bmatrix} + \begin{bmatrix} a_{11} & a_{12} \\ a_{21} & a_{22} \end{bmatrix} \cdot \begin{bmatrix} c_{11} & c_{12} \\ c_{21} & c_{22} \end{bmatrix}.$$

29. Show that

$$\left(\begin{bmatrix} b_{11} & b_{12} \\ b_{21} & b_{22} \end{bmatrix} + \begin{bmatrix} c_{11} & c_{12} \\ c_{21} & c_{22} \end{bmatrix} \right) \cdot \begin{bmatrix} a_{11} & a_{12} \\ a_{21} & a_{22} \end{bmatrix}$$

$$= \begin{bmatrix} b_{11} & b_{12} \\ b_{21} & b_{22} \end{bmatrix} \cdot \begin{bmatrix} a_{11} & a_{12} \\ a_{21} & a_{22} \end{bmatrix} + \begin{bmatrix} c_{11} & c_{12} \\ c_{21} & c_{22} \end{bmatrix} \cdot \begin{bmatrix} a_{11} & a_{12} \\ a_{21} & a_{22} \end{bmatrix}.$$

30. Show that $\begin{bmatrix} 0 & a \\ a & 0 \end{bmatrix}^2 = a^2 I$.

31. Show that $(A_{2 \times 2} \cdot B_{2 \times 2})^t = B_{2 \times 2}^t \cdot A_{2 \times 2}^t$.

32. Show that $A_{2 \times 2}^2 = (-A_{2 \times 2})^2$.

In Problems 33–39, prove the specified part of Theorem 9.2 for 2×2 *matrices.*

33. Part I   **34.** Part II   **35.** Part III   **36.** Part V

37. Part VI   **38.** Part VII   **39.** Part VIII

40. Prove that if A and B are 2×2 matrices and a is a real number, then we have $a(AB) = (aA)B = A(aB)$.

9.3   Solution of Linear Systems by Using Row-Equivalent Matrices

Elementary transforma-tions

An **elementary transformation** of a matrix $A_{n \times m}$ is a transformation that can be obtained in any one of the following ways:

1. Multiply the entries of any row of $A_{n \times m}$ by k, where $k \in R$, $k \neq 0$.
2. Interchange any two rows of $A_{n \times m}$.
3 Multiply the entries of any row of $A_{n \times m}$ by k, where $k \in R$, and add to the corresponding entries of any other row.

Notice that the inverse of an elementary transformation is an elementary transformation. That is, you can undo an elmentary transformation by means of an elementary transformation. For example, if A is transformed into B by interchanging two rows, then you can regain A from B by again interchanging the same two rows. It follows that if B results from performing a *succession* of elementary transformation on A, then A can similarly be obtained from B by performing the inverse operations in reverse order.

Row-equivalent matrices

Definition 9.11   *If B is a matrix resulting from a succession of a finite number of elementary transformations on a matrix A, then A and B are **row-equivalent** matrices. This is expressed by writing $A \sim B$ or $B \sim A$.*

Example

Show that $\begin{bmatrix} 1 & -2 & -1 \\ 1 & 0 & 2 \\ -4 & 3 & 1 \end{bmatrix} \sim \begin{bmatrix} 3 & -6 & -3 \\ 1 & 0 & 2 \\ -4 & 3 & 1 \end{bmatrix}$.

Solution

Multiplying each entry of the first row of the left-hand matrix by 3, we obtain the right-hand matrix.

Example

Show that $\begin{bmatrix} 1 & -2 & -1 \\ 1 & 0 & 2 \\ -4 & 3 & 1 \end{bmatrix} \sim \begin{bmatrix} -4 & 3 & 1 \\ 1 & 0 & 2 \\ 1 & -2 & -1 \end{bmatrix}$.

Solution

Interchanging the first and third row of the left-hand matrix, we obtain the right-hand matrix.

Sometimes it is convenient to make several elementary transformations on the same matrix.

Example

Show that $\begin{bmatrix} 1 & -2 & -1 \\ 1 & 0 & 2 \\ -4 & 3 & 1 \end{bmatrix} \sim \begin{bmatrix} 1 & -2 & -1 \\ 0 & 2 & 3 \\ 0 & -5 & -3 \end{bmatrix}$.

Solution

Multiplying the entries of the first row of the left-hand matrix by -1 and adding these products to the corresponding entries of the second row, and then multiplying the entries of the first row by 4 and adding these products to the corresponding entries of the third row, we obtain the right-hand matrix.

Most $n \times n$ matrices are row-equivalent to $I_{n \times n}$. These are the matrices that in Section 9.6 will be defined to be **nonsingular**.

Example   Show that $\begin{bmatrix} 1 & 2 & 1 \\ -1 & 1 & 0 \\ 1 & 0 & 1 \end{bmatrix} \sim \begin{bmatrix} 1 & 0 & 0 \\ 0 & 1 & 0 \\ 0 & 0 & 1 \end{bmatrix}$.

Solution   We first make appropriate transformations to obtain "0" elements in each entry (except for the principal diagonal) in columns 1, 2, and 3. Each reference to a row indicates the row of the preceding matrix.

$$\begin{bmatrix} 1 & 2 & 1 \\ -1 & 1 & 0 \\ 1 & 0 & 1 \end{bmatrix} \sim \begin{bmatrix} 1 & 2 & 1 \\ 0 & 3 & 1 \\ 0 & -2 & 0 \end{bmatrix} \begin{matrix} \\ \text{Row 2 + Row 1} \\ \text{Row 3 + [}(-1) \times \text{Row 1]} \end{matrix}$$

$$\sim \begin{bmatrix} 1 & 0 & 1 \\ 0 & 3 & 1 \\ 0 & 0 & \dfrac{2}{3} \end{bmatrix} \begin{matrix} \text{Row 1 + Row 3} \\ \\ \text{Row 3 + }\left[\left(\dfrac{2}{3}\right) \times \text{Row 2}\right] \end{matrix}$$

$$\sim \begin{bmatrix} 1 & 0 & 0 \\ 0 & 3 & 0 \\ 0 & 0 & \dfrac{2}{3} \end{bmatrix} \begin{matrix} \text{Row 1 + }\left[\left(-\dfrac{3}{2}\right) \times \text{Row 3}\right] \\ \text{Row 2 + }\left[\left(-\dfrac{3}{2}\right) \times \text{Row 3}\right] \\ \\ \end{matrix}$$

$$\sim \begin{bmatrix} 1 & 0 & 0 \\ 0 & 1 & 0 \\ 0 & 0 & 1 \end{bmatrix} \begin{matrix} \\ \left(\dfrac{1}{3}\right) \times \text{Row 2} \\ \left(\dfrac{3}{2}\right) \times \text{Row 3} \end{matrix}$$

Matrix
solution of
linear systems

In a linear system of the form

$$a_{11}x + a_{12}y + a_{13}z = c_1$$
$$a_{21}x + a_{22}y + a_{23}z = c_2$$
$$a_{31}x + a_{32}y + a_{33}z = c_3$$

the matrices

$$\begin{bmatrix} a_{11} & a_{12} & a_{13} \\ a_{21} & a_{22} & a_{23} \\ a_{31} & a_{32} & a_{33} \end{bmatrix} \quad \text{and} \quad \begin{bmatrix} a_{11} & a_{12} & a_{13} & c_1 \\ a_{21} & a_{22} & a_{23} & c_2 \\ a_{31} & a_{32} & a_{33} & c_3 \end{bmatrix}$$

are called the **coefficient matrix** and the **augmented matrix**, respectively. Similar definitions hold for a system of n linear equations.

Starting with the augmented matrix of a linear system, and generating a sequence of row-equivalent matrices, we can obtain a matrix from which the solution set of the system is evident simply by inspection. The validity of the method, which is illustrated by example below, follows from the fact that performing elementary row transformations on the augmented matrix of a system corresponds in this context to forming equivalent systems of equations.

Example   Solve $x + 2y - 3z = -4$
$$2x - y + z = 3$$
$$3x + 2y + z = 10$$

Solution   The augmented matrix of the system is

$$\begin{bmatrix} 1 & 2 & -3 & \vdots & -4 \\ 2 & -1 & 1 & \vdots & 3 \\ 3 & 2 & 1 & \vdots & 10 \end{bmatrix}.$$

The system of equations corresponding to each successive matrix is shown on the right of the matrix in the following solution.

$$\begin{matrix} & \\ \text{Row 2} + [-2 \times \text{Row 1}] \\ \text{Row 3} + [-3 \times \text{Row 1}] \end{matrix} \begin{bmatrix} 1 & 2 & -3 & \vdots & -4 \\ 0 & -5 & 7 & \vdots & 11 \\ 0 & -4 & 10 & \vdots & 22 \end{bmatrix} \begin{matrix} x + 2y - 3z = -4 \\ 0x - 5y + 7z = 11 \\ 0x - 4y + 10z = 22 \end{matrix}$$

$$\begin{matrix} \text{Row 1} + \left[\frac{2}{5} \times \text{Row 2}\right] \\ \\ \text{Row 3} + \left[-\frac{4}{5} \times \text{Row 2}\right] \end{matrix} \begin{bmatrix} 1 & 0 & -\frac{1}{5} & \vdots & \frac{2}{5} \\ 0 & -5 & 7 & \vdots & 11 \\ 0 & 0 & \frac{22}{5} & \vdots & \frac{66}{5} \end{bmatrix} \begin{matrix} x + 0y - \frac{1}{5}z = \frac{2}{5} \\ 0x - 5y + 7z = 11 \\ 0x + 0y + \frac{22}{5}z = \frac{66}{5} \end{matrix}$$

$$\begin{matrix} 5 \times \text{Row 1} \\ \\ \frac{5}{22} \times \text{Row 3} \end{matrix} \begin{bmatrix} 5 & 0 & -1 & \vdots & 2 \\ 0 & -5 & 7 & \vdots & 11 \\ 0 & 0 & 1 & \vdots & 3 \end{bmatrix} \begin{matrix} 5x + 0y - z = 2 \\ 0x - 5y + 7z = 11 \\ 0x + 0y + z = 3 \end{matrix}$$

$$\begin{matrix} \text{Row 1} + \text{Row 3} \\ \text{Row 2} + [-7 \times \text{Row 3}] \\ \cdot \end{matrix} \begin{bmatrix} 5 & 0 & 0 & \vdots & 5 \\ 0 & -5 & 0 & \vdots & -10 \\ 0 & 0 & 1 & \vdots & 3 \end{bmatrix} \begin{matrix} 5x + 0y + 0z = 5 \\ 0x - 5y + 0z = -10 \\ 0x + 0y + z = 3 \end{matrix}$$

$$\begin{matrix} \frac{1}{5} \times \text{Row 1} \\ \\ -\frac{1}{5} \times \text{Row 2} \\ \\ \end{matrix} \begin{bmatrix} 1 & 0 & 0 & \vdots & 1 \\ 0 & 1 & 0 & \vdots & 2 \\ 0 & 0 & 1 & \vdots & 3 \end{bmatrix} \begin{matrix} x + 0y + 0z = 1 \\ 0x + y + 0z = 2 \\ 0x + 0y + z = 3 \end{matrix}$$

The last system is equivalent to

$$x = 1$$
$$y = 2$$
$$z = 3.$$

From this, the solution set, $\{(1, 2, 3)\}$, for the given system is evident by inspection.

For any given $n \times n$ linear system with nonsingular coefficient matrix, there are many sequences of row operations which will transform the augmented matrix of a system equivalently to one of the form

$$\begin{bmatrix} 1 & 0 & 0 & \cdots & 0 & \vline & x_1 \\ 0 & 1 & 0 & & 0 & \vline & \cdot \\ 0 & 0 & 1 & & 0 & \vline & \cdot \\ \vdots & & & & \vdots & \vline & \cdot \\ 0 & 0 & 0 & \cdots & 1 & \vline & x_n \end{bmatrix},$$

from which the solution set, $\{(x_1, \ldots, x_n)\}$, of the original system is evident by inspection. Finding the most efficient sequence depends on experience and insight.

Exercise 9.3

Use row transformations of the augmented matrix to solve each system of equations. (Each coefficient matrix is nonsingular and is row-equivalent to $I_{n \times n}$.)

1. $x - 2y = 4$
    $x + 3y = -1$

2. $x + y = -1$
    $x - 4y = -14$

3. $3x - 2y = 13$
    $4x - y = 19$

4. $4x - 3y = 16$
    $2x + y = 8$

5. $x - 2y = 6$
    $3x + y = 25$

6. $x - y = -8$
    $x + 2y = 9$

7. $x + y - z = 0$
    $2x - y + z = -6$
    $x + 2y - 3z = 2$

8. $2x - y + 3z = 1$
    $x + 2y - z = -1$
    $3x + y + z = 2$

9. $2x - y = 0$
    $3y + z = 7$
    $2x + 3z = 1$

10. $3x - z = 7$
    $2x + y = 6$
    $3y - z = 7$

11. $2x - 5y + 3z = -1$
    $-3x - y + 2z = 11$
    $-2x + 7y + 5z = 9$

12. $2x + y + z = 4$
    $3x - z = 3$
    $2x + 3z = 13$

Show that each product is a matrix that is row-equivalent to $\begin{bmatrix} a & b \\ c & d \end{bmatrix}$ *if* $k \neq 0$.

13. $\begin{bmatrix} k & 0 \\ 0 & 1 \end{bmatrix} \begin{bmatrix} a & b \\ c & d \end{bmatrix}$

14. $\begin{bmatrix} 1 & 0 \\ 0 & k \end{bmatrix} \begin{bmatrix} a & b \\ c & d \end{bmatrix}$

15. $\begin{bmatrix} 0 & 1 \\ 1 & 0 \end{bmatrix} \begin{bmatrix} a & b \\ c & d \end{bmatrix}$

16. $\begin{bmatrix} 1 & k \\ 0 & 1 \end{bmatrix} \begin{bmatrix} a & b \\ c & d \end{bmatrix}$

17. $\begin{bmatrix} 1 & 0 \\ k & 1 \end{bmatrix} \begin{bmatrix} a & b \\ c & d \end{bmatrix}$

9.4   The Determinant Function

Associated with each square matrix A having real-number entries is a real number called the **determinant** of A and denoted by δA or $\delta(A)$ (read "the determinant of A"). Thus we have a function, δ (delta), with domain the set of all square matrices having real-number entries, and with range the set of all real numbers; $\delta(A_{n \times n})$ is called a determinant of **order** n.

Let us begin by examining δ over the set $S_{2 \times 2}$ of 2×2 matrices.

Definition 9.12   The **determinant** of the matrix

$$\begin{bmatrix} a_{11} & a_{12} \\ a_{21} & a_{22} \end{bmatrix}$$

is the number $a_{11}a_{22} - a_{12}a_{21}$.

The determinant of a square matrix is customarily displayed in the same form as the matrix, but with vertical bars in lieu of brackets. Thus,

$$\delta \begin{bmatrix} a_{11} & a_{12} \\ a_{21} & a_{22} \end{bmatrix} = \begin{vmatrix} a_{11} & a_{12} \\ a_{21} & a_{22} \end{vmatrix} = a_{11}a_{22} - a_{12}a_{21}.$$

Example
$$\delta \begin{bmatrix} 3 & 1 \\ -2 & 3 \end{bmatrix} = \begin{vmatrix} 3 & 1 \\ -2 & 3 \end{vmatrix} = (3)(3) - (1)(-2) = 9 + 2 = 11.$$

Turning next to 3×3 matrices, we have the following definition.

Definition 9.13   The **determinant** of the matrix

$$\begin{bmatrix} a_{11} & a_{12} & a_{13} \\ a_{21} & a_{22} & a_{23} \\ a_{31} & a_{32} & a_{33} \end{bmatrix}, \quad \text{denoted by} \quad \begin{vmatrix} a_{11} & a_{12} & a_{13} \\ a_{21} & a_{22} & a_{23} \\ a_{31} & a_{32} & a_{33} \end{vmatrix},$$

is the number

$$a_{11}a_{22}a_{33} - a_{11}a_{23}a_{32} + a_{12}a_{23}a_{31} - a_{12}a_{21}a_{33} + a_{13}a_{21}a_{32} - a_{13}a_{22}a_{31}.$$

Sign of a term in the expression of a determinant

An inspection of the subscripts of the factors of the products involved in this determinant (as well as those of the determinant of a 2×2 matrix) will show that each product is formed by taking one entry from each row and one entry from each column, with the restriction that no two factors be entries in the same row or column. The determinant consists of the sum of $\pm$ all such products as are possible. Whether a product or its negative is used in a determinant depends on the number of **inversions** in the second subscripts in the product when the first subscripts are in natural order, 1 2 3 .... An inversion occurs in a sequence of natural numbers each time a natural number is preceded by a greater natural number. For example, in the sequence 1 4 3 2, there are three inversions, because 4 precedes 3, 4 precedes 2, and 3 precedes 2. Now, if there is an odd number of inversions in the sequence formed by the *second subscripts of the factors in a product* when the first subscripts are in natural order, the negative of the product is used; otherwise, the product itself is used.

With this means of distinguishing products, we can generalize our definition of the determinant of a matrix.

Definition 9.14 *The **determinant** of the square matrix*

$$\begin{bmatrix} a_{11} & a_{12} & \cdots & a_{1n} \\ \vdots & \vdots & & \vdots \\ a_{n1} & a_{n2} & \cdots & a_{nn} \end{bmatrix}, \quad \text{denoted by} \quad \begin{vmatrix} a_{11} & a_{12} & \cdots & a_{1n} \\ \vdots & \vdots & & \vdots \\ a_{n1} & a_{n2} & \cdots & a_{nn} \end{vmatrix},$$

is equal to the sum of all products, $\pm a_{1j_1} a_{2j_2} a_{3j_3} \cdots a_{nj_n}$, where each j takes on all values from 1 to n, and no j's in the same product have the same subscript. For each term in the sum, the negative sign is used if the number of inversions in the sequence formed by the j's is odd; otherwise, the positive sign is used.

Value of a determinant

It is evident from this definition that the determinant of a matrix with real entries is a real number. We shall refer to this real number as the **value** of the determinant; writing the determinant as a sum is called **expanding** the determinant. Note that the arrays shown in the foregoing definition appear in print to be rectangular rather than square. The subscript on the lower right-hand entry, however, shows the dimension of the matrix or determinant. In any similar symbolism, this subscript should always be checked to determine the correct dimension. For example, in expanding the determinant of a 4×4 square matrix, one of the products is $a_{11}a_{23}a_{32}a_{44}$. The first subscripts are in natural order, 1 2 3 4, and the second subscripts are in the order 1 3 2 4. Since the only inversion is that 3 appears before 2, the negative of $a_{11}a_{23}a_{32}a_{44}$ is used in the expansion of the determinant.

Definition 9.15 *The **minor** M_{ij} of the element a_{ij} in a given determinant is the determinant that remains after the ith row and jth column in the given determinant have been deleted.*

For example, in the determinant

$$\begin{vmatrix} a_{11} & a_{12} & a_{13} \\ a_{21} & a_{22} & a_{23} \\ a_{31} & a_{32} & a_{33} \end{vmatrix}, \tag{1}$$

the minor of a_{11} is

$$M_{11} = \begin{vmatrix} a_{22} & a_{23} \\ a_{32} & a_{33} \end{vmatrix},$$

the minor of a_{23} is

$$M_{23} = \begin{vmatrix} a_{11} & a_{12} \\ a_{31} & a_{32} \end{vmatrix},$$

the minor of a_{31} is

$$M_{31} = \begin{vmatrix} a_{12} & a_{13} \\ a_{22} & a_{23} \end{vmatrix}.$$

Definition 9.16 *The **cofactor** A_{ij} of the element a_{ij} is the minor of a_{ij} if $i + j$ is an even integer, and the negative of the minor of a_{ij} if $i + j$ is an odd integer.*

For example, in the determinant (1),

the cofactor of a_{11} is $A_{11} = \begin{vmatrix} a_{22} & a_{23} \\ a_{32} & a_{33} \end{vmatrix}$, because $1 + 1$ is 2, an even integer;

the cofactor of a_{23} is $A_{23} = - \begin{vmatrix} a_{11} & a_{12} \\ a_{31} & a_{32} \end{vmatrix}$, because $2 + 3$ is 5, an odd integer;

the cofactor of a_{31} is $A_{31} = \begin{vmatrix} a_{12} & a_{13} \\ a_{22} & a_{23} \end{vmatrix}$, because $3 + 1$ is 4, an even integer; etc.

The following sign array is a convenient means of determining whether the cofactor of a given element equals the minor or whether it equals the negative of the minor.

$$
\begin{array}{cccccc}
+ & - & + & \cdots & \cdots & (-)^{n+1} \\
- & + & - & \cdots & \cdots & \cdots \\
+ & - & + & \cdots & \cdots & \cdots \\
\cdots & \cdots & \cdots & \cdots & \cdots & \cdots \\
\cdots & \cdots & \cdots & \cdots & \cdots & \cdots \\
(-)^{n+1} & \cdots & \cdots & \cdots & \cdots & +
\end{array}
$$

Expansion
by cofactors

With the definition of a cofactor in mind, let us look again at Definition 9.13 for the determinant of the matrix

$$
A = \begin{bmatrix} a_{11} & a_{12} & a_{13} \\ a_{21} & a_{22} & a_{23} \\ a_{31} & a_{32} & a_{33} \end{bmatrix}.
$$

The value of this determinant is

$$\delta(A) = a_{11}a_{22}a_{33} - a_{11}a_{23}a_{32} + a_{12}a_{23}a_{31} - a_{12}a_{21}a_{33} + a_{13}a_{21}a_{32} - a_{13}a_{22}a_{31}.$$

By suitably factoring pairs of terms in the right-hand member, we obtain

$$\delta(A) = a_{11}(a_{22}a_{33} - a_{23}a_{32}) + a_{12}(a_{23}a_{31} - a_{21}a_{33}) + a_{13}(a_{21}a_{32} - a_{22}a_{31}).$$

If the binomial factor in the middle term is rewritten $-(a_{21}a_{33} - a_{23}a_{31})$, we have

$$\delta(A) = a_{11}(a_{22}a_{33} - a_{23}a_{32}) + a_{12}[-(a_{21}a_{33} - a_{23}a_{31})] + a_{13}(a_{21}a_{32} - a_{22}a_{31}),$$

which is equal to

$$
a_{11} \begin{vmatrix} a_{22} & a_{23} \\ a_{32} & a_{33} \end{vmatrix} + a_{12} \left(- \begin{vmatrix} a_{21} & a_{23} \\ a_{31} & a_{33} \end{vmatrix} \right) + a_{13} \begin{vmatrix} a_{21} & a_{22} \\ a_{31} & a_{32} \end{vmatrix}.
$$

Accordingly, we have

$$\delta(A) = a_{11}A_{11} + a_{12}A_{12} + a_{13}A_{13}.$$

Thus, the determinant

$$
\delta(A) = \begin{vmatrix} a_{11} & a_{12} & a_{13} \\ a_{21} & a_{22} & a_{23} \\ a_{31} & a_{32} & a_{33} \end{vmatrix}
$$

is equal to the sum formed by multiplying each entry in the first row by its cofactor and then adding these products.

Similar methods can be used to show that this determinant is also equal to the sum formed by multiplying each entry in *any* row (or column) by its cofactor and then adding the products.

Although we shall not show it here, Definition 9.14 is logically equivalent to the following.

Definition 9.17   The **determinant** of the square matrix

$$\begin{bmatrix} a_{11} & a_{12} & \cdots & a_{1n} \\ a_{21} & a_{22} & \cdots & a_{2n} \\ \vdots & & & \vdots \\ a_{n1} & a_{n2} & \cdots & a_{nn} \end{bmatrix}$$

is the sum of the n products formed by multiplying each entry in any single row (or any single column) by its cofactor.

When this latter definition is used to rewrite a determinant, the determinant is said to be **expanded** about whatever row (or column) is chosen.

Example   If $\quad A = \begin{bmatrix} 3 & 2 & 1 \\ 0 & 1 & -2 \\ 1 & 3 & 4 \end{bmatrix}$, find $\delta(A)$ by expansion about the first column.

Solution   Noting that $a_{11} = 3$, $a_{21} = 0$, and $a_{31} = 1$, we have

$$\delta(A) = 3\begin{vmatrix} 1 & -2 \\ 3 & 4 \end{vmatrix} + 0\left(-\begin{vmatrix} 2 & 1 \\ 3 & 4 \end{vmatrix}\right) + 1\begin{vmatrix} 2 & 1 \\ 1 & -2 \end{vmatrix}$$

$$= 3(10) + 0 + (-5) = 25.$$

Exercise 9.4

$$\text{Let} \quad A = \begin{bmatrix} 2 & 1 & -2 & 0 \\ 1 & 0 & 3 & -1 \\ -2 & 1 & 2 & 2 \\ 1 & -1 & 3 & 1 \end{bmatrix}.$$

Each of the following products or its negative is a term in an expansion of $\delta(A)$. Determine the product, and write the product or its negative in accordance with the number of inversions in the second subscripts.

1. $a_{11}a_{22}a_{33}a_{44}$ 2. $a_{11}a_{23}a_{34}a_{42}$ 3. $a_{11}a_{24}a_{32}a_{43}$
4. $a_{12}a_{21}a_{33}a_{44}$ 5. $a_{13}a_{24}a_{32}a_{41}$ 6. $a_{14}a_{23}a_{32}a_{41}$

Let A be the matrix given above. Determine the minor M_{ij} and cofactor A_{ij} (in determinant form) of each of the following entries.

7. a_{11} 8. a_{13} 9. a_{23} 10. a_{41}

11. a_{31} 12. a_{33} 13. a_{44} 14. a_{14}

Evaluate each determinant.

Examples

a. $\begin{vmatrix} 2 & -3 \\ 1 & 4 \end{vmatrix}$ b. $\begin{vmatrix} 1 & 2 & 0 \\ 3 & -1 & 4 \\ -2 & 1 & 3 \end{vmatrix}$

Solutions

a. $\begin{vmatrix} 2 & -3 \\ 1 & 4 \end{vmatrix} = (2)(4) - (-3)(1) = 11.$

b. Expand about any row or column; the first row is used here:

$$\begin{vmatrix} 1 & 2 & 0 \\ 3 & -1 & 4 \\ -2 & 1 & 3 \end{vmatrix} = 1 \begin{vmatrix} -1 & 4 \\ 1 & 3 \end{vmatrix} - 2 \begin{vmatrix} 3 & 4 \\ -2 & 3 \end{vmatrix} + 0 \begin{vmatrix} 3 & -1 \\ -2 & 1 \end{vmatrix}$$

$$= 1[(-1)(3) - (4)(1)] - 2[(3)(3) - (4)(-2)] + 0$$

$$= -41.$$

15. $\begin{vmatrix} 1 & 0 \\ 2 & 1 \end{vmatrix}$ **16.** $\begin{vmatrix} 3 & -2 \\ 4 & 1 \end{vmatrix}$ **17.** $\begin{vmatrix} -3 & -1 \\ 3 & 1 \end{vmatrix}$ **18.** $\begin{vmatrix} -1 & 6 \\ 0 & -2 \end{vmatrix}$

19. $\begin{vmatrix} -4 & 2 \\ 1 & 7 \end{vmatrix}$ **20.** $\begin{vmatrix} 8 & -1 \\ 3 & 2 \end{vmatrix}$ **21.** $\begin{vmatrix} x & -1 \\ 1 & 3 \end{vmatrix}$ **22.** $\begin{vmatrix} -3 & x \\ -1 & 3 \end{vmatrix}$

23. $\begin{vmatrix} 2 & 0 & 1 \\ 1 & 1 & 2 \\ -1 & 0 & 1 \end{vmatrix}$ **24.** $\begin{vmatrix} 1 & 3 & 1 \\ -1 & 2 & 1 \\ 0 & 2 & 0 \end{vmatrix}$ **25.** $\begin{vmatrix} 1 & 2 & 3 \\ 3 & -1 & 2 \\ 2 & 0 & 2 \end{vmatrix}$

26. $\begin{vmatrix} 1 & 0 & 0 \\ 0 & 1 & 2 \\ 0 & 3 & 4 \end{vmatrix}$ **27.** $\begin{vmatrix} -1 & 0 & 2 \\ -2 & 1 & 0 \\ 0 & 1 & -3 \end{vmatrix}$ **28.** $\begin{vmatrix} 2 & 1 & 4 \\ 3 & 2 & 6 \\ 5 & -3 & 10 \end{vmatrix}$

29. $\begin{vmatrix} a & b & 1 \\ a & b & 1 \\ 1 & 1 & 1 \end{vmatrix}$ **30.** $\begin{vmatrix} a & a & a \\ 1 & 2 & 3 \\ 4 & 5 & 6 \end{vmatrix}$ **31.** $\begin{vmatrix} x & 0 & 0 \\ 0 & x & 0 \\ 0 & 0 & x \end{vmatrix}$ **32.** $\begin{vmatrix} 0 & 0 & x \\ 0 & x & 0 \\ x & 0 & 0 \end{vmatrix}$

Solve for x.

33. $\begin{vmatrix} x & -1 \\ 2 & 3 \end{vmatrix} = 17$ **34.** $\begin{vmatrix} 1 & -5 \\ x & 3 \end{vmatrix} = -7$

35. $\begin{vmatrix} x & 0 & 0 \\ 2 & 1 & 3 \\ 0 & 1 & 4 \end{vmatrix} = 3$ **36.** $\begin{vmatrix} x^2 & x & 1 \\ 0 & 2 & 1 \\ 3 & 1 & 4 \end{vmatrix} = 28$

Expand by cofactors and verify.

37. $\begin{vmatrix} 0 & 1 & 0 & 0 \\ 1 & 0 & 3 & 2 \\ 5 & -1 & 2 & 1 \\ 1 & 0 & 1 & 1 \end{vmatrix} = 5$ **38.** $\begin{vmatrix} 1 & 2 & 0 & -1 \\ 1 & 0 & -1 & 2 \\ 0 & 1 & 1 & 1 \\ 2 & -1 & 0 & 1 \end{vmatrix} = 17$

39. In accordance with Definition 9.17, the determinant of an $n \times n$ matrix is the sum of a certain number of products. What is this number for $n = 2$? For $n = 3$? For $n = 4$?

40. Generalize the results of Problem 39, making a conjecture about the number of such products in the determinant of an $n \times n$ matrix.

41. Show that for any 2×2 matrix A, $\delta(aA) = a^2\delta(A)$.

42. Show that for any 2×2 matrix A, $\delta(A^t) = \delta(A)$.

43. Show that for any 2×2 matrices A and B, $\delta(AB) = \delta(A) \cdot \delta(B)$.

9.5   *Properties of Determinants*

Determinants have some properties that are useful by virtue of the fact that they permit us to generate equal determinants with different and simpler configurations of entries. This, in turn, helps us find values for determinants.

Theorem 9.7   *If each entry in any row, or each entry in any column, of a determinant is 0, then the determinant is equal to 0.*

Proof By Definition 9.14, a determinant is a sum of products, each product having an entry from each row and each column of the determinant. This means that if any row or any column has only zero entries, then each product will contain 0 as a factor. Thus each product, and therefore the sum of all such products, is 0.

For example,

$$\begin{vmatrix} 0 & 0 \\ 1 & 2 \end{vmatrix} = 0, \quad \begin{vmatrix} 1 & 1 & 0 \\ 3 & 5 & 0 \\ 2 & 7 & 0 \end{vmatrix} = 0, \quad \text{and} \quad \begin{vmatrix} 0 & 1 & 0 & 0 \\ 1 & 0 & 0 & 0 \\ 0 & 0 & 0 & 1 \\ 0 & 0 & 0 & 1 \end{vmatrix} = 0.$$

Theorem 9.8   *If any two rows (or any two columns) of a determinant are interchanged, then the resulting determinant is the negative of the original determinant.*

Proof Let us look first at the case in which two *adjacent* rows of a determinant D of order n are interchanged. If the ith and $(i + 1)$st rows of D are interchanged to give us D' with entries a'_{ij}, then a_{ij} in D is equal to $a'_{i+1,j}$ in D' and the minors of a_{ij} in D are identical to those of $a'_{i+1,j}$ in D'. An expansion of D about the ith row leads to

$$D = a_{i1}A_{i1} + a_{i2}A_{i2} + \cdots + a_{in}A_{in},$$

and an expansion of D' about its $(i + 1)$st row leads to

$$D' = a'_{i+1,1}A'_{i+1,1} + a'_{i+1,2}A'_{i+1,2} + \cdots + a'_{i+1,n}A'_{i+1,n}.$$

Since $a_{ij} = a'_{i+1,j}$, and since, from the definition of cofactor, $A_{ij} = -A'_{i+1,j}$, it follows immediately that $D = -D'$. The same argument can be applied to an interchange of *adjacent* columns.

Now, let us turn to the case in which *any* two rows are interchanged. Let the interchange of the ith and $(i + k)$th rows of D lead to D'. This change can be viewed as the result of making k adjacent-row interchanges to bring the ith row into the $(i + k)$th position, and then $k - 1$ similar interchanges to bring the $(i + k)$th row into the ith position. Since each such interchange results in a sign reversal, and since $k + k - 1$ such reversals occur, $D' = (-1)^{2k-1} D$. But $2k - 1$ is an odd number for any natural number k, and hence $D' = -D$. A similar argument can be made for the interchange of the jth and $(j + k)$th columns.

For example,

$$\begin{vmatrix} 1 & 2 \\ 3 & 4 \end{vmatrix} = - \begin{vmatrix} 3 & 4 \\ 1 & 2 \end{vmatrix}, \quad \text{and} \quad \begin{vmatrix} 1 & 2 & 3 \\ 4 & 5 & 6 \\ 7 & 8 & 9 \end{vmatrix} = - \begin{vmatrix} 3 & 2 & 1 \\ 6 & 5 & 4 \\ 9 & 8 & 7 \end{vmatrix}.$$

In the first example, rows 1 and 2 were interchanged. In the second example, columns 1 and 3 were interchanged.

Theorem 9.9   *If two rows (or two columns) in a determinant have corresponding entries that are equal, the determinant is equal to 0.*

Proof   Let the determinant D have two rows (or columns) with corresponding entries equal. Then by Theorem 9.8, an interchange of these rows (or columns) will produce D', which is the negative of D. That is, $D' = -D$. Since, however, the interchanged rows (or columns) have identical elements, $D' = D$. Thus, $D = -D$, $2D = 0$, and D must be 0.

For example,

$$\begin{vmatrix} 1 & 1 \\ 3 & 3 \end{vmatrix} = 0, \quad \begin{vmatrix} 1 & 2 & 1 \\ 3 & 1 & 0 \\ 1 & 2 & 1 \end{vmatrix} = 0, \quad \text{and} \quad \begin{vmatrix} 1 & 2 & 3 & 4 \\ 5 & 6 & 7 & 8 \\ 0 & 0 & 1 & 0 \\ 1 & 2 & 3 & 4 \end{vmatrix} = 0.$$

Theorem 9.10   *If each of the entries of one row (or column) of a determinant is multiplied by k, the determinant is multiplied by k.*

Proof   Let the ith row of the determinant D be multipled by k to yield D'. Then an expansion of D' about the ith row leads to

$$\begin{aligned} D' &= ka_{i1}A_{i1} + ka_{i2}A_{i2} + \cdots + ka_{in}A_{in} \\ &= k(a_{i1}A_{i1} + a_{i2}A_{i2} + \cdots + a_{in}A_{in}) \\ &= kD. \end{aligned}$$

A similar argument holds for columns. Thus,

$$\begin{vmatrix} 1 & 0 & 0 \\ 2 & 1 & 3 \\ 1 \times 2 & 3 \times 2 & 4 \times 2 \end{vmatrix} = 2 \begin{vmatrix} 1 & 0 & 0 \\ 2 & 1 & 3 \\ 1 & 3 & 4 \end{vmatrix}, \quad \text{and} \quad \begin{vmatrix} 4 & 5 & 8 \\ 1 & 1 & 2 \\ 3 & 1 & 6 \end{vmatrix} = 2 \begin{vmatrix} 4 & 5 & 4 \\ 1 & 1 & 1 \\ 3 & 1 & 3 \end{vmatrix}.$$

Note that this process is different from that of the multiplication of a matrix by a real number. In the latter, each entry in the matrix is multiplied by the real number, rather than, as here, only the entries in a single row or column being so multiplied.

Theorem 9.11 *If each entry in a row (or column) of a determinant is written as the sum of two terms, the determinant can be written as the sum of two determinants as follows: If*

$$D = \begin{vmatrix} a_{11} & a_{12} & \cdots & a_{1n} \\ \vdots & \vdots & & \vdots \\ b_{i1} + c_{i1} & b_{i2} + c_{i2} & \cdots & b_{in} + c_{in} \\ \vdots & \vdots & & \vdots \\ a_{n1} & a_{n2} & \cdots & a_{nn} \end{vmatrix},$$

then

$$D = \begin{vmatrix} a_{11} & a_{12} & \cdots & a_{1n} \\ \vdots & \vdots & & \vdots \\ b_{i1} & b_{i2} & \cdots & b_{in} \\ \vdots & \vdots & & \vdots \\ a_{n1} & a_{n2} & \cdots & a_{nn} \end{vmatrix} + \begin{vmatrix} a_{11} & a_{12} & \cdots & a_{1n} \\ \vdots & \vdots & & \vdots \\ c_{i1} & c_{i2} & \cdots & c_{in} \\ \vdots & \vdots & & \vdots \\ a_{n1} & a_{n2} & \cdots & a_{nn} \end{vmatrix},$$

and if

$$D = \begin{vmatrix} a_{11} & \cdots & b_{1j} + c_{1j} & \cdots & a_{1n} \\ a_{21} & \cdots & b_{2j} + c_{2j} & \cdots & a_{2n} \\ \vdots & & \vdots & & \vdots \\ a_{n1} & \cdots & b_{nj} + c_{nj} & \cdots & a_{nn} \end{vmatrix},$$

then

$$D = \begin{vmatrix} a_{11} & \cdots & b_{1j} & \cdots & a_{1n} \\ a_{21} & \cdots & b_{2j} & \cdots & a_{2n} \\ \vdots & & \vdots & & \vdots \\ a_{n1} & \cdots & b_{nj} & \cdots & a_{nn} \end{vmatrix} + \begin{vmatrix} a_{11} & \cdots & c_{1j} & \cdots & a_{1n} \\ a_{21} & \cdots & c_{2j} & \cdots & a_{2n} \\ \vdots & & \vdots & & \vdots \\ a_{n1} & \cdots & c_{nj} & \cdots & a_{nn} \end{vmatrix}.$$

Proof Let

$$D = \begin{vmatrix} a_{11} & \cdots & a_{1n} \\ \vdots & & \vdots \\ b_{i1} + c_{i1} & \cdots & b_{in} + c_{in} \\ \vdots & & \vdots \\ a_{n1} & \cdots & a_{nn} \end{vmatrix},$$

and expand about the ith row. This leads to

$$\begin{aligned} D &= (b_{i1} + c_{i1})A_{i1} + (b_{i2} + c_{i2})A_{i2} + \cdots + (b_{in} + c_{in})A_{in} \\ &= b_{i1}A_{i1} + c_{i1}A_{i1} + b_{i2}A_{i2} + c_{i2}A_{i2} + \cdots + b_{in}A_{in} + c_{in}A_{in} \\ &= b_{i1}A_{i1} + b_{i2}A_{i2} + \cdots + b_{in}A_{in} + c_{i1}A_{i1} + c_{i2}A_{i2} + \cdots + c_{in}A_{in} \\ &= \begin{vmatrix} a_{11} & \cdots & a_{1n} \\ \vdots & & \vdots \\ b_{i1} & \cdots & b_{in} \\ \vdots & & \vdots \\ a_{n1} & \cdots & a_{nn} \end{vmatrix} + \begin{vmatrix} a_{11} & \cdots & a_{1n} \\ \vdots & & \vdots \\ c_{i1} & \cdots & c_{in} \\ \vdots & & \vdots \\ a_{n1} & \cdots & a_{nn} \end{vmatrix}. \end{aligned}$$

A similar argument proves the result for columns.

For example,

$$\begin{vmatrix} 1 & 3 \\ 2 & 5 \end{vmatrix} = \begin{vmatrix} 1 & 1 \\ 2 & 4 \end{vmatrix} + \begin{vmatrix} 1 & 2 \\ 2 & 1 \end{vmatrix}$$

and

$$\begin{vmatrix} 4 & 0 & 0 \\ 0 & 4 & 0 \\ 0 & 0 & 4 \end{vmatrix} = \begin{vmatrix} 4 & 0 & 0 \\ 0 & 2 & 0 \\ 0 & 0 & 4 \end{vmatrix} + \begin{vmatrix} 4 & 0 & 0 \\ 0 & 2 & 0 \\ 0 & 0 & 4 \end{vmatrix}.$$

Theorem 9.12   *If each entry of one row (or column) of a determinant is multiplied by a real number k and the resulting product is added to the corresponding entry in another row (or column, respectively) in the determinant, the resulting determinant is equal to the original determinant.*

Proof Let

$$D = \begin{vmatrix} a_{11} & \cdots & a_{1n} \\ \vdots & & \vdots \\ a_{n1} & \cdots & a_{nn} \end{vmatrix}.$$

Then, if the ith row of D is multiplied by k and added to another row, say the pth row, we have

$$\begin{vmatrix} a_{11} & \cdots & a_{1n} \\ \vdots & & \vdots \\ a_{i1} & \cdots & a_{in} \\ \vdots & & \vdots \\ a_{p1} + ka_{i1} & \cdots & a_{pn} + ka_{in} \\ \vdots & & \vdots \\ a_{n1} & \cdots & a_{nn} \end{vmatrix},$$

which, by Theorem 9.11, is equal to

$$\begin{vmatrix} a_{11} & \cdots & a_{1n} \\ \vdots & & \vdots \\ a_{i1} & \cdots & a_{in} \\ \vdots & & \vdots \\ a_{p1} & \cdots & a_{pn} \\ \vdots & & \vdots \\ a_{n1} & \cdots & a_{nn} \end{vmatrix} + \begin{vmatrix} a_{11} & \cdots & a_{1n} \\ \vdots & & \vdots \\ a_{i1} & \cdots & a_{in} \\ \vdots & & \vdots \\ ka_{i1} & \cdots & ka_{in} \\ \vdots & & \vdots \\ a_{n1} & \cdots & a_{nn} \end{vmatrix}.$$

The last determinant in this sum can, by Theorem 9.10, be written

$$k \begin{vmatrix} a_{11} & \cdots & a_{1n} \\ \vdots & & \vdots \\ a_{i1} & \cdots & a_{in} \\ \vdots & & \vdots \\ a_{i1} & \cdots & a_{in} \\ \vdots & & \vdots \\ a_{n1} & \cdots & a_{nn} \end{vmatrix},$$

and since this determinant contains two rows that have their corresponding entries equal, it follows from Theorem 9.9 that it is 0. Hence, the sum is D, and the theorem is proved for rows. A similar argument shows that it is true also for columns.

For example,

$$\begin{vmatrix} 1 & 1 \\ 2 & 1 \end{vmatrix} = \begin{vmatrix} 1 & 1 \\ 2 + 3(1) & 1 + 3(1) \end{vmatrix} = \begin{vmatrix} 1 & 1 \\ 5 & 4 \end{vmatrix}$$

and

$$\begin{vmatrix} 1 & 2 & 3 \\ 4 & 5 & 6 \\ 7 & 8 & 9 \end{vmatrix} = \begin{vmatrix} 1+2(3) & 2 & 3 \\ 4+2(6) & 5 & 6 \\ 7+2(9) & 8 & 9 \end{vmatrix} = \begin{vmatrix} 7 & 2 & 3 \\ 16 & 5 & 6 \\ 25 & 8 & 9 \end{vmatrix}$$

Evaluation of determinants

The preceding theorems can be used to write sequences of equal determinants, leading from one form of a determinant to another and more useful form.

Example

Evaluate

$$D = \begin{vmatrix} 2 & -1 & 1 & -3 \\ 1 & 3 & -4 & 2 \\ 1 & 0 & -2 & 1 \\ 3 & -1 & 5 & 2 \end{vmatrix}.$$

Solution

As a step toward evaluating the determinant, we shall use Theorem 9.12 to produce an equal determinant with a row or a column containing zero entries in all but one place. Let us arbitrarily select the second column for this role, because one entry is already zero. Multiplying a_{1j} by 3 and adding the result to a_{2j}, we obtain

$$D = \begin{vmatrix} 2 & -1 & 1 & -3 \\ 1+3(2) & 3+3(-1) & -4+3(1) & 2+3(-3) \\ 1 & 0 & -2 & 1 \\ 3 & -1 & 5 & 2 \end{vmatrix} = \begin{vmatrix} 2 & -1 & 1 & -3 \\ 7 & 0 & -1 & -7 \\ 1 & 0 & -2 & 1 \\ 3 & -1 & 5 & 2 \end{vmatrix}.$$

Next, multiplying a_{1j} by -1 and adding the result to a_{4j}, we find that

$$D = \begin{vmatrix} 2 & -1 & 1 & -3 \\ 7 & 0 & -1 & -7 \\ 1 & 0 & -2 & 1 \\ 3-1(2) & -1-1(-1) & 5-1(1) & 2-1(-3) \end{vmatrix} = \begin{vmatrix} 2 & -1 & 1 & -3 \\ 7 & 0 & -1 & -7 \\ 1 & 0 & -2 & 1 \\ 1 & 0 & 4 & 5 \end{vmatrix}.$$

If we now expand the determinant about the second column, we have

$$D = \begin{vmatrix} 2 & -1 & 1 & -3 \\ 7 & 0 & -1 & -7 \\ 1 & 0 & -2 & 1 \\ 1 & 0 & 4 & 5 \end{vmatrix} = -(-1)\begin{vmatrix} 7 & -1 & -7 \\ 1 & -2 & 1 \\ 1 & 4 & 5 \end{vmatrix} + 0A_{22} + 0A_{32} + 0A_{42}.$$

From this point, we can reduce the third-order determinant to a second-order determinant by a similar procedure or, alternatively, expand directly about the elements in any row or column. Expanding about the elements of the first row, we obtain

$$D = \begin{vmatrix} 7 & -1 & -7 \\ 1 & -2 & 1 \\ 1 & 4 & 5 \end{vmatrix} = 7\begin{vmatrix} -2 & 1 \\ 4 & 5 \end{vmatrix} - (-1)\begin{vmatrix} 1 & 1 \\ 1 & 5 \end{vmatrix} + (-7)\begin{vmatrix} 1 & -2 \\ 1 & 4 \end{vmatrix},$$

from which

$$D = 7(-14) + (4) - 7(6) = -98 + 4 - 42 = -136.$$

Exercise 9.5

Without evaluating, state why each statement is true. Verify selected examples by expansion.

1. $\begin{vmatrix} 2 & 3 & 1 \\ 0 & 0 & 0 \\ -1 & 2 & 0 \end{vmatrix} = 0$

2. $\begin{vmatrix} 3 & 1 & 3 \\ 0 & 1 & 0 \\ 1 & 2 & 1 \end{vmatrix} = 0$

3. $\begin{vmatrix} 2 & 3 & 1 & 1 \\ 2 & 0 & 1 & 2 \\ 2 & 3 & 1 & 1 \\ 0 & 1 & 2 & 0 \end{vmatrix} = 0$

4. $\begin{vmatrix} 7 & 3 & 2 & 0 \\ 2 & 1 & 2 & 0 \\ 4 & 1 & 1 & 0 \\ 0 & 2 & 1 & 0 \end{vmatrix} = 0$

5. $\begin{vmatrix} 4 & 2 & 1 \\ 0 & -1 & -2 \\ 1 & 0 & 2 \end{vmatrix} = - \begin{vmatrix} 4 & 2 & 1 \\ 0 & 1 & 2 \\ 1 & 0 & 2 \end{vmatrix}$

6. $\begin{vmatrix} -2 & 3 & 1 \\ -1 & 0 & 1 \\ -2 & 1 & 0 \end{vmatrix} = - \begin{vmatrix} 2 & 3 & 1 \\ 1 & 0 & 1 \\ 2 & 1 & 0 \end{vmatrix}$

7. $2 \begin{vmatrix} 1 & 0 & 2 \\ -1 & 2 & 0 \\ 1 & 1 & 1 \end{vmatrix} = \begin{vmatrix} 1 & 0 & 2 \\ -1 & 2 & 0 \\ 2 & 2 & 2 \end{vmatrix}$

8. $\begin{vmatrix} 3 & -4 & 2 \\ 1 & -2 & 0 \\ 0 & 8 & 1 \end{vmatrix} = -2 \begin{vmatrix} 3 & 2 & 2 \\ 1 & 1 & 0 \\ 0 & -4 & 1 \end{vmatrix}$

9. $\begin{vmatrix} 1 & 2 \\ 3 & 4 \end{vmatrix} = \begin{vmatrix} 1+2 & 2 \\ 3+4 & 4 \end{vmatrix}$

10. $\begin{vmatrix} 1 & 2 \\ 3 & 4 \end{vmatrix} = \begin{vmatrix} 1+4 & 2 \\ 3+8 & 4 \end{vmatrix}$

11. $\begin{vmatrix} 1 & 2 & 1 \\ 0 & 2 & 3 \\ 2 & -1 & 2 \end{vmatrix} = \begin{vmatrix} 1 & 2 & 1 \\ 0 & 2 & 3 \\ 0 & -5 & 0 \end{vmatrix}$

12. $\begin{vmatrix} -1 & 1 & 0 \\ 2 & 3 & -1 \\ 2 & 1 & 2 \end{vmatrix} = \begin{vmatrix} 0 & 1 & 0 \\ 5 & 3 & -1 \\ 3 & 1 & 2 \end{vmatrix}$

Theorem 9.12 was used on the left-hand member of each of the following equalities to produce the elements in the right-hand member. Complete the entries.

13. $\begin{vmatrix} 1 & 3 \\ 2 & 2 \end{vmatrix} = \begin{vmatrix} 1 & 3 \\ 0 & \end{vmatrix}$

14. $\begin{vmatrix} 2 & -1 \\ 3 & 1 \end{vmatrix} = \begin{vmatrix} & 0 \\ 3 & 1 \end{vmatrix}$

15. $\begin{vmatrix} 1 & -2 & 1 \\ 3 & 1 & 4 \\ 0 & 2 & 1 \end{vmatrix} = \begin{vmatrix} 1 & -2 & 1 \\ 0 & 7 & \\ 0 & 2 & 1 \end{vmatrix}$

16. $\begin{vmatrix} 3 & -1 & 0 \\ 1 & 2 & 1 \\ 2 & 3 & 1 \end{vmatrix} = \begin{vmatrix} 3 & -1 & 0 \\ 1 & 2 & 1 \\ 1 & & 0 \end{vmatrix}$

17. $\begin{vmatrix} 2 & 3 & 1 & 4 \\ 0 & 2 & 1 & 2 \\ 1 & 1 & 2 & 3 \\ 0 & 1 & 1 & 1 \end{vmatrix} = \begin{vmatrix} 0 & 1 & & -2 \\ 0 & 2 & 1 & 2 \\ 1 & 1 & 2 & 3 \\ 0 & 1 & 1 & 1 \end{vmatrix}$

18. $\begin{vmatrix} 2 & 1 & 1 & 0 \\ 1 & 2 & 0 & 2 \\ 3 & 1 & 0 & 3 \\ 2 & 1 & 4 & 2 \end{vmatrix} = \begin{vmatrix} 2 & 1 & 1 & 0 \\ 1 & 2 & 0 & 2 \\ 3 & 1 & 0 & 3 \\ & -3 & 0 & 2 \end{vmatrix}$

First reduce each determinant to an equal 2×2 determinant and then evaluate.

19. $\begin{vmatrix} 2 & 1 & 0 \\ 3 & 2 & 1 \\ -1 & 2 & 0 \end{vmatrix}$

20. $\begin{vmatrix} 1 & 2 & 1 \\ 2 & -1 & 2 \\ 0 & 1 & 0 \end{vmatrix}$

21. $\begin{vmatrix} 1 & 0 & 3 \\ 2 & -1 & 1 \\ 1 & 2 & 1 \end{vmatrix}$

22. $\begin{vmatrix} 1 & 2 & -1 \\ 2 & 1 & 3 \\ 0 & 1 & 2 \end{vmatrix}$

23. $\begin{vmatrix} 1 & 2 & 1 \\ -1 & 2 & 3 \\ 2 & -1 & 1 \end{vmatrix}$

24. $\begin{vmatrix} 3 & -1 & 2 \\ 1 & 2 & 1 \\ -2 & 1 & 3 \end{vmatrix}$

25. $\begin{vmatrix} 11 & -10 & 32 \\ 12 & -11 & 35 \\ 12 & -6 & 33 \end{vmatrix}$

26. $\begin{vmatrix} 9 & 31 & 16 \\ 11 & 38 & 19 \\ 10 & 34 & 21 \end{vmatrix}$

27. $\begin{vmatrix} 28 & 27 & 25 \\ 31 & 30 & 26 \\ 36 & 35 & 30 \end{vmatrix}$

28. $\begin{vmatrix} 26 & 29 & 29 \\ 25 & 27 & 30 \\ 25 & 26 & 28 \end{vmatrix}$

29. $\begin{vmatrix} 13 & 16 & 19 \\ 28 & 34 & 40 \\ 27 & 33 & 39 \end{vmatrix}$

30. $\begin{vmatrix} 19 & 54 & 17 \\ 20 & 57 & 18 \\ 19 & 55 & 19 \end{vmatrix}$

31. $\begin{vmatrix} 0 & 0 & 1 & 2 \\ 6 & 0 & 0 & 1 \\ 6 & 1 & 0 & -1 \\ 6 & 1 & 0 & 2 \end{vmatrix}$

32. $\begin{vmatrix} 4 & 2 & 0 & 2 \\ -1 & 0 & 2 & 1 \\ 3 & 0 & -1 & 1 \\ 0 & 0 & 2 & 1 \end{vmatrix}$

33. $\begin{vmatrix} 0 & 1 & 0 & 2 \\ 0 & 2 & 0 & 3 \\ 2 & -1 & 1 & 0 \\ 0 & 0 & 8 & 8 \end{vmatrix}$

34. $\begin{vmatrix} 0 & 2 & -1 & 3 \\ 0 & 0 & 2 & 1 \\ 3 & 0 & 1 & 0 \\ -6 & 6 & 0 & 0 \end{vmatrix}$

35. $\begin{vmatrix} 1 & 2 & 3 & -1 \\ 0 & 4 & 8 & 4 \\ -2 & 0 & 1 & 1 \\ 2 & 1 & 0 & 1 \end{vmatrix}$

36. $\begin{vmatrix} 1 & 2 & 1 & 1 \\ 2 & -1 & 0 & 1 \\ 0 & 6 & 3 & 9 \\ 2 & 0 & -1 & 1 \end{vmatrix}$

37. Show that $\begin{vmatrix} x & y & 1 \\ x_1 & y_1 & 1 \\ x_2 & y_2 & 1 \end{vmatrix} = 0$ represents an equation of the line through

the points (x_1, y_1) and (x_2, y_2).

38. Use the results in Problem 37 to find an equation of the line through the points $(3, -1)$ and $(-2, 5)$.

39. Show that $\begin{vmatrix} 1 & a & a^2 \\ 1 & b & b^2 \\ 1 & c & c^2 \end{vmatrix} = (b - c)(c - a)(a - b)$.

40. Show that $\begin{vmatrix} a_{11} & a_{12} & a_{13} & a_{14} \\ a_{21} & a_{22} & a_{23} & a_{24} \\ 0 & 0 & a_{33} & a_{34} \\ 0 & 0 & a_{43} & a_{44} \end{vmatrix} = \begin{vmatrix} a_{11} & a_{12} \\ a_{21} & a_{22} \end{vmatrix} \cdot \begin{vmatrix} a_{33} & a_{34} \\ a_{43} & a_{44} \end{vmatrix}$.

9.6   The Inverse of a Square Matrix

In the field of real numbers, every element a except 0 has a multiplicative inverse $1/a$ with the property that $a \cdot 1/a = 1$. The question should (and does) arise, "Does every square matrix A have a multiplicative inverse A^{-1}?"

Definition 9.18   *For a given square matrix A of order n, if there is a square matrix A^{-1} of order n such that*

$$AA^{-1} = I \quad and \quad A^{-1}A = I,$$

where I is the multiplicative identity matrix of order n, then A^{-1} is the **multiplicative inverse** *of A.*

To answer the question about the existence of a multiplicative inverse for a matrix, we shall begin by considering the simple case of 2×2 matrices. If we let

$$A = \begin{bmatrix} a_{11} & a_{12} \\ a_{21} & a_{22} \end{bmatrix},$$

then we must see whether or not there exists a 2×2 matrix A^{-1} such that $AA^{-1} = I$. If so, let $A^{-1} = \begin{bmatrix} b & c \\ d & e \end{bmatrix}$. We wish to have

$$\begin{bmatrix} a_{11} & a_{12} \\ a_{21} & a_{22} \end{bmatrix} \begin{bmatrix} b & c \\ d & e \end{bmatrix} = \begin{bmatrix} 1 & 0 \\ 0 & 1 \end{bmatrix}.$$

This leads to

$$\begin{bmatrix} a_{11}b + a_{12}d & a_{11}c + a_{12}e \\ a_{21}b + a_{22}d & a_{21}c + a_{22}e \end{bmatrix} = \begin{bmatrix} 1 & 0 \\ 0 & 1 \end{bmatrix},$$

which is true if and only if

$$a_{11}b + a_{12}d = 1, \quad a_{11}c + a_{12}e = 0,$$
$$a_{21}b + a_{22}d = 0, \quad a_{21}c + a_{22}e = 1. \tag{1}$$

Solving these equations for b, c, d, and e, we have

$$(a_{11}a_{22} - a_{12}a_{21})b = a_{22}, \quad (a_{11}a_{22} - a_{12}a_{21})c = -a_{12},$$
$$(a_{11}a_{22} - a_{12}a_{21})d = -a_{21}, \quad (a_{11}a_{22} - a_{12}a_{21})e = a_{11},$$

from which

$$b = \frac{a_{22}}{a_{11}a_{22} - a_{12}a_{21}}, \quad c = \frac{-a_{12}}{a_{11}a_{22} - a_{12}a_{21}},$$

$$d = \frac{-a_{21}}{a_{11}a_{22} - a_{12}a_{21}}, \quad e = \frac{a_{11}}{a_{11}a_{22} - a_{12}a_{21}},$$

provided $a_{11}a_{22} - a_{12}a_{21} \neq 0$. Now the denominator of each of these fractions is just $\delta(A)$, so that

$$A^{-1} = \begin{bmatrix} b & c \\ d & e \end{bmatrix} = \begin{bmatrix} \dfrac{a_{22}}{\delta(A)} & \dfrac{-a_{12}}{\delta(A)} \\ \dfrac{-a_{21}}{\delta(A)} & \dfrac{a_{11}}{\delta(A)} \end{bmatrix} = \frac{1}{\delta(A)} \begin{bmatrix} a_{22} & -a_{12} \\ -a_{21} & a_{11} \end{bmatrix}.$$

By direct multiplication, it can be verified not only that

$$AA^{-1} = I,$$

but also (surprisingly, since matrix multiplication is not always commutative) that

$$A^{-1}A = I.$$

Inverse of a
2 × 2 matrix   Thus, to write the inverse of a 2×2 square matrix A for which $\delta(A) \neq 0$, we interchange the entries on the principal diagonal, replace each of the other two entries with its negative, and multiply the result by $1/\delta(A)$.

Example   If $A = \begin{bmatrix} 1 & 3 \\ 2 & -1 \end{bmatrix}$, find A^{-1}.

Solution   We first observe that $\delta(A) = (1)(-1) - (3)(2) = -7$. Hence,

$$A^{-1} = -\frac{1}{7} \begin{bmatrix} -1 & -3 \\ -2 & 1 \end{bmatrix} = \begin{bmatrix} \dfrac{1}{7} & \dfrac{3}{7} \\ \dfrac{2}{7} & -\dfrac{1}{7} \end{bmatrix}.$$

It is a good idea always to check the result when finding A^{-1}, because there is much room for blundering in the process of determining the inverse. In the present example, we have

$$A^{-1}A = -\frac{1}{7} \begin{bmatrix} -1 & -3 \\ -2 & 1 \end{bmatrix} \begin{bmatrix} 1 & 3 \\ 2 & -1 \end{bmatrix} = -\frac{1}{7} \begin{bmatrix} -7 & 0 \\ 0 & -7 \end{bmatrix} = \begin{bmatrix} 1 & 0 \\ 0 & 1 \end{bmatrix}.$$

Matrices with no inverse   We have now arrived at a position where we can answer the question, "Does every 2×2 square matrix A have an inverse?" The answer is "No," for if $\delta(A)$ is 0, then the foregoing equations (1) for b, c, d, e would have no solution.

Example   The matrix $\begin{bmatrix} 3 & 5 \\ 6 & 10 \end{bmatrix}$ has no inverse because

$$\delta(A) = 3(10) - 6(5) = 0.$$

More generally, and without proving it, we have the following result.

Theorem 9.13   *If*

$$A = \begin{bmatrix} a_{11} & a_{12} & \cdots & a_{1n} \\ a_{21} & a_{22} & \cdots & a_{2n} \\ \vdots & \vdots & & \vdots \\ a_{n1} & a_{n2} & \cdots & a_{nn} \end{bmatrix},$$

and if $\delta(A) \neq 0$, then

$$A^{-1} = \frac{1}{\delta(A)} \begin{bmatrix} A_{11} & A_{21} & \cdots & A_{n1} \\ A_{12} & A_{22} & \cdots & A_{n2} \\ \vdots & \vdots & & \vdots \\ A_{1n} & A_{2n} & \cdots & A_{nn} \end{bmatrix},$$

where A_{ij} is the cofactor of a_{ij} in A. If $\delta(A) = 0$, then A has no inverse.

Square matrices A for which $\delta(A) = 0$ are called **singular matrices**. Those for which $\delta(A) \neq 0$ are **nonsingular**. Thus by Theorem 9.13, A has an inverse if and only if A is nonsingular.

Inverse of an $n \times n$ matrix   Observe that A^{-1} is the matrix having as its entries the cofactors of the entries in A multiplied by $1/\delta(A)$, but that the cofactors of the *row* entries in A are the *column* entries in A^{-1}. One way to obtain A^{-1} is to replace each entry in A with its cofactor and multiply the *transpose* of the resulting matrix by $1/\delta(A)$.

Example   If $A = \begin{bmatrix} 1 & 0 & 1 \\ 2 & 1 & 0 \\ 1 & -1 & 1 \end{bmatrix}$,   find A^{-1}.

Solution   We first observe that $\delta(A) = -2$, and since $\delta(A)$ is not zero, A has an inverse. Next, replacing each entry in A with its cofactor, we obtain the matrix

$$\begin{bmatrix} 1 & -2 & -3 \\ -1 & 0 & 1 \\ -1 & 2 & 1 \end{bmatrix}, \quad \text{whose transpose is} \quad \begin{bmatrix} 1 & -1 & -1 \\ -2 & 0 & 2 \\ -3 & 1 & 1 \end{bmatrix},$$

so that

$$A^{-1} = -\frac{1}{2} \begin{bmatrix} 1 & -1 & -1 \\ -2 & 0 & 2 \\ -3 & 1 & 1 \end{bmatrix}.$$

As a check, we have

$$A^{-1}A = -\frac{1}{2} \begin{bmatrix} 1 & -1 & -1 \\ -2 & 0 & 2 \\ -3 & 1 & 1 \end{bmatrix} \begin{bmatrix} 1 & 0 & 1 \\ 2 & 1 & 0 \\ 1 & -1 & 1 \end{bmatrix}$$

$$= -\frac{1}{2} \begin{bmatrix} -2 & 0 & 0 \\ 0 & -2 & 0 \\ 0 & 0 & -2 \end{bmatrix} = \begin{bmatrix} 1 & 0 & 0 \\ 0 & 1 & 0 \\ 0 & 0 & 1 \end{bmatrix}.$$

Theorem 9.13 is applicable to $n \times n$ square matrices, although, clearly, the process of actually determining A^{-1} by the formula given in that theorem becomes very laborious for matrices much larger than 3×3.

Properties of matrices and their inverses   There are a number of useful properties associated with matrices and their inverses. The following theorem gives one example.

Theorem 9.14   *If A and B are $n \times n$ nonsingular square matrices, then AB has an inverse, namely*

$$(AB)^{-1} = B^{-1}A^{-1}.$$

Proof If we right-multiply AB by $B^{-1}A^{-1}$, and apply the associative law for the multiplication of matrices, we have

$$AB \cdot B^{-1}A^{-1} = A \cdot I \cdot A^{-1} = A \cdot A^{-1} = I.$$

Moreover, if we left-multiply AB by $B^{-1}A^{-1}$, we have

$$B^{-1}A^{-1} \cdot AB = B^{-1} \cdot I \cdot B = B^{-1} \cdot B = I.$$

Thus, since $(AB)(B^{-1}A^{-1}) = (B^{-1}A^{-1})(AB) = I$, by the definition of the inverse of a matrix, we have

$$(AB)^{-1} = B^{-1}A^{-1}.$$

This theorem can be used to find the inverse of products of any number of non-singular matrices. For example, if there are three factors A, B, and C in a product,

$$(ABC)^{-1} = [(AB)C]^{-1} = C^{-1}(AB)^{-1} = C^{-1}B^{-1}A^{-1}.$$

Exercise 9.6

Find the inverse of each matrix if one exists.

Example $B = \begin{bmatrix} 1 & 0 & -1 \\ 1 & 3 & 1 \\ 0 & 1 & 2 \end{bmatrix}$

Solution   The determinant $\delta(B)$ is given by

$$\delta \begin{bmatrix} 1 & 0 & -1 \\ 1 & 3 & 1 \\ 0 & 1 & 2 \end{bmatrix} = 1(5) - 0 - 1(1) = 4.$$

Replacing each entry of B with its cofactor gives

$$\begin{bmatrix} 5 & -2 & 1 \\ -1 & 2 & -1 \\ 3 & -2 & 3 \end{bmatrix}; \quad \begin{bmatrix} 5 & -2 & 1 \\ -1 & 2 & -1 \\ 3 & -2 & 3 \end{bmatrix}^t = \begin{bmatrix} 5 & -1 & 3 \\ -2 & 2 & -2 \\ 1 & -1 & 3 \end{bmatrix};$$

$$B^{-1} = \frac{1}{\delta(B)} \begin{bmatrix} \text{each } b_{ij} \text{ of} \\ B \text{ replaced} \\ \text{by } B_{ij} \end{bmatrix}^t = \frac{1}{4} \begin{bmatrix} 5 & -1 & 3 \\ -2 & 2 & -2 \\ 1 & -1 & 3 \end{bmatrix}$$

$$= \begin{bmatrix} \dfrac{5}{4} & -\dfrac{1}{4} & \dfrac{3}{4} \\ -\dfrac{2}{4} & \dfrac{2}{4} & -\dfrac{2}{4} \\ \dfrac{1}{4} & -\dfrac{1}{4} & \dfrac{3}{4} \end{bmatrix}$$

1. $\begin{bmatrix} 1 & 2 \\ 1 & 3 \end{bmatrix}$ 2. $\begin{bmatrix} 3 & 1 \\ 2 & -1 \end{bmatrix}$ 3. $\begin{bmatrix} 2 & -3 \\ 1 & 1 \end{bmatrix}$

4. $\begin{bmatrix} 3 & -2 \\ 2 & 1 \end{bmatrix}$ 5. $\begin{bmatrix} -2 & -1 \\ 4 & 2 \end{bmatrix}$ 6. $\begin{bmatrix} 3 & 1 \\ 9 & 3 \end{bmatrix}$

7. $\begin{bmatrix} 5 & 7 \\ 3 & 4 \end{bmatrix}$ 8. $\begin{bmatrix} 5 & -4 \\ 4 & -3 \end{bmatrix}$ 9. $\begin{bmatrix} 7 & 4 \\ -4 & -2 \end{bmatrix}$

10. $\begin{bmatrix} -9 & 5 \\ -4 & 2 \end{bmatrix}$ 11. $\begin{bmatrix} -2 & -6 \\ -3 & -9 \end{bmatrix}$ 12. $\begin{bmatrix} 21 & 7 \\ 9 & 3 \end{bmatrix}$

13. $\begin{bmatrix} 1 & -1 & 2 \\ 2 & 1 & 3 \\ 0 & 0 & 2 \end{bmatrix}$ 14. $\begin{bmatrix} 0 & 4 & 2 \\ 1 & 0 & 2 \\ 0 & -1 & 1 \end{bmatrix}$ 15. $\begin{bmatrix} 2 & -1 & 1 \\ 3 & 0 & 1 \\ 2 & 2 & 1 \end{bmatrix}$

16. $\begin{bmatrix} 1 & 2 & 1 \\ 0 & 2 & 1 \\ -2 & 2 & 3 \end{bmatrix}$ **17.** $\begin{bmatrix} 2 & 1 & 1 \\ 1 & 0 & 2 \\ 4 & 2 & 2 \end{bmatrix}$ **18.** $\begin{bmatrix} -3 & 1 & -6 \\ 2 & 1 & 4 \\ 2 & 0 & 4 \end{bmatrix}$

19. $\begin{bmatrix} 1 & 2 & -3 \\ 3 & -1 & 0 \\ 5 & 3 & -6 \end{bmatrix}$ **20.** $\begin{bmatrix} 2 & 4 & -1 \\ 1 & 6 & 2 \\ 5 & 14 & 0 \end{bmatrix}$ **21.** $\begin{bmatrix} 2 & -1 & -5 \\ 1 & 3 & 4 \\ 0 & 1 & 2 \end{bmatrix}$

22. $\begin{bmatrix} 2 & 1 & -8 \\ 1 & 1 & -2 \\ 1 & 2 & 3 \end{bmatrix}$ **23.** $\begin{bmatrix} 0 & 0 & 1 \\ 0 & 1 & 0 \\ 1 & 0 & 0 \end{bmatrix}$ **24.** $\begin{bmatrix} 1 & 0 & 1 \\ 0 & 1 & 0 \\ 1 & 0 & 0 \end{bmatrix}$

25. Verify that

$$\left(\begin{bmatrix} 2 & 3 \\ 1 & -1 \end{bmatrix} \cdot \begin{bmatrix} 0 & 1 \\ 3 & 1 \end{bmatrix} \right)^{-1} = \begin{bmatrix} 0 & 1 \\ 3 & 1 \end{bmatrix}^{-1} \cdot \begin{bmatrix} 2 & 3 \\ 1 & -1 \end{bmatrix}^{-1}.$$

26. Verify that

$$\left(\begin{bmatrix} 1 & 2 \\ -1 & 0 \end{bmatrix} \cdot \begin{bmatrix} 1 & 1 \\ 2 & 0 \end{bmatrix} \cdot \begin{bmatrix} 2 & -1 \\ 0 & 1 \end{bmatrix} \right)^{-1} = \begin{bmatrix} 2 & -1 \\ 0 & 1 \end{bmatrix}^{-1} \cdot \begin{bmatrix} 1 & 1 \\ 2 & 0 \end{bmatrix}^{-1} \cdot \begin{bmatrix} 1 & 2 \\ -1 & 0 \end{bmatrix}^{-1}.$$

27. Verify that

$$\left(\begin{bmatrix} 3 & 0 & 1 \\ 2 & 1 & 0 \\ 0 & 1 & 2 \end{bmatrix} \cdot \begin{bmatrix} 2 & 1 & 0 \\ 1 & 1 & 2 \\ 0 & 1 & 0 \end{bmatrix} \right)^{-1} = \begin{bmatrix} 2 & 1 & 0 \\ 1 & 1 & 2 \\ 0 & 1 & 0 \end{bmatrix}^{-1} \cdot \begin{bmatrix} 3 & 0 & 1 \\ 2 & 1 & 0 \\ 0 & 1 & 2 \end{bmatrix}^{-1}.$$

28. Show that $[A^t]^{-1} = [A^{-1}]^t$ for each nonsingular 2×2 matrix.

29. Show that $\delta(A^{-1}) = 1/\delta(A)$ for each nonsingular 2×2 matrix.

30. Prove that if a and b are real numbers, then $\delta(aA^2 + bA) = \delta(aA + bI) \times \delta(A)$ for all 2×2 matrices A.

31. Prove that $\delta(B^{-1}AB) = \delta(A)$ for all nonsingular 2×2 matrices A and B.

32. Prove that if A is a 2×2 matrix and a, b, and c are real numbers, with $c \neq 0$, and if $aA^2 + bA + cI = 0$, then A has an inverse.

9.7   *Solution of Linear Systems Using Inverses of Matrices*

In Section 9.3 we solved linear systems using row-equivalent matrices. The solution for a linear system can also be found by using the inverse of a matrix.

We first verify the matrix product equation

$$\begin{bmatrix} a_{11} & a_{12} & \cdots & a_{1n} \\ \vdots & \vdots & & \vdots \\ a_{n1} & a_{n2} & \cdots & a_{nn} \end{bmatrix} \begin{bmatrix} x_1 \\ \vdots \\ x_n \end{bmatrix} = \begin{bmatrix} a_{11}x_1 + a_{12}x_2 + \cdots + a_{1n}x_n \\ \vdots & \vdots & & \vdots \\ a_{n1}x_1 + a_{n2}x_2 + \cdots + a_{nn}x_n \end{bmatrix},$$

and hence note that the linear system

$$\begin{aligned}
a_{11}x_1 + a_{12}x_2 + \cdots + a_{1n}x_n &= c_1 \\
a_{21}x_1 + a_{22}x_2 + \cdots + a_{2n}x_n &= c_2 \\
\vdots \qquad \vdots \qquad\qquad \vdots \qquad \vdots \\
a_{n1}x_1 + a_{n2}x_2 + \cdots + a_{nn}x_n &= c_n
\end{aligned} \tag{1}$$

can be written as the matrix equation

$$\begin{bmatrix} a_{11} & a_{12} & \cdots & a_{1n} \\ \vdots & \vdots & & \vdots \\ a_{n1} & a_{n2} & \cdots & a_{nn} \end{bmatrix} \begin{bmatrix} x_1 \\ \vdots \\ x_n \end{bmatrix} = \begin{bmatrix} c_1 \\ \vdots \\ c_n \end{bmatrix},$$

where the first factor in the left-hand member is the coefficient matrix for the system. In more concise notation, this latter equation can be written

$$AX = B,$$

where A is an $n \times n$ square matrix, and X and B are $n \times 1$ column matrices.

Solution of
systems

If A in the foregoing equation is nonsingular, we can left-multiply both members of this equation by A^{-1} to obtain the equivalent matrices

$$A^{-1}AX = A^{-1}B,$$
$$IX = A^{-1}B,$$
$$X = A^{-1}B,$$

where $A^{-1}B$ is an $n \times 1$ column matrix. Since X and $A^{-1}B$ are equal, each entry in X is equal to the corresponding entry in $A^{-1}B$, and hence these latter entries constitute the components of the solution of the given linear system. If A is a singular matrix, then of course it has no inverse, and either the system has no solution or the solution is not unique.

Example

Use matrices to find the solution set of

$$\begin{aligned}
2x + y + z &= 1 \\
x - 2y - 3z &= 1 \\
3x + 2y + 4z &= 5.
\end{aligned}$$

Solution

We first write this as a matrix equation of the form $AX = B$, thus:

$$\begin{bmatrix} 2 & 1 & 1 \\ 1 & -2 & -3 \\ 3 & 2 & 4 \end{bmatrix} \begin{bmatrix} x \\ y \\ z \end{bmatrix} = \begin{bmatrix} 1 \\ 1 \\ 5 \end{bmatrix}.$$

We next determine $\delta(A)$, obtaining

$$\delta \begin{bmatrix} 2 & 1 & 1 \\ 1 & -2 & -3 \\ 3 & 2 & 4 \end{bmatrix} = 2(-2) - 1(13) + 1(8) = -9,$$

(Solution continued overleaf)

observe that A is nonsingular, and then find A^{-1} by the method of Section 9.6, as shown below.

$$A^{-1} = \begin{bmatrix} 2 & 1 & 1 \\ 1 & -2 & -3 \\ 3 & 2 & 4 \end{bmatrix}^{-1} = -\frac{1}{9}\begin{bmatrix} -2 & -2 & -1 \\ -13 & 5 & 7 \\ 8 & -1 & -5 \end{bmatrix}$$

As a matter of routine, we check the latter by verifying that $A^{-1}A = I$.

$$A^{-1}A = -\frac{1}{9}\begin{bmatrix} -2 & -2 & -1 \\ -13 & 5 & 7 \\ 8 & -1 & -5 \end{bmatrix}\begin{bmatrix} 2 & 1 & 1 \\ 1 & -2 & -3 \\ 3 & 2 & 4 \end{bmatrix}$$

$$= -\frac{1}{9}\begin{bmatrix} -9 & 0 & 0 \\ 0 & -9 & 0 \\ 0 & 0 & -9 \end{bmatrix}$$

$$= \begin{bmatrix} 1 & 0 & 0 \\ 0 & 1 & 0 \\ 0 & 0 & 1 \end{bmatrix}.$$

Now, since $X = A^{-1}B$, we have

$$\begin{bmatrix} x \\ y \\ z \end{bmatrix} = -\frac{1}{9}\begin{bmatrix} -2 & -2 & -1 \\ -13 & 5 & 7 \\ 8 & -1 & -5 \end{bmatrix}\begin{bmatrix} 1 \\ 1 \\ 5 \end{bmatrix} = -\frac{1}{9}\begin{bmatrix} -9 \\ 27 \\ -18 \end{bmatrix} = \begin{bmatrix} 1 \\ -3 \\ 2 \end{bmatrix}.$$

Hence, $x = 1$, $y = -3$, and $z = 2$, and the solution set of the system is $\{(1, -3, 2)\}$.

The computation of A^{-1} is laborious when A is a square matrix containing many rows and columns. The foregoing method is not always the easiest to use in solving systems, but it is most valuable for theoretical developments.

Exercise 9.7

Find the solution set of the given system by means of matrices. If the system has no unique solution, so state.

1. $2x - 3y = -1$
    $x + 4y = 5$

2. $3x - 4y = -2$
    $x - 2y = 0$

3. $3x - 4y = -2$
    $6x + 12y = 36$

4. $2x - 4y = 7$
    $x - 2y = 1$

5. $2x - 3y = 0$
    $2x + y = 16$

6. $2x + 3y = 3$
    $3x - 4y = 0$

7. $x + y = 2$
    $2x - z = 1$
    $2y - 3z = -1$

8. $2x - 6y + 3z = -12$
    $3x - 2y + 5z = -4$
    $4x + 5y - 2z = 10$

9. $x - 2y + z = -1$
    $3x + y - 2z = 4$
    $y - z = 1$

10. $2x + 5z = 9$
     $4x + 3y = -1$
     $3y - 4z = -13$

11. $2x + 2y + z = 1$
     $x - y + 6z = 21$
     $3x + 2y - z = -4$

12. $4x + 8y + z = -6$
     $2x - 3y + 2z = 0$
     $x + 7y - 3z = -8$

13. $x + y + z = 0$
     $2x - y - 4z = 15$
     $x - 2y - z = 7$

14. $x + y - 2z = 3$
     $3x - y + z = 5$
     $3x + 3y - 6z = 9$

9.8   Cramer's Rule

In section 9.7, we obtained the solution set of the linear system (1) on page 241 with nonsingular coefficient matrix by first expressing the system in the matrix form $AX = B$ and then left-multiplying both members of the equation by A^{-1} to obtain

$$A^{-1}AX = X = A^{-1}B.$$

If, now, this technique is viewed in terms of determinants, we arrive at a general solution for such systems.

If the coefficient matrix A is nonsingular, then its inverse, A^{-1}, is

$$A^{-1} = \frac{1}{\delta(A)}\begin{bmatrix} A_{11} & A_{21} & \cdots & A_{n1} \\ \vdots & \vdots & & \vdots \\ A_{1n} & A_{2n} & \cdots & A_{nn} \end{bmatrix}.$$

Now, since $B = \begin{bmatrix} c_1 \\ c_2 \\ \vdots \\ c_n \end{bmatrix}$, we have

$$X = A^{-1}B = \frac{1}{\delta(A)}\begin{bmatrix} c_1A_{11} + c_2A_{21} + \cdots + c_nA_{n1} \\ c_1A_{12} + c_2A_{22} + \cdots + c_nA_{n2} \\ \vdots \qquad \vdots \qquad \qquad \vdots \\ c_1A_{1n} + c_2A_{2n} + \cdots + c_nA_{nn} \end{bmatrix}.$$

Each entry in $X = A^{-1}B$ can be seen to be of the form

$$\frac{c_1A_{1j} + c_2A_{2j} + \cdots + c_nA_{nj}}{\delta(A)}.$$

But $c_1A_{1j} + c_2A_{2j} + \cdots + c_nA_{nj}$ is just the expansion of the determinant

$$\begin{array}{c} j\text{th} \\ \text{column} \\ \downarrow \end{array}$$

$$\begin{vmatrix} a_{11} & a_{12} & \cdots & c_1 & \cdots & a_{1n} \\ a_{21} & a_{22} & \cdots & c_2 & \cdots & a_{2n} \\ \vdots & \vdots & & \vdots & & \vdots \\ a_{n1} & a_{n2} & \cdots & c_n & \cdots & a_{nn} \end{vmatrix}$$

about the jth column, which has entries $c_1, c_2, c_3, \ldots, c_n$ in place of $a_{1j}, a_{2j}, \ldots,$ a_{nj}. Thus, each entry x_j in the matrix $X = \begin{bmatrix} x_1 \\ x_2 \\ \vdots \\ x_n \end{bmatrix} = A^{-1}B$ is

$$x_j = \frac{\delta(A_j)}{\delta(A)} = \frac{\begin{vmatrix} a_{11} & a_{12} & \cdots & c_1 & \cdots & a_{1n} \\ a_{21} & a_{22} & \cdots & c_2 & \cdots & a_{2n} \\ \vdots & \vdots & & \vdots & & \vdots \\ a_{n1} & a_{n2} & \cdots & c_n & \cdots & a_{nn} \end{vmatrix}}{\begin{vmatrix} a_{11} & a_{12} & & \cdots & & a_{1n} \\ a_{21} & a_{22} & & \cdots & & a_{2n} \\ \vdots & \vdots & & & & \vdots \\ a_{n1} & a_{n2} & & \cdots & & a_{nn} \end{vmatrix}}.$$

Application of Cramer's rule

This relationship expresses **Cramer's rule**. Cramer's rule is the assertion that if the determinant of the coefficient matrix of an $n \times n$ linear system *is not* 0, then the equations are consistent (the system has a solution) and have a unique solution which can be found as follows.

To find x_j in solving the matrix equation $AX = B$:

1. Write the determinant of the coefficient matrix for the system.
2. Replace each entry in the *j*th column of the coefficient matrix A with the corresponding entry from the column matrix B, and find the determinant of the resulting matrix.
3. Divide the result in Step 2 by the result in Step 1.

Hence, in a nonsingular 3×3 system:

$$x = \frac{\delta(A_x)}{\delta(A)}, \quad y = \frac{\delta(A_y)}{\delta(A)}, \quad \text{and} \quad z = \frac{\delta(A_z)}{\delta(A)}.$$

Example

Use Cramer's rule to solve the system

$$-4x + 2y - 9z = 2$$
$$3x + 4y + z = 5$$
$$x - 3y + 2z = 8.$$

Solution

By inspection,

$$\delta(A) = \begin{vmatrix} -4 & 2 & -9 \\ 3 & 4 & 1 \\ 1 & -3 & 2 \end{vmatrix},$$

$$= -4(11) - 2(5) - 9(-13)$$

$$= -44 - 10 + 117 = 63.$$

Replacing the entries in the first column of A with corresponding constants 2, 5, and 8, we have

$$\delta(A_x) = \begin{vmatrix} 2 & 2 & -9 \\ 5 & 4 & 1 \\ 8 & -3 & 2 \end{vmatrix}$$

$$= 2(11) - 2(2) - 9(-47)$$

$$= 22 - 4 + 423 = 441.$$

Hence, by Cramer's rule,

$$x = \frac{\delta(A_x)}{\delta(A)} = \frac{441}{63} = 7.$$

Similarly, by replacing, in turn, the entries of the second and third columns of A with the corresponding constants, 2, 5, and 8, we have

$$\delta(A_y) = \begin{vmatrix} -4 & 2 & -9 \\ 3 & 5 & 1 \\ 1 & 8 & 2 \end{vmatrix} \quad \text{and} \quad \delta(A_z) = \begin{vmatrix} -4 & 2 & 2 \\ 3 & 4 & 5 \\ 1 & -3 & 8 \end{vmatrix}.$$

Now,

$$\delta(A_y) = -4(2) - 2(5) - 9(19)$$
$$= -8 - 10 - 171 = -189$$

and

$$\delta(A_z) = -4(47) - 2(19) + 2(-13)$$
$$= -188 - 38 - 26 = -252,$$

so that

$$y = \frac{\delta(A_y)}{\delta(A)} = \frac{-189}{63} = -3$$

and

$$z = \frac{\delta(A_z)}{\delta(A)} = \frac{-252}{63} = -4,$$

and the solution set of the system is $\{(7, -3, -4)\}$.

Test for consistency

If $\delta(A) = 0$ for a linear system, then the system either has infinitely many members in its solution set (the equations are consistent and one of them can be obtained from the others by linear combinations) or has an empty solution set (the equations are inconsistent). The distinction can be determined as follows: Consider the matrix of coefficients

$$\begin{bmatrix} a_{11} & \cdots & a_{1n} \\ \vdots & & \vdots \\ a_{n1} & \cdots & a_{nn} \end{bmatrix}$$

and the augmented matrix

$$\begin{bmatrix} a_{11} & \cdots & a_{1n} & c_1 \\ \vdots & & \vdots & \vdots \\ a_{n1} & \cdots & a_{nn} & c_n \end{bmatrix},$$

and in each find a determinant (obtained by striking out certain rows and columns) of order as great as possible with value not 0. The order of such a nonvanishing determinant is called the **rank** of the matrix. The rank of the augmented matrix is either the same as, or 1 greater than, that of the matrix of coefficients. The equations are consistent if and only if the two ranks are the same.

For example, the coefficient matrix C for the system

$$x + 2y + 3z = 2$$
$$2x + 4y + 2z = -1$$
$$x + 2y - 2z = 5$$

is given by

$$C = \begin{bmatrix} 1 & 2 & 3 \\ 2 & 4 & 2 \\ 1 & 2 & -2 \end{bmatrix},$$

while the augmented matrix C_A for the system is given by

$$C_A = \begin{bmatrix} 1 & 2 & 3 & 2 \\ 2 & 4 & 2 & -1 \\ 1 & 2 & -2 & 5 \end{bmatrix}.$$

Since $\delta(C) = 0$, the rank of C is not 3, and we therefore know that the system does not have a unique solution. If the first column and third row are deleted, the remaining determinant

$$\begin{vmatrix} 2 & 3 \\ 4 & 2 \end{vmatrix}$$

is not zero (check this), so C has rank 2.

Now, if the first column of C_A is deleted, the remaining entries form the determinant

$$\delta \begin{bmatrix} 2 & 3 & 2 \\ 4 & 2 & -1 \\ 2 & -2 & 5 \end{bmatrix},$$

which is also not zero (check this). Hence C_A has rank 3. Therefore, the equations in the system are inconsistent.

Exercise 9.8

Find the solution set of each of the following systems by Cramer's rule. If $\delta(A) = 0$ in any of the systems, use the ranks of the coefficient matrix and the augmented matrix to determine whether or not the equations in the system are consistent.

1. $x - y = 2$
  $x + 4y = 5$

2. $x + y = 4$
  $x - 2y = 0$

3. $3x - 4y = -2$
  $x + y = 6$

4. $2x - 4y = 7$
  $x - 2y = 1$

5. $\dfrac{1}{3}x - \dfrac{1}{2}y = 0$
  $\dfrac{1}{2}x + \dfrac{1}{4}y = 4$

6. $\dfrac{2}{3}x + y = 1$
  $x - \dfrac{4}{3}y = 0$

7. $x - 2y = 6$
  $\dfrac{2}{3}x - \dfrac{4}{3}y = 6$

8. $\dfrac{1}{2}x + y = 3$
  $-\dfrac{1}{4}x - y = -3$

9. $x - 3y = 1$
  $y = 1$

10. $2x - 3y = 12$
  $x = 4$

11. $ax + by = 1$
  $bx + ay = 1$

12. $x + y = a$
  $x - y = b$

13. $x - 2y + z = -1$
  $3x + y - 2z = 4$
  $y - z = 1$

14. $2x + 5z = 9$
  $4x + 3y = -1$
  $3y - 4z = -13$

15. $2x + 2y + z = 1$
  $x - y + 6z = 21$
  $3x + 2y - z = -4$

16. $4x + 8y + z = -6$
  $2x - 3y + 2z = 0$
  $x + 7y - 3z = -8$

17. $x + y + z = 0$
  $2x - y - 4z = 15$
  $x - 2y - z = 7$

18. $x + y - 2z = 2$
  $3x - y + z = 5$
  $3x + 3y - 6z = 6$

19. $x - 2y - 2z = 3$
  $2x - 4y + 4z = 1$
  $3x - 3y - 3z = 4$

20. $3x - 2y + 5z = 6$
  $4x - 4y + 3z = 0$
  $5x - 4y + z = -5$

21.
$$x - 4z = -1$$
$$3x + 3y = 2$$
$$3x + 4z = 5$$

22.
$$2x - \frac{2}{3}y + z = 2$$
$$\frac{1}{2}x - \frac{1}{3}y - \frac{1}{4}z = 0$$
$$4x + 5y - 3z = -1$$

23.
$$x + y + z = 0$$
$$w + 2y - z = 4$$
$$2w - y + 2z = 3$$
$$-2w + 2y - z = -2$$

24.
$$x + y + z = 0$$
$$x + z + w = 0$$
$$x + y + w = 0$$
$$y + z + w = 0$$

25. For the system
$$a_1 x + b_1 y + c_1 = 0$$
$$a_2 x + b_2 y + c_2 = 0,$$

show that if both $\delta(A_y) = 0$ and $\delta(A_x) = 0$, and if c_1 and c_2 are not both 0, then $\delta(A) = 0$, and the equations are consistent. *Hint:* Show that the first two determinant equations imply that $a_1 c_2 = a_2 c_1$ and $b_1 c_2 = b_2 c_1$ and that the rest follows from the formation of a proportion with these equations.

26. Show that if $\delta(A) = 0$ and $\delta(A_x) = 0$, and if a_1 and a_2 are not both 0, then $\delta(A_y) = 0$, where $\delta(A)$ is the determinant of the coefficient matrix of the system in Problem 25.

Chapter Review

[9.1] *Write each sum or difference as a single matrix.*

1. $\begin{bmatrix} 4 & -7 \\ 2 & 1 \end{bmatrix} + \begin{bmatrix} -3 & 6 \\ -1 & 0 \end{bmatrix}$

2. $\begin{bmatrix} 3 & -1 & 7 \\ 6 & 2 & 5 \end{bmatrix} + \begin{bmatrix} -1 & 6 & -9 \\ 8 & -3 & 7 \end{bmatrix}$

3. $\begin{bmatrix} 2 & -4 & 3 \\ 6 & 1 & 7 \\ 2 & 8 & 0 \end{bmatrix} - \begin{bmatrix} 4 & -1 & 2 \\ 3 & 8 & 1 \\ 7 & 6 & -5 \end{bmatrix}$

4. $\begin{bmatrix} -11 & 2 & -6 \\ 7 & 1 & 2 \\ -3 & 4 & 8 \end{bmatrix} - \begin{bmatrix} -3 & 5 & 0 \\ 1 & 4 & 2 \\ 6 & -1 & 3 \end{bmatrix}$

[9.2] *Write each product as a single matrix.*

5. $-7 \begin{bmatrix} 3 & -1 \\ 2 & 0 \\ 1 & 1 \end{bmatrix}$

6. $\begin{bmatrix} 3 & -1 & 2 \end{bmatrix} \begin{bmatrix} 4 \\ -1 \\ 0 \end{bmatrix}$

7. $\begin{bmatrix} 3 & -1 \\ 6 & 5 \end{bmatrix} \cdot \begin{bmatrix} -4 & 2 \\ 1 & 3 \end{bmatrix}$

8. $\begin{bmatrix} -1 & 7 & 6 \\ 3 & 1 & 2 \\ 1 & 0 & 1 \end{bmatrix} \cdot \begin{bmatrix} 1 & -1 & 2 \\ 1 & 0 & 3 \\ 2 & 1 & 1 \end{bmatrix}$

[9.3] *Use row transformation on the augmented matrix to solve each system of equations.*

9.
$$2x - y = 5$$
$$x + 3y = -1$$

10.
$$2x - y + z = 4$$
$$x + 3y - z = 4$$
$$x + 2y + z = 5$$

[9.4]   *Evaluate each determinant.*

11. $\begin{vmatrix} -3 & 0 \\ 2 & 1 \end{vmatrix}$ **12.** $\begin{vmatrix} 1 & 5 \\ -1 & 2 \end{vmatrix}$ **13.** $\begin{vmatrix} 3 & 1 & 0 \\ 2 & 0 & 1 \\ 1 & 2 & -1 \end{vmatrix}$ **14.** $\begin{vmatrix} 3 & 1 & -2 \\ -1 & 2 & 1 \\ 1 & -2 & 1 \end{vmatrix}$

[9.5]   *Reduce each determinant to an equal* 2×2 *determinant and evaluate.*

15. $\begin{vmatrix} 3 & -1 & 2 \\ 1 & -2 & 0 \\ 2 & 1 & -1 \end{vmatrix}$ **16.** $\begin{vmatrix} 1 & 7 & -11 \\ 12 & 10 & 15 \\ 5 & 11 & 14 \end{vmatrix}$

[9.6]   *Find the inverse of each nonsingular matrix.*

17. $\begin{bmatrix} -4 & 2 \\ 11 & 3 \end{bmatrix}$ **18.** $\begin{bmatrix} 1 & -1 & 2 \\ 3 & 1 & 0 \\ 2 & 1 & 1 \end{bmatrix}$

[9.7]   *Use matrices to solve each system.*

19. $x - y = -3$ **20.** $2x + z = 7$
        $2x + 3y = -1$ $y + 2z = 1$
                                        $3x + y + z = 9$

[9.8]   *Use Cramer's rule to solve each system.*

21. $3x - y = -5$ **22.** $x - y + 2z = 3$
        $x + 2y = -6$ $2x + y - z = 3$
                                $x - 2y + 2z = 4$

10
Complex Numbers and Vectors

10.1   *Definitions and Their Consequences*

Some real-number equations do not have solutions in the set of real numbers. Thus, if $b > 0$ then $x^2 = -b$ has no real-number solution, since there is no real number whose square is negative. Therefore, the expression $\sqrt{-b}$, for $b \in R$, $b > 0$, is undefined in the set of real numbers. In this chapter, we wish to consider a set of numbers containing members whose squares are negative real numbers and also containing a subset of members that can be identified with the set of real numbers. We shall see that this new set of numbers, called the set C of **complex numbers**, provides solutions for all polynomial equations in one variable with real coefficients. Furthermore, we shall see that the field postulates F-1 through F-11 are satisfied by the complex numbers and thus that C constitutes a field.

Complex numbers as ordered pairs

To begin, consider the set of all ordered pairs of real numbers. This set is, of course, $R \times R$, but for our purposes let us use C to denote it, and let us also use z to represent an unspecified element of C. That is, $z \in C$ is a complex number.

Definition 10.1 $C = R \times R = R^2 = \{z \mid z = (a, b),\ a \in R \text{ and } b \in R\}$.

Equality properties in the set C

Definition 10.2   *Let* $z_1 = (a_1, b_1) \in C$ *and* $z_2 = (a_2, b_2) \in C$. *Then* $z_1 = z_2$ *if and only if* $a_1 = a_2$ *and* $b_1 = b_2$.

Definition 10.2 gives meaning to **equality** in C. Since the equality postulates E-1 through E-4 hold in the set R of real numbers, it follows from Definition 10.2 that they hold also in the set C. Formal verification is left as an exercise.

Next, let us define sums and products of complex numbers. Although these definitions may appear arbitrary and unusual to start with (particularly in the case of products), we shall see in Section 10.3 that they relate to the corresponding operations in R in a direct and useful way.

Definition 10.3   *If $z_1 = (a_1, b_1) \in C$, and $z_2 = (a_2, b_2) \in C$, then*

$$\text{I} \quad z_1 + z_2 = (a_1, b_1) + (a_2, b_2) = (a_1 + a_2, b_1 + b_2),$$

$$\text{II} \quad z_1 z_2 = (a_1, b_1) \cdot (a_2, b_2) = (a_1 a_2 - b_1 b_2, a_1 b_2 + a_2 b_1).$$

Examples

a. $(3, 5) + (7, 9) = (3 + 7, 5 + 9) = (10, 14)$

b. $(3, 5) \cdot (7, 9) = (3 \cdot 7 - 5 \cdot 9, 3 \cdot 9 + 5 \cdot 7)$

$$= (21 - 45, 27 + 35)$$

$$= (-24, 62)$$

*Addition
properties
in C*

With the operations of addition and multiplication thus defined, we can now prove that the set C constitutes a field. The proof is divided into two parts, the first dealing with properties of addition and the second with those of multiplication.

Theorem 10.1   *If $z_1, z_2, z_3 \in C$, then:*

F-1 $z_1 + z_2 \in C$.   *Closure law for addition.*

F-2 $(z_1 + z_2) + z_3 = z_1 + (z_2 + z_3)$.   *Associative law for addition.*

F-3   *There exists an element*   *Additive-identity law.*
$z_0 = (0, 0) \in C$ *such that*

$$z + z_0 = z \quad and \quad z_0 + z = z$$

for all z in C.

F-4   *For each $z \in C$ there exists an*   *Additive-inverse law.*
element $-z \in C$ *such that*

$$z + (-z) = z_0 \quad and \quad (-z) + z = z_0 .$$

F-5 $z_1 + z_2 = z_2 + z_1$.   *Commutative law for addition.*

Proof of F-1   Let $z_1 = (a_1, b_1)$ and $z_2 = (a_2, b_2)$. By definition 10.3-I,

$$z_1 + z_2 = (a_1, b_1) + (a_2, b_2) = (a_1 + a_2, b_1 + b_2),$$

and since the operation of addition is closed in the field R of real numbers, we have $(a_1 + a_2)$ and $(b_1 + b_2) \in R$. Hence, by Definition 10.1, $(a_1 + a_2, b_1 + b_2) \in C$; that is, $z_1 + z_2 \in C$.

The proof of F-2 is left as an exercise. It depends on Definition 10.3-I and the fact that addition is associative in the field R of real numbers.

To give meaning to the expressions $z_1 + z_2 + z_3$ and $z_1 z_2 z_3$, we make the following statement.

Definition 10.4   *If $z_1, z_2, z_3 \in C$, then*

$$z_1 + z_2 + z_3 = (z_1 + z_2) + z_3 \quad and \quad z_1 z_2 z_3 = (z_1 z_2) z_3 .$$

This definition can be extended to cover $z_1 + z_2 + z_3 + z_4$, etc. According to Property F-2, it is, of course, immaterial whether $z_1 + z_2 + z_3$ is considered in accordance with Definition 10.4 or as $z_1 + (z_2 + z_3)$.

Proof of F-3   Let $z = (a, b) \in C$, and consider the element $z_0 = (0, 0)$. Then $z_0 \in C$. By Definition 10.3-I,

$$z + z_0 = (a, b) + (0, 0) = (a + 0, b + 0) = (a, b) = z$$

and

$$z_0 + z = (0, 0) + (a, b) = (0 + a, 0 + b) = (a, b) = z.$$

Here, of course, we have used the fact that 0 is the identity element for addition in the field R of real numbers. We say that $z_0 = (0, 0)$ is the **identity element for addition**, or the **zero element**, in the set C.

Proof of F-4   Let $z = (a, b) \in C$, and consider the element $-z = (-a, -b)$. Then $-z \in C$. We have

$$z + (-z) = (a, b) + (-a, -b) = (a + (-b), b + (-b)) = (0, 0),$$

and similarly

$$(-z) + z = (-a, -b) + (a, b) = ((-a) + a, (-b) + b) = (0, 0).$$

We say that $-z = (-a, -b)$ is the **additive inverse**, or the **negative**, of $z = (a, b)$.
    The proof of F-5 is also left as an exercise.

Multiplication properties in C   Turning now to properties of multiplication, we have the following result.

Theorem 10.2   If $z_1, z_2, z_3 \in C$, then:

F-6 $z_1 z_2 \in C$.   *Closure law for multiplication.*

F-7 $(z_1 z_2) z_3 = z_1 (z_2 z_3)$.   *Associative law for multiplication.*

F-8 $z_1 (z_2 + z_3) = z_1 z_2 + z_1 z_3$ *and*   *Distributive law.*
      $(z_2 + z_3) z_1 = z_2 z_1 + z_3 z_1$.

F-9   *There exists an element*   *Multiplicative-identity law.*
      $z_I = (1, 0) \in C$ *such that for all* $z \in C$
          $zz_I = z$ *and* $z_I z = z$.

F-10 $z_1 z_2 = z_2 z_1$.   *Commutative law for multiplication.*

F-11 *For each* $z \in C$ *other than the zero*   *Multiplicative-inverse law.*
      element z_0, *there exists an element*
      $z^{-1} \in C$ *such that*
          $zz^{-1} = z_I$ *and* $z^{-1} z = z_I$.

The proof of F-6 is analogous to that of F-1, with multiplication in place of addition. It is left as an exercise, as are the proofs of F-7 and F-8. For efficiency, let us prove F-10 before turning to F-9 and F-11.

Proof of F-10   Let $z_1 = (a_1, b_1)$ and $z_2 = (a_2, b_2)$; then by Definition 10.3-II,

$$z_1 z_2 = (a_1, b_1)(a_2, b_2) = (a_1 a_2 - b_1 b_2, a_1 b_2 + a_2 b_1)$$

and

$$z_2 z_1 = (a_2, b_2)(a_1, b_1) = (a_2 a_1 - b_2 b_1, a_2 b_1 + a_1 b_2).$$

But since multiplication and addition are commutative in the field R of real numbers, we have

$$a_1 a_2 - b_1 b_2 = a_2 a_1 - b_2 b_1 \quad \text{and} \quad a_1 b_2 + a_2 b_1 = a_2 b_1 + a_1 b_2,$$

from which

$$z_1 z_2 = z_2 z_1.$$

Proof of F-9   Let $z = (a, b)$, and consider the element $z_1 = (1, 0) \in C$. We have

$$z z_I = (a, b)(1, 0) = (a \cdot 1 - b \cdot 0, a \cdot 0 + 1 \cdot b) = (a, b) = z.$$

Similarly,

$$z_I z = (1, 0)(a, b) = (1 \cdot a - 0 \cdot b, 1 \cdot b + a \cdot 0) = (a, b) = z.$$

though this follows equally well from the former result together with F-10.

We say that $(1, 0)$ is the **identity element for multiplication** in the set C.

Proof of F-11   Let $z = (a, b)$. Since $a \in R$ and $b \in R$ and $(a, b) \neq (0, 0)$, we have $a^2 + b^2 \neq 0$. Now consider the element

$$z^{-1} = \left(\frac{a}{a^2 + b^2}, \frac{-b}{a^2 + b^2} \right) \in C. \tag{1}$$

By direct computation, we obtain

$$z z^{-1} = (a, b) \left(\frac{a}{a^2 + b^2}, \frac{-b}{a^2 + b^2} \right) = \left(\frac{a^2 + b^2}{a^2 + b^2}, \frac{-ab + ab}{a^2 + b^2} \right) = (1, 0),$$

as desired. By F-10, we observe also that

$$z^{-1} z = (1, 0),$$

thus completing the proof of F-11.

We say that z^{-1}, given by (1), is the **multiplicative inverse** of $z = (a, b)$. You might wonder how the expression (1) entered the picture. Actually, it can be found by solving the equation

$$(a, b)(x, y) = (1, 0)$$

for (x, y), a task that is left as an exercise.

The field
properties
for C

Let us recapitulate. We started with the set C of ordered pairs of real numbers, defined equality and the operations of addition and multiplication on the elements of this set, and then, one-by-one, established as theorems the properties F-1 through F-11 that characterize a field. Thus we have shown that the elements of R^2, when viewed as the set of complex numbers, are the elements of a field.

Exercise 10.1

Write each sum as an ordered pair (a, b).

1. $(3, 6) + (2, 1)$	**2.** $(7, 1) + (3, -5)$	**3.** $(-6, -2) + (0, 1)$
4. $(3, -2) + (-2, 0)$	**5.** $(0, 7) + (3, 0)$	**6.** $(-2, -1) + (2, 1)$
7. $(2, 3) + (1, 1)$	**8.** $(4, 5) + (0, 0)$	

Write each product as an ordered pair (a, b).

9. $(1, 1) \cdot (1, 1)$ **10.** $(1, 0) \cdot (2, 1)$ **11.** $(2, 3) \cdot (4, 1)$

12. $(3, 1) \cdot (0, 2)$ **13.** $(2, 2) \cdot (3, 4)$ **14.** $(-2, 1) \cdot (1, 3)$

15. $(3, -2) \cdot (1, -1)$ **16.** $(0, 1) \cdot (1, 0)$ **17.** $(3, 4) \cdot (1, 1)$

18. $(0, 1) \cdot (0, 2)$

19. Write the sum $(a_1, 0) + (a_2, 0)$ as an ordered pair. Write the product $(a_1, 0) \cdot (a_2, 0)$ as an ordered pair.

20. Use Definition 10.1 to show that since the equality postulates E-1 through E-4 of Chapter 1 hold in the set R of real numbers, they hold also in the set C.

21. Make a chart showing F-1 through F-11 for the set R of real numbers compared with the corresponding properties of the set C of ordered pairs (a, b). Start the chart thus:

If $a, b \in R$, then: If $(a, b), (c, d) \in C$, then

F-1 $a + b \in R$ $(a, b) + (c, d) \in C$ Closure law for addition.

Let $z_1, z_2, z_3 \in C$.

22. Show that $(z_1 + z_2) + z_3 = z_1 + (z_2 + z_3)$.

23. Show that $z_1 + z_2 = z_2 + z_1$.

24. Show that $z_1 z_2 \in C$.

25. Show that $(z_1 \cdot z_2) \cdot z_3 = z_1 \cdot (z_2 \cdot z_3)$.

26. Show that $z_1 \cdot (z_2 + z_3) = z_1 \cdot z_2 + z_1 \cdot z_3$ and $(z_2 + z_3) \cdot z_1 = z_2 \cdot z_1 + z_3 \cdot z_1$.

27. Solve the equation $(a, b) \cdot (x, y) = (1, 0)$ for (x, y), given that $(a, b) \neq (0, 0)$.

28. Show that if $z_1 \cdot z_2 = (0, 0)$, then either $z_1 = (0, 0)$, $z_2 = (0, 0)$, or both.

10.2   *Additional Properties of Complex Numbers*

Subtraction and division in C

We can extend the work of the preceding section to explore a few of the properties of complex numbers. First, let us define two more operations, subtraction and division, for the field C, in terms of addition and multiplication, respectively.

Definition 10.5   *If* $z_1, z_2 \in C$, *then*

$$z_1 - z_2 = z_1 + (-z_2).$$

You will recall that if $z_2 = (a, b)$, then $-z_2 = (-a, -b)$. The number $z_1 - z_2$ is the **difference** of z_1 and z_2 and is viewed as the result of **subtracting** z_2 from z_1.

Definition 10.6   *If* $z_1, z_2 \in C$ *and* $z_2 \neq (0, 0)$, *then*

$$\frac{z_1}{z_2} = z_1 \cdot z_2^{-1}.$$

You will recall from Equation (1) in Section 10.1 that if $z_2 = (a, b)$, then

$$z_2^{-1} = \left(\frac{a}{a^2 + b^2}, \frac{-b}{a^2 + b^2} \right).$$

The number z_1/z_2 is called the **quotient** of z_1 and z_2, and is viewed as the result of **dividing** z_1 by z_2.

Examples
    a. $(2, 3) - (5, 6) = (2, 3) + (-5, -6) = (-3, -3)$

    b. $\dfrac{(2, 3)}{(5, 6)} = (2, 3)\left(\dfrac{5}{25 + 36}, \dfrac{-6}{25 + 36} \right)$

            $= \left(\dfrac{2 \cdot 5 + 3 \cdot 6}{61}, \dfrac{-2 \cdot 6 + 5 \cdot 3}{61} \right) = \left(\dfrac{28}{61}, \dfrac{3}{61} \right)$

It might be noted that the foregoing definitions of subtraction and division in the field C are analogous to the respective definitions given in Chapter 1 for subtraction and division in the field R of real numbers. Indeed, these definitions might be extended to *any* field, since in any field each element has an additive inverse, and each element other than the zero element has a multiplicative inverse.

The structure of algebraic systems
  We can now obtain a considerable advantage from our structural study of algebraic systems. In Section 1.4, many properties of the field R of real numbers were established. Since, however, the theorems stated there were derived exclusively from the field postulates along with the equality postulates, and did not otherwise depend on the fact that we were dealing with real numbers, it follows that the results are valid in any field—in particular, in the field C. Thus from Theorem 1.3 and Problem 22 of Exercise 1.4, we conclude that for $z \in C$, the additive inverse and the multiplicative inverse (except for the zero element z_0) are unique.

The remaining results of Section 1.4 also extend, of course, to the field C. As an example, the analogue of Theorem 1.10, the *fundamental principle of fractions*, is stated as follows for elements of C.

Theorem 10.3   *If z_1, z_2, and z_3 are elements of C, and z_2 and z_3 are not the zero element z_0 then,*

$$\frac{z_1}{z_2} = \frac{z_1 z_3}{z_2 z_3}.$$

In applying this theorem to quotients, it is helpful to have the following result available, as well as the succeeding definition.

Theorem 10.4   *If $(a, b) \in C$ and $c \neq 0$, then*

$$\frac{(a, b)}{(c, 0)} = \left(\frac{a}{c}, \frac{b}{c} \right).$$

Proof From Definition 10.6 and Equation (1) in Section 10.1,

$$\frac{(a, b)}{(c, 0)} = (a, b)(c, 0)^{-1} = (a, b)\left(\frac{c}{c^2 + 0^2}, \frac{0}{c^2 + 0^2}\right) = (a, b)\left(\frac{1}{c}, 0\right),$$

from which, by Definition 10.3, we obtain

$$\frac{(a, b)}{(c, 0)} = \left(\frac{a}{c}, \frac{b}{c}\right).$$

For example,

$$\frac{(13, -13)}{(26, 0)} = \left(\frac{13}{26}, \frac{-13}{26}\right) = \left(\frac{1}{2}, -\frac{1}{2}\right).$$

Definition 10.7   *The **conjugate** of $z = (a, b) \in C$, denoted by $\bar{z}$, is*

$$\bar{z} = (a, -b).$$

Thus, the conjugate of $(5, 1)$ is $(5, -1)$, and if $z = (-2, -7)$, then $\bar{z} = (-2, 7)$.

Now, for $(c, d) \neq (0, 0)$, we already know that the quotient $(a, b)/(c, d)$ is an element of C. This quotient can be written as an ordered pair by using Theorem 10.3 to multiply the numerator and the denominator of $(a, b)/(c, d)$ by the conjugate of the denominator $(c, -d)$, and then applying Theorem 10.4.

Example   Write the quotient $\dfrac{(3, -2)}{(5, 1)}$ as an ordered pair, by first using Theorem 10.3 and then using Theorem 10.4.

Solution   By Theorem 10.3 and Definition 10.3, we have

$$\frac{(3, -2)}{(5, 1)} = \frac{(3, -2)(5, -1)}{(5, 1)(5, -1)} = \frac{(15 - 2, -3 - 10)}{(25 + 1, 0)} = \frac{(13, -13)}{(26, 0)},$$

from which we obtain, by Theorem 10.4,

$$\frac{(13, -13)}{(26, 0)} = \left(\frac{13}{26}, \frac{-13}{26}\right) = \left(\frac{1}{2}, -\frac{1}{2}\right).$$

Exercise 10.2

Write each difference as an ordered pair.

1. $(4, 2) - (1, 1)$ **2.** $(-3, 4) - (0, 5)$ **3.** $(-6, 1) - (3, 0)$

4. $(0, 1) - (6, 6)$ **5.** $(4, -4) - (-4, 4)$ **6.** $(0, 0) - (2, -3)$

Write each quotient as an ordered pair. Use Definition 10.6, $\dfrac{(a, b)}{(c, d)} = (a, b) \cdot (c, d)^{-1}$.

7. $\dfrac{(4, 3)}{(2, 2)}$ **8.** $\dfrac{(6, 1)}{(1, 3)}$ **9.** $\dfrac{(2, 1)}{(-1, 3)}$

10. $\dfrac{(6, -3)}{(4, 1)}$ **11.** $\dfrac{(-2, -2)}{(1, 1)}$ **12.** $\dfrac{(0, 0)}{(3, 3)}$

13. Write the product $\dfrac{(4,\,1)}{(1,\,2)} \cdot \dfrac{(-1,\,3)}{(6,\,1)}$ as an ordered pair.

14. Write the product $\dfrac{(0,\,1)}{(3,\,-1)} \cdot \dfrac{(2,\,0)}{(2,\,1)}$ as an ordered pair.

Write each quotient as an ordered pair by first multiplying the numerator and denominator by the conjugate of the denominator.

15. $\dfrac{(4,\,1)}{(-1,\,2)}$ **16.** $\dfrac{(6,\,-1)}{(0,\,4)}$ **17.** $\dfrac{(2,\,3)}{(-1,\,-1)}$

18. $\dfrac{(0,\,1)}{(2,\,-3)}$ **19.** $\dfrac{(4,\,-4)}{(2,\,-2)}$ **20.** $\dfrac{(0,\,0)}{(-2,\,3)}$

21. Show that $\dfrac{(a,\,b)}{(a,\,b)} = (1,\,0)$ for every nonzero ordered pair $(a,\,b)$.

22. Show that if z_1, z_2, and z_3 are elements of C, and z_2 and z_3 are not $(0,\,0)$, then

$$\frac{z_1}{z_2} = \frac{z_1 z_3}{z_2 z_3}.$$

23. Show that $\overline{z_1 + z_2} = \overline{z_1} + \overline{z_2}$.   **24.** Show that $\overline{z_1 \cdot z_2} = \overline{z_1} \cdot \overline{z_2}$.

25. Show that the set of ordered pairs of the form $(a,\,0)$, $a \in R$, with the operations of addition and multiplication in C, constitutes a field. (Show that the field postulates are all satisfied in this set.)

10.3   Subsets of C; $a + bi$ Form

The set R as a subset of C   Let us turn now to a consideration of the relationship between complex numbers and real numbers. First, recall that, graphically, the set of ordered pairs of real numbers $(a,\,b)$ is in one-to-one correspondence with the points in the geometric $(x,\,y)$-plane, just as the set R of real numbers a is in one-to-one correspondence with the points on the x-axis. Thus, the subset $\{(a,\,0)\,|\,a \in R\}$ of C can be considered as corresponding in a one-to-one way to the set R. Moreover, from Definitions 10.3, 10.5, 10.6, and Theorem 10.4, we have the following results:

$$\left.\begin{aligned}
(a_1,\,0) + (a_2,\,0) &= (a_1 + a_2,\,0), \\
(a_1,\,0) \cdot (a_2,\,0) &= (a_1 a_2,\,0), \\
(a_1,\,0) - (a_2,\,0) &= (a_1 - a_2,\,0) \\
\frac{(a_1,\,0)}{(a_2,\,0)} &= \left(\frac{a_1}{a_2},\,0\right), \quad a_2 \neq 0.
\end{aligned}\right\} \quad (1)$$

The behavior exhibited under the four basic operations by the first components a of the numbers $(a,\,0) \in C$, and by the numbers $a \in R$, is identical. Let us, then, identify $(a,\,0)$ and a by the following definition.

Definition 10.8   *The element* $(a, 0)$ *in the set* C *is identified with the element* a
in the set R, *and we write*

$$a = (a, 0).$$

It should be noted that a and $(a, 0)$ actually are *conceptually* different, but the identification is useful and should not be confusing. Under this convention, we consider that $R \subset C$.

The set I of elements (a, b) of C in which $b \neq 0$ is called the set of **imaginary numbers**; thus $I \subset C$.

Now, consider the subset of I consisting of all ordered pairs of the form $(0, b)$. In particular, observe that if we square $(0, b)$, we obtain

$$(0, b)^2 = (0, b) \cdot (0, b) = (-b^2, 0).$$

Since we have agreed to identify $(-b^2, 0)$ with the real number $-b^2$, and since $b^2 > 0$ for every real number $b \neq 0$, we have $-b^2 < 0$ for every such number. Thus, we see that the set C provides us with a square root, $(0, b)$, for each negative real number $-b^2$! Moreover, since

$$(0, -b)^2 = (0, -b) \cdot (0, -b) = (-b^2, 0),$$

C provides us with *two* such square roots. That is, $(0, b)$ and $(0, -b)$ are square roots of $(-b^2, 0)$, or $-b^2$. The square roots of nonnegative real numbers are in $\{(a, 0) | a \in R\}$, and those of negative real numbers are in $\{(0, b) | b \in R, b \neq 0\}$. As a special case, the square roots of $(-1, 0)$, or -1, are $(0, 1)$ and $(0, -1)$.

The foregoing discussion implies that the set C of complex numbers can be partitioned as follows:

1. $\{(a, 0) | a \in R\}$ or $\{a | a \in R\}$, the set of real numbers R;

2. $\{(a, b) | a, b \in R, b \neq 0\}$, the set of imaginary numbers I.

The set $\{(0, b) | b \in R, b \neq 0\}$, a subset of I, is called the set of **pure imaginary numbers**.

*Complex
numbers in
the form
a + bi*

Historically, and for practical reasons, complex numbers are not ordinarily represented in ordered-pair form. To obtain an alternative representation, let us begin by adopting the following convention.

Definition 10.9   *In the set* C, $i = (0, 1)$.

With this definition, and with the convention $b = (b, 0)$ of Definition 10.8,

$$bi = (b, 0)(0, 1) = (0, b).$$

In accordance with this definition, we can extend our symbolism $\sqrt{b^2} = |b|$, which was first introduced on page 64.

Definition 10.10   *In the set* C, *for* $b \in R$, $\sqrt{-b^2} = |b| i$.

In particular, for $b = 1$, we have

$$\sqrt{-1} = 1 \cdot i = i.$$

Any ordered pair (a, b) can now be represented by the sum $(a, 0) + (0, b)$, so that, in our new notation, for $a, b \in R$,

$$(a, b) = a + bi = a + b\sqrt{-1}.$$

Both the expressions $a + bi$ and $a + ib$ are **standard forms** for a complex number in this notation.

The number i is called the **imaginary unit**, a is called the **real part** of $a + bi$, and b is called its **imaginary part**. Definitions 10.2, 10.3, 10.5, and 10.6 can now be expressed using $a + bi$.

Definition 10.11   *If $a_1, b_1, a_2, b_2 \in R$ and $i = (0, 1) = \sqrt{-1}$, then*

I $a_1 + b_1 i = a_2 + b_2 i$ *if and only if* $a_1 = a_2$ *and* $b_1 = b_2$,

II $(a_1 + b_1 i) + (a_2 + b_2 i) = (a_1 + a_2) + (b_1 + b_2)i$,

III $(a_1 + b_1 i) \cdot (a_2 + b_2 i) = (a_1 a_2 - b_1 b_2) + (a_1 b_2 + a_2 b_1)i$,

IV $(a_1 + b_1 i) - (a_2 + b_2 i) = (a_1 - a_2) + (b_1 - b_2)i$,

V $\dfrac{a_1 + b_1 i}{a_2 + b_2 i} = \dfrac{a_1 a_2 + b_1 b_2}{a_2^2 + b_2^2} + \dfrac{b_1 a_2 - a_1 b_2}{a_2^2 + b_2^2} i$ (a_2, b_2 *not both* 0).

From the definitions we now have, we may rewrite expressions involving complex numbers in the form $a + bi$ in the same way that we rewrote real polynomial expressions, except that i^2 is replaced with -1.

Examples   a. $(2 + 3i) + (6 - 2i) = (2 + 6) + (3 - 2)i = 8 + i$

b. $(2 - i)(1 + 3i) = 2 + 6i - i - 3i^2 = 5 + 5i$

c. $i - (2 + 3i) = (0 - 2) + (1 - 3)i = -2 - 2i$

d. $\dfrac{4 - i}{1 + i} = \dfrac{(4 - i)(1 - i)}{(1 + i)(1 - i)} = \dfrac{4 - 5i + i^2}{1 - i^2} = \dfrac{3 - 5i}{2} = \dfrac{3}{2} - \dfrac{5}{2}i$

Observe that, in example (d) above, Theorem 10.3 was applied to multiply both $4 - i$ and $1 + i$ by $1 - i$, the conjugate of $1 + i$.

The symbol
$\sqrt{-b}, b > 0$

In accordance with Definition 10.10, $\sqrt{-b} = i\sqrt{b}, b > 0$. The symbol $\sqrt{-b}, b > 0$, should be used with care since certain relationships involving the square root symbol that are valid for real numbers are not valid when the symbol does not represent a real number. For instance,

$$\sqrt{2}\sqrt{3} = \sqrt{6} = \sqrt{(2)(3)},$$

but

$$\sqrt{-2}\sqrt{-3} = (i\sqrt{2})(i\sqrt{3}) = i^2\sqrt{6} = -\sqrt{6} \neq \sqrt{(-2)(-3)}.$$

To avoid difficulty with this point, you should rewrite all expressions of the form $\sqrt{-b}, b > 0$, in the form $i\sqrt{b}$ or $\sqrt{b}i$ before rewriting expressions involving complex numbers.

Examples   a. $3 + \sqrt{-5} = 3 + i\sqrt{5}$

b. $(2 + \sqrt{-3})(2 - \sqrt{-3}) = (2 + i\sqrt{3})(2 - i\sqrt{3}) = 4 - 3i^2 = 7$

c. $\dfrac{1}{1 + \sqrt{-4}} = \dfrac{1(1 - 2i)}{(1 + 2i)(1 - 2i)} = \dfrac{1 - 2i}{1 - 4i^2} = \dfrac{1}{5} - \dfrac{2}{5}i$

The *absolute value*, or *modulus*, of a complex number is given by the following definition.

Definition 10.12   *The **absolute value**, or **modulus**, of* (a, b) *is*

$$|(a, b)| = |a + bi| = \sqrt{a^2 + b^2}.$$

For the number $(a, 0)$, we have

$$|(a, 0)| = |a| = \sqrt{a^2},$$

so that the definition of $|(a, 0)|$ is consistent with that of $|a|$.

Graphs of complex numbers

If $b = 0$, then $(a, b) = a$, a real number. Hence, $\{(a, 0) \,|\, a \in R\}$ can be associated with the points on the horizontal axis in the plane and $\{(0, b) \,|\, b \neq 0,\ b \in R\}$ can be associated with points other than the origin on the vertical axis. Thus, the graphical counterpart [or Argand plane as it is sometimes called, after the French mathematician Jean Argand (1768–1822)] of the complex-number system appears as in Figure 10.1. The point with Cartesian coordinates (a, b) is given the complex designation $a + bi$; for brevity, it is called the point (a, b), or the point $a + bi$.

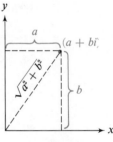

Figure 10.1

Note that $|a + bi|$, defined as $\sqrt{a^2 + b^2}$, is the distance from the origin to the point $a + bi$.

Exercise 10.3

Write each ordered pair (complex number) in the form $a + bi$.

1. $(2, 6)$	**2.** $(-3, 4)$	**3.** $(5, -2)$	**4.** $(0, 6)$
5. $(-7, -3)$	**6.** $(-3, 2)$	**7.** $(4, 0)$	**8.** $(0, 0)$

Write each complex number as an ordered pair.

9. $2 + 3i$	**10.** $4 - 2i$	**11.** $-3 + i$	**12.** $-6 - 3i$
13. $4i$	**14.** 0	**15.** 7	**16.** $-i$

17–24.   Write the conjugate of each complex number in Problems 9–16 in the form $a + bi$.

Find real numbers x and y for which the following statements are true.

Example $(x - 2i)^2 = yi$

Solution   Write each member in the form $a + bi$.
$$x^2 - 4xi + 4i^2 = yi$$
$$(x^2 - 4) - 4xi = yi$$

For equality,
$$x^2 - 4 = 0 \quad \text{and} \quad -4x = y.$$

Therefore, $x = 2$ or -2. If $x = 2$, then $y = -8$; and if $x = -2$, then $y = 8$.
The desired real numbers are 2 and -8, and -2 and 8.

25. $2x - yi = 3 + 2i$ 26. $-2i = 3x + yi$ 27. $4 + xi = x^2 - yi$
28. $x + 9i = y + y^2i$ 29. $(x + 3i)^2 = 2yi$ 30. $(x - 2i)^2 = 3x + yi$

Write each of the following expressions in the form $a + bi$.

31. $(2 + 4i) + (3 + i)$ 32. $(4 - i) - (6 - 2i)$ 33. $(2 - i) + (3 - 2i)$
34. $(2 + i) - (4 - 2i)$ 35. $3 - (4 + 2i)$ 36. $(2 - 6i) - 3$
37. $\dfrac{2}{1 - i}$ 38. $\dfrac{6}{3 + 2i}$ 39. $\dfrac{2 + i}{1 - 3i}$
40. $\dfrac{3 - i}{2i}$ 41. $(1 - 3i)^2$ 42. $(2 + i)^2$
43. $(1 - i)^2(1 + i)$ 44. $(3 - 4i)^2(1 - i)^2$ 45. $4 - \sqrt{-7}$
46. $2 + \sqrt{-1}$ 47. $5 - \sqrt{-4}$ 48. $\sqrt{-9}$
49. $\sqrt{-7}\sqrt{-1}$ 50. $\sqrt{-4}\sqrt{-5}$ 51. $\dfrac{2}{1 + \sqrt{-1}}$
52. $\dfrac{3}{1 - \sqrt{-4}}$ 53. $\dfrac{2 + \sqrt{-1}}{3 - \sqrt{-4}}$ 54. $\dfrac{\sqrt{-3}}{1 - \sqrt{-7}}$

Write each of the following expressions without using absolute-value notation.

Examples   a. $|-3|$ b. $|(-3, 2)|$

Solutions   a. $|-3| = \sqrt{(-3)^2} = 3$ b. $|(-3, 2)| = \sqrt{(-3)^2 + (2)^2} = \sqrt{13}$

55. $|4|$ 56. $|-2|$ 57. $|(2, 0)|$ 58. $|(3, -5)|$
59. $|-2 + i|$ 60. $|3i|$ 61. $|(-2, -1)|$ 62. $|(-7, -1)|$

Graph each complex number, its negative, and its conjugate.

63. (4, 2)   **64.** (−3, 1)   **65.** 2 − 3*i*   **66.** −4 − *i*

67. Show that $|z| = |\bar{z}|$.

68. Show that $|z_1 \cdot z_2| = |z_1| \cdot |z_2|$.

69. Prove that the sum and product of two conjugate complex numbers are real numbers.

10.4   *Quadratic Equations and Complex Numbers*

In Chapter 4, we examined the general quadratic equation

$$ax^2 + bx + c = 0 \tag{1}$$

for real numbers a, b, and c, with $a \neq 0$, and found any real solutions that exist. We saw that the discriminant, $b^2 - 4ac$, determines whether such solutions exist. If $b^2 - 4ac > 0$, then $ax^2 + bx + c = 0$ has two real solutions; if $b^2 - 4ac = 0$, then there is one real solution; and if $b^2 - 4ac < 0$, then there are no real solutions.

With the field of complex numbers at our disposal, we can take a further look at Equation (1). By the field properties, the solution set is still given by

$$\left\{ \frac{-b + \sqrt{b^2 - 4ac}}{2a}, \frac{-b - \sqrt{b^2 - 4ac}}{2a} \right\},$$

but it is no longer necessary to require $b^2 - 4ac \geq 0$ in order for a solution to exist.

Example   Find the solution set of $x^2 + x + 1 = 0$.

Solution   Substituting in the quadratic formula with $a = 1$, $b = 1$, and $c = 1$ gives

$$x = \frac{-1 \pm \sqrt{1^2 - 4(1)(1)}}{2(1)} = \frac{-1 \pm \sqrt{-3}}{2} = \frac{-1 \pm i\sqrt{3}}{2},$$

and the solution set is $\left\{ -\dfrac{1}{2} + \dfrac{\sqrt{3}}{2} i, \ -\dfrac{1}{2} - \dfrac{\sqrt{3}}{2} i \right\}$.

Every quadratic equation with real coefficients has a solution in the field C of complex numbers. Moreover, it can be shown that every quadratic equation with complex coefficients also has a solution in the field of complex numbers.

Exercise 10.4

Solve each linear equation.

Example $4ix = 8 + 2ix$

(*Solution overleaf*)

Solution $4ix - 2ix = 8$

                  $2ix = 8$

                  $x = \dfrac{8}{2i} = \dfrac{4i}{i^2} = -4i$

The solution set is $\{-4i\}$.

1. $3ix = 2 - i$ 2. $4ix - 7 = i$ 3. $2ix + x = 6$

4. $ix - 3x = i$ 5. $\dfrac{2ix}{5} = 3 + i$ 6. $\dfrac{ix - 2i}{2} = i$

Solve each quadratic equation by factoring.

Example $9x^2 + 2 = 0$

Solution $9x^2 + 2 = (3x + i\sqrt{2})(3x - i\sqrt{2}) = 0$

                  $x = -\dfrac{\sqrt{2}}{3}i \quad \text{or} \quad x = \dfrac{\sqrt{2}}{3}i$

The solution set is $\left\{-\dfrac{\sqrt{2}}{3}i, \dfrac{\sqrt{2}}{3}i\right\}$.

7. $x^2 + 4 = 0$ 8. $4x^2 + 9 = 0$ 9. $x^2 + 3 = 0$
10. $9x^2 + 2 = 0$ 11. $(x + 2)^2 + 1 = 0$ 12. $(x - 3)^2 + 5 = 0$
13. $x^2 + 10x + 41 = 0$ *Hint:* First rewrite the left-hand member as $(x + 5)^2 + 16$.
14. $x^2 + 6x + 13 = 0$ 15. $x^2 + 4x + 7 = 0$ 16. $x^2 - 2x + 6 = 0$

Solve each quadratic equation by using the quadratic formula.

17. $x^2 - 2x + 3 = 0$ 18. $x^2 + x + 5 = 0$
19. $3x^2 + x = -7$ 20. $2x^2 = 5x - 10$
21. $x^2 + (10 - i)x - i = 0$ 22. $x^2 + 3ix + 4 = 0$
23. $2x^2 - (4 + i)x = 1$ 24. $ix^2 + 2ix = 4$
25. $2ix^2 - 3x + 4 + i = 0$ 26. $3ix^2 - 2x - 4 + i = 0$

Find equations having the given numbers as solutions. Hint: Recall that if x_1 is a solution of $P(x) = 0$, then $x - x_1$ is a factor of $P(x)$.

27. $2 - i, 2 + i$ 28. $3 + i, 3 - i$ 29. $\sqrt{2}i, -\sqrt{2}i$
30. $2 + \sqrt{3}i, 2 - \sqrt{3}i$ 31. $-3, i$ 32. $2i, i$
33. $2 + i, 1 - 2i$ 34. $3 + 2i, 3 + 2i$ 35. $3, i, -i$
36. $0, 2i, -2i$ 37. $2, 3, i$ 38. $i, -i, 2i$

10.5   *Two-Dimensional Vectors*

Ordered pairs as vectors   In Chapter 9, we referred to a $1 \times n$ matrix $[a_1 \ a_2 \ \cdots \ a_n]$ as a row vector, and to its transpose $[a_1 \ a_2 \ \cdots \ a_n]^t$ as a column vector. We shall now consider an ordered pair of real numbers as a vector.

Definition 10.13   *A **two-dimensional vector** **v** is an ordered pair of real numbers. That is, if $a, b \in R$, then*

$$\mathbf{v} = (a, b)$$

is a vector.

The set of all two-dimensional vectors is the set

$$V = \{(a, b) \,|\, a, b \in R\}.$$

Note that we use **boldface** type to denote vectors. In handwritten form, vectors are frequently identified by arrows drawn above symbols. Thus, $\vec{v}$ denotes a vector. Since we shall be dealing only with two-dimensional vectors, that is, vectors with only two components, we shall refer to these simply as *vectors*. Most of the concepts developed here, however, extend to sets of ordered triples, ordered quadruples, or, indeed, to ordered *n*-tuples of real numbers. These, in turn, can be considered as sets of vectors of three, four, or *n* dimensions.

Geometric representation   Because every ordered pair (a, b) can be associated with a directed line segment, or **geometric vector**, originating (having **initial point**) at the origin and terminating (having **terminal point**) at the point in the plane corresponding to (a, b), a geometric interpretation (similar to the one made for complex numbers) can be made of the vector-space algebra of ordered pairs. See Figure 10.2.

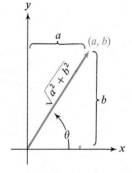

Definition 10.14   *For each vector $\mathbf{v} = (a, b)$, the **norm** or **magnitude**, of **v** is the real number*

$$\|\mathbf{v}\| = \sqrt{a^2 + b^2}.$$

Figure 10.2

It is evident from the Pythagorean theorem that the norm of a vector is just the length of the associated geometric vector.

Definition 10.15   *If $\mathbf{v}_1 = (a_1, b_1)$ and $\mathbf{v}_2 = (a_2, b_2)$, then*

$$\mathbf{v}_1 = \mathbf{v}_2 \quad \text{if and only if} \quad a_1 = a_2 \text{ and } b_1 = b_2.$$

Accordingly, vectors are equal if and only if they correspond to the same geometric vector.

Operations on vectors   Now, let us define two operations, as follows.

Definition 10.16   *If $\mathbf{v}_1 = (a_1, b_1)$, $\mathbf{v}_2 = (a_2, b_2)$, and $c \in R$, then*

    I $\mathbf{v}_1 + \mathbf{v}_2 = (a_1, b_1) + (a_2, b_2) = (a_1 + a_2, b_1 + b_2),$

    II $c\mathbf{v}_1 = c(a_1, b_1) = (ca_1, cb_1).$

From the vector viewpoint of ordered pairs, real numbers, such as c in the fore-going definition, are called **scalars**, and the operation defined in II is called **the multiplication of a vector by a scalar**. Graphically, operations I and II can be interpreted in terms of geometric vectors, as shown in Figures 10.3 and 10.4.

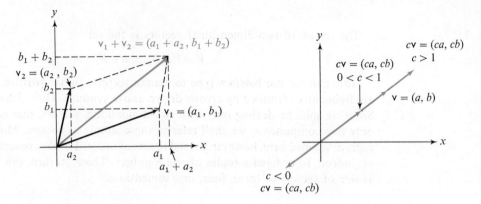

Figure 10.3   **Figure 10.4**

In Figure 10.3, the vectors $\mathbf{v}_1$ and $\mathbf{v}_2$ are said to be **noncollinear**, because they correspond to noncollinear geometric vectors. It is evident that the sum of two such noncollinear vectors $\mathbf{v}_1$ and $\mathbf{v}_2$ corresponds to the diagonal of the parallelogram with adjacent sides corresponding to $\mathbf{v}_1$ and $\mathbf{v}_2$. This statement can be considered valid, in a limiting way, even if the geometric vectors corresponding to $\mathbf{v}_1$ and $\mathbf{v}_2$ are collinear. Accordingly, we say that vectors are added according to the **parallelogram law**.

In Figure 10.4, the vectors $\mathbf{v}_1$ and $c\mathbf{v}_1$ are said to be collinear, since they correspond to geometric vectors on the same line. For $c > 0$, the product $c\mathbf{v}$ corresponds to a geometric vector with the *same direction* as the one corresponding to $\mathbf{v}$. If $0 < c < 1$, then $\|c\mathbf{v}\| < \|\mathbf{v}\|$. If $c < 0$, then $c\mathbf{v}$ corresponds to a geometric vector of *direction opposite* to the geometric vector corresponding to $\mathbf{v}$.

Definition 10.17   *The **zero vector**, $\mathbf{0}$, or **identity vector for addition**, is*

$$\mathbf{0} = (0, 0).$$

The zero vector has the property that $\mathbf{v} + \mathbf{0} = \mathbf{0} + \mathbf{v} = \mathbf{v}$ for each vector $\mathbf{v}$. The norm of $\mathbf{0}$ is, of course, 0. No particular direction is assigned to the geometric vector corresponding to $\mathbf{0}$; for this reason, *any and every direction might be assigned to it*, as suits our convenience.

Definition 10.18   *If $\mathbf{v} = (a, b)$ is a vector, then the **negative** of $\mathbf{v}$ is*

$$-\mathbf{v} = (-a, -b).$$

With Definitions 10.15–10.18 we now have exactly the same algebraic properties, as regards addition and the multiplication by a real-number scalar, for the set of real-number pairs (a, b), whether they are regarded as 1×2 matrices, as 2×1

matrices, or as vectors. These properties are listed, for $m \times n$ matrices with m and n fixed, in Theorems 9.1 and 9.2 on pages 212 and 215. Mathematical systems possessing these properties are called **vector spaces** over the field R of real-number scalars. We can assert this fact for vectors $\mathbf{v} = (a, b)$ as follows:

Theorem 10.5   *The set V of vectors $\mathbf{v} = (a, b)$, $a, b \in R$, with the set R of real numbers as scalar multipliers, is a vector space over the field R of real numbers. That is, if $\mathbf{v}_1$, $\mathbf{v}_2$, and $\mathbf{v}_3$ are vectors, and c and d are real scalars, then*

I $\mathbf{v}_1 + \mathbf{v}_2 \in V$,   VIII $(c + d)\mathbf{v}_1 = c\mathbf{v}_1 + d\mathbf{v}_1$,

II $(\mathbf{v}_1 + \mathbf{v}_2) + \mathbf{v}_3 = \mathbf{v}_1 + (\mathbf{v}_2 + \mathbf{v}_3)$,   IX $c(\mathbf{v}_1 + \mathbf{v}_2) = c\mathbf{v}_1 + c\mathbf{v}_2$,

III $\mathbf{v}_1 + \mathbf{0} = \mathbf{v}_1$ and $\mathbf{0} + \mathbf{v}_1 = \mathbf{v}_1$,   X $1 \cdot \mathbf{v}_1 = \mathbf{v}_1$,

IV $\mathbf{v}_1 + (-\mathbf{v}_1) = \mathbf{0}$,   XI $(-1)\mathbf{v}_1 = -\mathbf{v}_1$,

V $\mathbf{v}_1 + \mathbf{v}_2 = \mathbf{v}_2 + \mathbf{v}_1$,   XII $0 \cdot \mathbf{v}_1 = \mathbf{0}$,

VI $c\mathbf{v}_1 \in V$,   XIII $c \cdot \mathbf{0} = \mathbf{0}$.

VII $c(d\mathbf{v}_1) = (cd)\mathbf{v}_1$,

The proofs of some of these were considered in Chapter 9 for the vector spaces of 2×1 or 1×2 matrices. The exercise set following this section also asks for proofs of a few of them, but this time in terms of vectors.

In accord with Definition 10.18 we can also define the *difference* of two vectors.

Definition 10.19   *If $\mathbf{v}_1$ and $\mathbf{v}_2$ are vectors, then the **difference** $\mathbf{v}_1 - \mathbf{v}_2$ is*

$$\mathbf{v}_1 - \mathbf{v}_2 = \mathbf{v}_1 + (-\mathbf{v}_2),$$

and is viewed as the result of subtracting $\mathbf{v}_2$ from $\mathbf{v}_1$.

The geometric vector associated with $\mathbf{v}_1 - \mathbf{v}_2$ is shown in Figure 10.5.

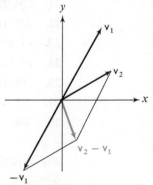

Figure 10.5

Exercise 10.5

Write each sum or difference in the form (a, b). Represent the vector operation graphically and find $\|(a, b)\|$.

Example $(5, 3) - (2, -1)$

Solution $(5, 3) - (2, -1) = (5, 3) + (-2, 1)$

$$= (3, 4)$$

$$\|(3, 4)\| = \sqrt{3^2 + 4^2} = 5$$

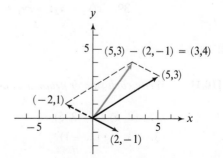

1.	$(2, 4) + (6, 1)$	**2.**	$(3, -1) + (4, 2)$
3.	$(-3, 4) + (2, -1)$	**4.**	$(-2, -4) + (7, -2)$
5.	$(0, 3) + (5, 6)$	**6.**	$(5, 2) + (3, 0)$
7.	$(2, 0) + (3, 5) + (3, 1)$	**8.**	$(-2, 4) + (3, 0) + (4, 3)$
9.	$(6, -1) + (2, 0) + (0, 2)$	**10.**	$(0, 4) + (2, -1) + (5, 4)$
11.	$(5, 3) - (2, 2)$	**12.**	$(7, 8) - (3, 4)$
13.	$(-2, 4) - (2, -1)$	**14.**	$(8, -3) - (2, -1)$
15.	$(2, 4) + (0, 3) - (2, -1)$	**16.**	$(-3, 0) + (2, 1) - (-1, 0)$
17.	$(0, 2) - (3, 1) + (2, 0)$	**18.**	$(4, 2) - (0, -2) + (4, 4)$

Example $2(3, 5) + 3(1, -2)$

Solution $2(3, 5) + 3(1, -2) = (6, 10) + (3, -6)$

$$= (9, 4)$$

$$\|(9, 4)\| = \sqrt{9^2 + 4^2} = \sqrt{97}$$

19.	$3(1, 1) + 4(2, 2)$
20.	$3(1, -4) + 2(3, 5)$
21.	$5(0, 2) + 3(-2, -3)$
22.	$2(3, 1) + 5(-6, 0)$
23.	$(5, 6) - 2(1, -2)$
24.	$(6, 2) - 3(-1, 2)$

25.	$3(0, 4) - 4(2, 0)$	**26.**	$7(5, 0) - 2(-3, 0)$
27.	$2(1, 3) - 4(-2, 1) + 2(0, 1)$	**28.**	$3(0, 2) + 5(1, -2) - 3(0, 1)$
29.	$3(4, -1) - 2(0, 1) + (3, 0)$	**30.**	$(4, 0) - 2(1, 3) + 3(0, -2)$

For Problems 31–40, let $\mathbf{v}_1 = (a_1, b_1)$, $\mathbf{v}_2 = (a_2, b_2)$, and $\mathbf{v}_3 = (a_3, b_3)$; $c, d \in R$. Show that each of the following statements is true, using appropriate definitions and properties of real numbers.

31.	$\mathbf{v}_1 + \mathbf{v}_2 \in V.$	**32.**	$\mathbf{v}_1 + \mathbf{v}_2 = \mathbf{v}_2 + \mathbf{v}_1.$
33.	$(\mathbf{v}_1 + \mathbf{v}_2) + \mathbf{v}_3 = \mathbf{v}_1 + (\mathbf{v}_2 + \mathbf{v}_3)$	**34.**	$\mathbf{v}_1 + \mathbf{0} = \mathbf{v}_1$
35.	$\mathbf{v}_1 + (-\mathbf{v}_1) = \mathbf{0}$	**36.**	$c\mathbf{v}_1 \in V$
37.	$c(d\mathbf{v}_1) = (cd)\mathbf{v}_1$	**38.**	$(c + d)\mathbf{v}_1 = c\mathbf{v}_1 + d\mathbf{v}_1$
39.	$c(\mathbf{v}_1 + \mathbf{v}_2) = c\mathbf{v}_1 + c\mathbf{v}_2$	**40.**	$1 \cdot \mathbf{v}_1 = \mathbf{v}_1$

Chapter Review

[10.1]   *Write each sum or product as an ordered pair.*

1.	$(2, 6) + (-3, 5)$	**2.**	$(-6, 5) + (2, -1)$
3.	$(3, 1) \cdot (2, -1)$	**4.**	$(4, -2) \cdot (1, -3)$

[10.2] *Write each difference or quotient as an ordered pair.*

5. $(6, 3) - (-2, 4)$

6. $(-8, 2) - (-3, -1)$

7. $\dfrac{(3, 2)}{(2, 1)}$

8. $\dfrac{(-3, 1)}{(2, -4)}$

[10.3] *Find real numbers x and y for which each statement is true.*

9. $3x + yi = 2 - 3i$

10. $(x - 2i)^2 = 5 + 2yi$

Write each expression in the form a + bi.

11. $(2 + 3i) + (3 - 2i)$

12. $(3 - i) - (2 + 4i)$

13. $(5 - 2i)^2$

14. $\dfrac{2 - i}{2 - i}$

15. $\sqrt{-2}\sqrt{-7}$

16. $\dfrac{\sqrt{-4}}{1 + \sqrt{-1}}$

Write each expression without using absolute-value notation.

17. $|(2, -4)|$

18. $|3 + 2i|$

Graph each complex number, its negative, and its conjugate.

19. $(3, -2)$

20. $-3 + i$

[10.4] *Solve each linear equation.*

21. $2 + 4ix = i$

22. $2ix - 4 = 3i$

Solve each quadratic equation.

23. $9x^2 + 16 = 0$

24. $x^2 + 2x + 3 = 0$

25. $x^2 + 2ix - 3 = 0$

26. $ix^2 + 3x - 2 - i = 0$

[10.5] *Write each sum or difference in the form (a, b) and find $\|(a, b)\|$.*

27. $(3, 0) + (-2, 1) - (2, 0)$

28. $(2, -1) - (0, 2) + (3, -4)$

29. $2(3, 4) + 3(-1, 2)$

30. $(-2, -1) - 4(-3, 2)$

11

Theory of Equations

In Section 2.1, we defined a polynomial over the field R of real numbers as one in which each of the coefficients is a real number. The polynomial equation

$$P(x) = a_0 x^n + a_1 x^{n-1} + \cdots + a_n, \quad a_0 \neq 0,$$

can also be used to define a function of the real or complex variable x over the field C of complex numbers as well as over the field R of real numbers, depending on whether we take the $a_j \in C$ or $a_j \in R$. In this chapter we shall study some characteristics of both.

11.1   *Synthetic Division and the Factor Theorem*

In Section 2.5, a synthetic process was developed for the division of $P(x)$ by $(x - r)$, where $P(x)$ was a real polynomial over R, and r was any real number. Then, in Section 6.4, the remainder theorem was developed, and these two concepts were combined to help us find ordered pairs $(x, P(x))$ in real polynomial functions.

Now that we have the field C of complex numbers available, we can reexamine these ideas for polynomials over C. To begin with, let us examine the problem of finding $P(r)$ for any given real or complex number r by the method that we employed in Section 6.4.

Remainder theorem

Because no properties other than the field properties are necessary for its validity, the remainder theorem also holds in the set of complex numbers.

> **Theorem 11.1**   *If $P(x)$ is a polynomial over the field C of complex numbers, and $c \in C$, then there exists a unique polynomial $Q(x)$ and a complex number r, such that*
>
> $$P(x) = (x - c)Q(x) + r,$$
>
> *and $r = P(c)$.*

268

Example   If $P(x) = 2x^4 - x^3 + 2x^2 - x + 1$,   find $P(1 - i)$.

Solution   Using synthetic division in the field of complex numbers, we obtain:

$$
\begin{array}{r|ccccc}
1 - i & 2 & -1 & 2 & -1 & 1 \\
      &   & 2 - 2i & -1 - 3i & -2 - 4i & -7 - i \\
\hline
      & 2 & 1 - 2i & 1 - 3i & -3 - 4i & -6 - i
\end{array}
$$

Thus $P(1 - i) = -6 - i$.

Factor theorem   Another theorem, called the **factor theorem**, follows immediately from the remainder theorem.

Theorem 11.2   *If $P(x)$ is a polynomial over the field C of complex numbers, and $P(r) = 0$, then $(x - r)$ is a factor of $P(x)$.*

Proof If $P(r) = 0$, then, by the remainder theorem,

$$P(x) = (x - r)Q(x) + P(r) = (x - r)Q(x) + 0,$$

and by the definition of a factor, $(x - r)$ is a factor of $P(x)$.

Converse of factor theorem   It might be noted, as a converse of the factor theorem, that if $(x - r)$ is a factor of $P(x)$, so that $P(x) = (x - r)Q(x)$, then $P(r) = 0$, since

$$P(r) = (r - r)Q(r) = 0 \cdot Q(r) = 0.$$

Example   Show that $(x - 2i)$ is a factor of $P(x) = x^3 - x^2 + 4x - 4$.

Solution   If $P(2i) = 0$, then, by the factor theorem, $x - 2i$ must be a factor of $P(x)$. Dividing $P(x)$ by $x - 2i$ synthetically, we have

$$
\begin{array}{r|cccc}
2i & 1 & -1 & 4 & -4 \\
   &   & 2i & -4 - 2i & 4 \\
\hline
   & 1 & -1 + 2i & -2i & 0
\end{array}
$$

and, since the remainder is 0, we have shown that $(x - 2i)$ is a factor of $P(x)$.

As a last observation, note that, in the synthetic division of $P(x)$ by $x - r$, the terms occurring in the bottom row of the division process are the coefficients in the polynomial $Q(x)$, except that the final one is $P(r)$. Thus, in the preceding example, we showed that $x - 2i$ is a factor of $P(x) = x^3 - x^2 + 4x - 4$. By looking at the last row in the synthetic-division process, we see that, since $P(2i) = 0$. Hence, the polynomial $P(x)$ can be expressed as

$$P(x) = (x - 2i)[x^2 + (-1 + 2i)x - 2i].$$

Exercise 11.1

Find the value of the polynomial for the given value of x.

Example $P(x) = 3x^3 + 2x^2 - x + 1;\ 2i$

Solution   Using synthetic division, we obtain

$$
\begin{array}{r|rrrr}
2i & 3 & 2 & -1 & 1 \\
   &   & 6i & -12 + 4i & -8 - 26i \\
\hline
   & 3 & 2 + 6i & -13 + 4i & -7 - 26i
\end{array}.
$$

By the remainder theorem, $P(2i) = -7 - 26i.$

1. $P(x) = 4x^3 - x^2 + 5x + 2;\ 2$
2. $Q(x) = 3x^3 + 2x^2 - x - 1;\ -3$
3. $Q(x) = x^3 + 4x^2 - 2x - 5;\ -3i$
4. $G(x) = 3x^3 - 2x^2 + 7x - 1;\ i$
5. $P(x) = x^4 - 3x^3 + x - 2;\ -2$
6. $P(x) = 2x^4 - 3x^2 - x;\ 4$
7. $Q(x) = x^5 - x^3 + 3x^2 + 1;\ 2i$
8. $Q(x) = 3x^5 - x^4 + 2x - 1;\ -i$
9. $P(x) = x^3 + 3x^2 - x + 1;\ 1 - i$
10. $P(x) = 2x^3 - x^2 + 3x - 5;\ 2 + i$

For each expression, find the quotient and the remainder.

Example $(3x^3 - x^2 + 1) \div (x + 2)$

Solution   Using synthetic division, we obtain

$$
\begin{array}{r|rrrr}
-2 & 3 & -1 & 0 & 1 \\
   &   & -6 & 14 & -28 \\
\hline
   & 3 & -7 & 14 & -27
\end{array}.
$$

The quotient is $3x^2 - 7x + 14$ and the remainder is $-27.$

11. $(2x^3 - x^2 + 3x + 4) \div (x - 3)$
12. $(x^3 - 4x^2 - 2x - 2) \div (x + 1)$
13. $(x^3 + x^2 - 3x + 3) \div (x - 2i)$
14. $(2x^3 - 2x^2 + 4x + 1) \div (x + i)$
15. $(x^4 - 3x^2 + x - 1) \div (x + i)$
16. $(x^4 - x^3 + 2x^2 + 2) \div (x - 2i)$
17. $(x^3 - 2x^2 + x + 1) \div [x - (1 + i)]$
18. $(2x^3 + x^2 - 3x) \div [x + (1 - i)]$

19. Show that $x - 2$ is a factor of $x^3 + 2x^2 - 5x - 6.$
20. Show that $x + 1$ is a factor of $x^4 - 5x^3 - 13x^2 + 53x + 60.$
21. Show that $-2i$ is a root of $x^3 - x^2 + 4x - 4 = 0.$
22. Show that $2 + 4i$ is a root of $x^4 - 4x^3 + 18x^2 + 8x - 40 = 0.$
23. Find $Q(x)$ if $x^3 - 3x^2 + x - 3 = (x - i)Q(x).$
24. Find $Q(x)$ if $2x^3 - 3x^2 - 4x + 4 = (x - 2)Q(x).$

Find a value for k so that the second polynomial is a factor of the first.

25. $x^3 - 5x^2 - 16x + k;\ x - 5$
26. $x^3 - 9x + 14x + k;\ x + 1$
27. $x^4 + 2x^3 - 21x^2 + kx + 40;\ x - 4$
28. $3x^4 - 40x^3 + 130x^2 + kx + 27;\ x - 9$

11.2   Complex Zeros of Polynomial Functions

Conjugate complex roots

Recalling from Section 10.2 that the conjugate of the complex number $z = a + bi$ is $\bar{z} = a - bi$, where $a, b, \in R$, we state without proof an important property of a polynomial over R.

Theorem 11.3   *If $P(z)$ is a polynomial over the field R of real numbers, and $P(z) = 0$ for some $z \in C$, then $P(\bar{z}) = 0$.*

One implication this theorem has for the zeros of a function defined by a polynomial with *real coefficients* is that *complex zeros always occur in conjugate pairs.*

Example

Given that $2 - i$ is a zero of

$$\{(x, P(x)) \mid P(x) = x^3 - 6x^2 + 13x - 10\},$$

find all zeros of P.

Solution

The zeros of P are the solutions of $P(x) = 0$. By synthetic division, we have

$$
\begin{array}{r|rrrr}
2-i & 1 & -6 & 13 & -10 \\
 & & 2-i & -9+2i & 10 \\
\hline
 & 1 & -4-i & 4+2i & 0
\end{array},
$$

and the quotient is $x^2 + (-4 - i)x + 4 + 2i$. By Theorem 11.3, we know that $2 + i$ must also be a root of $P(x) = 0$. We can now write

$$P(x) = [x - (2 - i)][x^2 + (-4 - i)x + 4 + 2i] = 0,$$

and since, by inspection, $2 + i$ is not a root of $x - (2 - i) = 0$, it must be a root of

$$x^2 + (-4 - i)x + 4 + 2i = 0.$$

Using synthetic division again, we have

$$
\begin{array}{r|rrr}
2+i & 1 & -4-i & 4+2i \\
 & & 2+i & -4-2i \\
\hline
 & 1 & -2 & 0
\end{array},
$$

where the quotient is $x - 2$. Then

$$P(x) = [x - (2 - i)][x - (2 + i)](x - 2) = 0$$

and, by inspection, the solutions of this equation—and hence the zeros of P—are $2 - i$, $2 + i$, and 2.

Fundamental theorem of algebra

When Theorem 11.3 is coupled with the following theorem, which is called the **fundamental theorem of algebra**, a great deal of information relative to the zeros of polynomial functions becomes readily available.

Theorem 11.4   *Every polynomial function of degree $n \geq 1$ over the field C of complex numbers has at least one real or complex zero.*

The proof of this theorem involves concepts beyond those available to us, and is omitted.

As an extension of Theorem 11.4, we have the following consequent result.

Theorem 11.5   *Every polynomial of degree $n \geq 1$ over the field C can be expressed as a product of a constant and n linear factors of the form $x - x_j$.*

Proof   Let $P(x) = a_0 x^n + a_1 x^{n-1} + \cdots + a_n$, $a_i \in C$, $a_0 \neq 0$, $x \in C$, and $n \in N$. Then, since this equation defines a polynomial function, by the fundamental theorem of algebra there is at least one real or complex number x, say x_1, such that

$$P(x_1) = a_0 x_1^n + a_1 x_1^{n-1} + \cdots + a_n = 0.$$

By the factor theorem,

$$P(x) = (x - x_1)Q_{n-1}(x),$$

where $Q_{n-1}(x)$ is of degree $n - 1$. Again, by the fundamental theorem, if $n - 1 \geq 1$, there must exist an $x_2 \in C$ such that $Q_{n-1}(x_2) = 0$. Hence, we can write

$$P(x) = (x - x_1)(x - x_2)Q_{n-2}(x),$$

where $Q_{n-2}(x)$ is a polynomial of degree $n - 2$. If this factoring process is performed n times, the result is

$$P(x) = (x - x_1)(x - x_2) \cdots (x - x_n)Q_0(x),$$

where $Q_0(x)$ consists solely of a_0, and the theorem is proved.

An nth-degree polynomial function has n zeros

If a factor $(x - x_i)$ occurs k times in such a linear factorization of $P(x)$, then x_i is said to be a zero of *multiplicity* k. With this agreement, Theorem 11.5 shows that every polynomial function defined by a polynomial $P(x)$ of degree n with complex coefficients has exactly n zeros.

Note that any theorem stated in terms of zeros of polynomial functions applies to solutions of polynomial equations and vice versa; a *zero* of

$$\{(x, P(x)) \mid P(x) = a_0 x^n + a_1 x^{n-1} + \cdots + a_n\}$$

is a *solution* of $P(x) = 0$.

Exercise 11.2

In Problems 1–8, one or more zeros are given for each of the polynomial functions; find the other zeros. Verify by synthetic division.

1. $\{(x, P(x)) \mid P(x) = x^2 + 4\}$; $2i$ is one zero.
2. $\{(x, P(x)) \mid P(x) = 3x^2 + 27\}$; $-3i$ is one zero.
3. $\{(x, P(x)) \mid P(x) = x^3 - 3x^2 + x - 3\}$; 3 and i are zeros.

4. $\{(x, Q(x)) \mid Q(x) = x^3 - 5x^2 + 7x + 13\}$; -1 and $3 - 2i$ are zeros.

5. $\{(x, Q(x)) \mid Q(x) = x^4 + 5x^2 + 4\}$; $-i$ and $2i$ are zeros.

6. $\{(x, P(x)) \mid P(x) = x^4 + 11x^2 + 18\}$; $3i$ and $\sqrt{2}\,i$ are zeros.

7. $\{(x, Q(x)) \mid Q(x) = x^4 + 3x^3 + 4x^2 + 27x - 45\}$; $-3i$ is a zero.

8. $\{(x, Q(x)) \mid Q(x) = x^5 - 2x^4 + 8x^3 - 16x^2 + 16x - 32\}$; $2i$ (multiplicity 2) and 2 are zeros.

9. A cubic equation with real coefficients has roots -2 and $1 + i$. What is the third root? Write the equation in the form $P(x) = 0$, given that the leading coefficient (the coefficient of the highest power of x) is 1.

10. A cubic equation with real coefficients has roots 4 and $2 - i$. What is the third root? Write the equation in the form $P(x) = 0$, given that the leading coefficient is 1.

11. One zero of $P(x) = 2x^3 - 11x^2 + 28x - 24$ is $2 - 2i$. Factor $P(x)$ over the complex numbers.

12. One zero of $Q(x) = 3x^3 - 10x^2 + 7x + 10$ is $2 + i$. Factor $Q(x)$ over the complex numbers.

13. One root of $x^4 - 10x^3 + 35x^2 - 50x + 34 = 0$ is $4 - i$. Find the remaining roots.

14. One root of $5x^4 + 34x^3 + 40x^2 - 78x + 51 = 0$ is $-4 - i$. Find the remaining roots.

15. Argue that every polynomial equation with real coefficients and of odd degree has at least one real root.

16. One zero of a polynomial P is i. Is $-i$ necessarily a zero of P? Why or why not? The number i is a zero of $P(x) = x^3 + i$. Determine by synthetic division whether or not $-i$ is a zero.

17. Determine the three cube roots of -1. *Hint:* Consider the roots of the equation $x^3 + 1 = 0$.

18. Show that there are four fourth roots of -1. *Hint:* Consider the roots of the equation $x^4 + 1 = 0$, that is, of $(x^2 + i)(x^2 - i) = 0$.

11.3   *Real Zeros of Polynomial Functions*

Since we have adopted the convention that $R \subset C$, if $\{z_1, z_2, z_3, \ldots z_n\}$ is the set of zeros for a complex polynomial function of degree n, any one or more of these zeros, or, indeed, all of them, may be real numbers. Now we wish to find some ways of identifying real zeros of these polynomials and we shall, for the time being, restrict replacements for x in $P(x)$ to real numbers.

Continuity   For real-number replacements for x, we know that for the $a_j \in R$, the graph of

$$\{(x, P(x)) \mid P(x) = a_0 x^n + a_1 x^{n-1} + \cdots + a_n\}$$

lies entirely in the real plane (see Section 5.1). Moreover, although we shall not prove it here, we have the following theorem on continuity.

Theorem 11.6   If $P(x) = a_0 x^n + a_1 x^{n-1} + \cdots + a_n$, a_j, $x \in R$, and if $k \in R$ is between $P(x_1)$ and $P(x_2)$, then there exists at least one $c \in R$ between x_1 and x_2 such that $P(c) = k$.

More precisely, we can show that there are no breaks or jumps in the values of $P(x)$, so we know that $P(x)$ must assume all values between any two of its values. Thus, the graph of P must be a continuous, unbroken curve.

Descartes'
Rule of
Signs

Another useful theorem, which again we shall not prove here, is that which establishes **Descartes' Rule of Signs.**

Theorem 11.7   *If $P(x)$ is a polynomial over the field R of real numbers, then the number of positive real solutions of $P(x) = 0$ either is equal to the number of variations in sign occurring in the coefficients of $P(x)$, or else is less than this number by an even natural number. Moreover, the number of negative real solutions of $P(x) = 0$ either is equal to the number of variations in sign occurring in $P(-x)$, or else is less than this number by an even natural number.*

Variations
in sign

A **variation in sign** occurs in a polynomial with real coefficients if, as the polynomial is viewed from left to right, successive coefficients are opposite in sign. For example, in the polynomial

$$P(x) = 3x^5 - 2x^4 - 2x^2 + x - 1,$$

there are three variations in sign, and, in

$$P(-x) = -3x^5 - 2x^4 - 2x^2 - x - 1,$$

there are no variations in sign.

Example

Find an upper bound on the number of positive real solutions and an upper bound on the number of negative real solutions for the equation

$$3x^4 + 3x^3 - 2x^2 + x + 1 = 0. \tag{1}$$

Solution

Since $P(x) = 3x^4 + 3x^3 - 2x^2 + x + 1$ has but two variations in sign, the equation can have no more than two positive real solutions. Since

$$P(-x) = 3x^4 - 3x^3 - 2x^2 - x + 1$$

has two variations in sign, the number of negative real solutions of (1) cannot exceed two.

Isolation of
real zeros

The following theorem is sometimes helpful in isolating real zeros of a polynomial function with real coefficients.

Theorem 11.8   *Let $P(x)$ be a polynomial over the field R of real numbers.*

I   *If $r_1 \geq 0$, and the coefficients of the terms in $Q(x)$ and the term $P(r_1)$ are all of the same sign in the right-hand member of*

$$P(x) = (x - r_1)Q(x) + P(r_1),$$

then $P(x) = 0$ can have no solution greater than r_1.

II   *If* $r_2 \leq 0$, *and the coefficients of the terms in* $Q(x)$ *and the term* $P(r_2)$ *alternate in sign (zero suitably denoted by* $+0$ *or* -0*) in the right-hand member of*

$$P(x) = (x - r_2)Q(x) + P(r_2),$$

then $P(x) = 0$ *can have no solution less than* r_2.

Proof   We shall prove only the first part of the theorem here. The second part follows from the first by considering $P(-x) = (-x - r_2)Q(-x) + P(r_2)$.

For all $x > r_1$, $(x - r_1) > 0$. Moreover, if all the coefficients in $Q(x)$ are of the same sign, say positive, then, since $x > r_1 \geq 0$, we have $Q(x) > 0$ and $(x - r_1)Q(x) > 0$. Since $P(r_1) \geq 0$ by hypothesis, we have

$$P(x) = (x - r_1)Q(x) + P(r_1) > 0,$$

and the first part is proved.

This theorem permits us to place upper and lower bounds on the set of real zeros of the polynomial function

$$P = \{(x, P(x)) \,|\, P(x) = a_0 x^n + \cdots + a_n\},$$

and consequently on the real members of the solution set of $P(x) = 0$.

Example   Show that 2 and -2 are upper and lower bounds, respectively, for the location of the zeros of

$$\{(x, P(x)) \,|\, P(x) = 18x^3 - 12x^2 - 11x + 10\}.$$

Solution   Using synthetic division to divide $P(x)$ by $x - 2$, we have

$$
\begin{array}{r|rrrr}
2 & 18 & -12 & -11 & 10 \\
  &   & 36 & 48 & 74 \\
\hline
  & 18 & 24 & 37 & 84
\end{array}.
$$

Thus $Q(x) = 18x^2 + 24x + 37$ and $P(2) = 84 > 0$. Hence, by Theorem 11.8-I, 2 is an upper bound for the zeros of P. Next dividing $P(x)$ by $x + 2$, we have

$$
\begin{array}{r|rrrr}
-2 & 18 & -12 & -11 & 10 \\
   &   & -36 & 96 & -170 \\
\hline
   & 18 & -48 & 85 & -160
\end{array}.
$$

Here $Q(x) = 18x^2 - 48x + 85$ and $P(-2) = -160$. Therefore, by Theorem 11.8-II, -2 is a lower bound for the zeros of P.

Example   Find the least nonnegative integer and the greatest nonpositive integer that are, by Theorem 11.8, upper and lower bounds, respectively, for the real zeros of

$$\{(x, P(x)) \,|\, P(x) = x^4 - x^3 - 10x^2 - 2x + 12\}.$$

Solution   We shall first seek an upper bound by dividing $P(x)$ successively by $x - 1$, $x - 2$, and so on. Each row after the first in the following array is the bottom row in the respective synthetic division involved.

(*Solution continued overleaf*)

	1	−1	−10	−2	12
1	1	0	−10	−12	0
2	1	1	−8	−18	−24
3	1	2	−4	−14	−30
4	1	3	2	6	36

Since the numbers in the last row are all positive, 4 is an upper bound. Next, we divide by $x + 1$, $x + 2$, and so on, in search of a lower bound.

	1	−1	−10	−2	12
−1	1	−2	−8	6	6
−2	1	−3	−4	6	0
−3	1	−4	2	−8	36

Since the numbers in the row following −3 alternate from positive to negative, etc., −3 is a lower bound. (Had the numbers in the row been 1, 0, 2, −8, 36, then the sign "−" could arbitrarily have been assigned to 0 to give the desired pattern of alternating signs.)

The location theorem

To narrow the search for real zeros of a polynomial function still further, we have the following **location theorem**.

Theorem 11.9   *Let $P(x)$ be a polynomial over the field R of real numbers. If $x_1, x_2 \in R$, with $x_1 < x_2$, and if $P(x_1)$ and $P(x_2)$ are opposite in sign, then there exists at least one $c \in R$, $x_1 < c < x_2$, such that $P(c) = 0$.*

Proof   First, let $P(x_1) < 0 < P(x_2)$, and $x_1 < x_2$. Then, by Theorem 11.6, there must exist $c \in R$, $x_1 < c < x_2$, such that $P(c) = 0$. Next, let $P(x_2) < 0 < P(x_1)$ and $x_1 < x_2$. Then, also by Theorem 11.6, there must exist $c \in R$, $x_1 < c < x_2$, such that $P(c) = 0$. Since x_1 and x_2 are arbitrary, the theorem is proved.

This theorem expresses the fact that if the graphs of $(x_1, P(x_1))$ and $(x_2, P(x_2))$ are on opposite sides of the x-axis, then the graph of $y = P(x)$ must cross the x-axis at some (at least one) point c between x_1 and x_2.

Example

Discuss the possibilities for real zeros of

$$\{(x, P(x)) \mid P(x) = 32x^4 - 8x^3 - 148x^2 + 162x - 45\}.$$

Solution

We observe first that, by Descartes' Rule of Signs, Theorem 11.7, we can have at most three positive and one negative real zero. Next, we apply synthetic division to obtain the following array:

	32	−8	−148	162	−45
0	32	−8	−148	162	−45
1	32	24	−124	38	−7
2	32	56	−36	90	135
3	32	88	116	510	1485
−1	32	−40	−108	270	−315
−2	32	−72	−4	170	−385
−3	32	−104	164	−330	945

Examining this array, we find that, by Theorem 11.8, 3 is an upper and −3 is a lower bound for real zeros. By the remainder theorem, $P(1) = -7$ and $P(2) = 135$, and since these are of opposite sign, Theorem 11.9 assures us that there is at least one real zero between 1 and 2. Similarly, we find there is at least one zero between −2 and −3, because $P(-2) = -385$ while $P(-3) = 945$. Furthermore, by Theorem 11.5, P must have precisely 4 real or complex zeros, not necessarily all distinct, because $P(x)$ is of degree four. By Theorem 11.3, any nonreal (imaginary) complex roots must occur in conjugate pairs, so we can assert, as a conclusion, the following possibilities:

1. P has two real zeros, one between 1 and 2, and one between −2 and −3;

2. P has four real zeros:
   a. three between 1 and 2 and one between −2 and −3, or vice versa,
   b. two between 2 and 3, one between 1 and 2, and one between −2 and −3,
   c. two between −2 and 1, one between 1 and 2, and one between −2 and −3.

As a matter of fact, the zeros of P are $1/2$, $3/4$, $3/2$, and $-5/2$, so that case 2(c) is the true situation; but in order to have detected this, we would have had either to find these zeros or to conduct a rather extensive search for sign changes in $P(x)$ over the interval $0 \le x \le 1$.

Exercise 11.3

Use Theorem 11.7 to discuss the nature of the roots of each polynomial equation.

1. $x^4 - 2x^3 + 2x + 1 = 0$
2. $3x^4 + 3x^3 + 2x^2 - x + 1 = 0$
3. $2x^5 + 3x^3 + 2x + 1 = 0$
4. $4x^5 - 2x^3 - 3x - 2 = 0$
5. $3x^4 + 1 = 0$
6. $2x^5 - 1 = 0$

Find an upper bound and a lower bound for the real roots of each polynomial equation.

7. $x^3 + 2x^2 - 7x - 8 = 0$
8. $x^3 - 8x + 5 = 0$
9. $x^4 - 2x^3 - 7x^2 + 10x + 10 = 0$
10. $x^3 - 4x^2 - 4x + 12 = 0$
11. $x^5 - 3x^3 + 24 = 0$
12. $x^5 - 3x^4 - 1 = 0$
13. $2x^5 + x^4 - 2x - 1 = 0$
14. $2x^5 - 2x^2 + x - 2 = 0$

Use Theorem 11.9 to verify each statement in Problems 15–20.

15. $\{(x, f(x))|f(x) = x^3 - 3x + 1\}$ has a zero between 0 and 1.

16. $\{(x, f(x))|f(x) = 2x^3 + 7x^2 + 2x - 6\}$ has a zero between -2 and -1.

17. $\{(x, g(x))|g(x) = x^4 - 2x^2 + 12x - 17\}$ has a zero between -3 and -2.

18. $\{(x, g(x))|g(x) = 2x^4 + 3x^3 - 14x^2 - 15x + 9\}$ has a zero between -2 and -1.

19. $\{(x, P(x))|P(x) = 2x^2 + 4x - 4\}$ has one zero between -3 and -2, and one between 0 and 1.

20. $\{(x, P(x))|P(x) = x^3 - x^2 - 2x + 1\}$ has one zero between -2 and -1, one between 0 and 1, and one between 1 and 2.

11.4   *Rational Zeros of Polynomial Functions*

Prime and composite numbers

The results of the present section depend on the notion of a **prime number**. If a is an element of the set N of natural numbers, and $a \neq 1$, then a is a prime number if and only if a has no prime factor in N other than itself and 1; otherwise, a is a **composite number**. For example, 2 and 3 are prime numbers, and $6 = 2 \cdot 3$ is composite; but 1 is considered to be neither prime nor composite.

The **unique-factorization theorem**, or **fundamental theorem of arithmetic** (which we shall not prove), follows.

> **Theorem 11.10**   *If a is a composite number, then a is the product of only one set of prime factors.*

Two integers a and b are said to be **relatively prime** if and only if they have no prime factors in common. For example, -6 and 35 are relatively prime, since $-6 = -2 \cdot 3$ and $35 = 5 \cdot 7$; but 6 and 8 are not, since they have the prime factor 2 in common. The fraction a/b is said to express a rational number in **lowest terms** if and only if a and b are relatively prime.

Identifying possible rational zeros

If all the coefficients of the defining equation

$$P(x) = a_0 x^n + a_1 x^{n-1} + \cdots + a_n$$

of a polynomial function P are integers, then we can identify all possible rational zeros of P by means of the following theorem.

> **Theorem 11.11**   *If the rational number p/q, in lowest terms, is a solution of*
>
> $$P(x) = a_0 x^n + a_1 x^{n-1} + \cdots + a_n = 0,$$
>
> *where $a_j \in J$, then p is an integral factor of a_n and q is an integral factor of a_0.*

Proof   Since p/q is a solution of $P(x) = 0$, we have

$$a_0 \left(\frac{p}{q}\right)^n + a_1 \left(\frac{p}{q}\right)^{n-1} + \cdots + a_n = 0,$$

and we can multiply each member here by q^n to obtain

$$a_0 p^n + a_1 p^{n-1}q + \cdots + a_n q^n = 0.$$

Adding $-a_n q^n$ to each member and factoring p from each term in the left-hand member of the resulting equation, we have

$$p(a_0 p^{n-1} + a_1 p^{n-2}q + \cdots + a_{n-1}q^{n-1}) = -a_n q^n.$$

Since J is closed with respect to addition and multiplication, the expression in parentheses in the left-hand member here represents an integer, say r, so that we have

$$pr = -a_n q^n,$$

where pr is an integer having p as a factor. Hence, p is a factor of $-a_n q^n$. But, by Theorem 11.10, p and q^n have no factor in common, because, by hypothesis, p/q is in lowest terms; hence p must be a factor of a_n. In a similar manner, by writing the equation

$$a_0 p^n + a_1 p^{n-1}q + \cdots + a_n q^n = 0$$

in the form

$$-a_0 p^n = a_1 p^{n-1}q + \cdots + a_n q^n,$$

we can factor q from each term in the right-hand member and show that q must be a factor of a_0.

Example

List all possible rational zeros of

$$\{(x,\ P(x))\,|\,P(x) = 2x^3 - 4x^2 + 3x + 9\}.$$

Solution

Rational zeros, p/q, must, by Theorem 11.11, be such that p is an integral factor of 9 and q is an integral factor of 2. Hence

$$p \in \{-9,\ -3,\ -1,\ 1,\ 3,\ 9\}, \quad q \in \{-2,\ -1,\ 1,\ 2\}.$$

and the set of possible rational zeros of P is

$$\left\{-9,\ -\frac{9}{2},\ -3,\ -\frac{3}{2}, -1,\ -\frac{1}{2}, \frac{1}{2},\ 1, \frac{3}{2}, 3, \frac{9}{2}, 9\right\}.$$

Test for rational zeros

It is important to observe that Theorem 11.11 does not assure us that a polynomial function with integral coefficients indeed has a rational zero; it simply enables us to identify possibilities for rational zeros. These can then be checked by synthetic division. The identification of the zeros of P in the previous example is left as an exercise.

As a special case of Theorem 11.11, it is evident that if a function P is defined by the equation

$$P(x) = x^n + a_1 x^{n-1} + \cdots + a_n,$$

in which $a_i \in J$ and $a_0 = 1$, then any rational zero of P must be an integer, and, moreover, must be an integral factor of a_n.

Example   Find all rational zeros of $\{(x, P(x)) \mid P(x) = x^3 - 4x^2 + x + 6\}$.

Solution   The only possible rational zeros of P are -6, -3, -2, -1, 1, 2, 3, and 6. Using synthetic division, we set up the following array:

	1	-4	1	6
1	1	-3	-2	4
-1	1	-5	6	0

We can cease our trials with -1, since by the remainder theorem, -1 is a zero of P, and the remaining zeros can be obtained from the equation $x^2 - 5x + 6 = 0$, whose coefficients are taken from the last line in the array. We can write the equation as $(x - 2)(x - 3) = 0$ and observe that 2 and 3 are also zeros. Note that it is often useful to start with possible roots of *least* absolute value, since the synthetic-division testing might reveal an upper or lower bound and thus eliminate the necessity of testing some of the numbers.

Exercise 11.4

Find all integral zeros of each function.

1. $\{(x, f(x)) \mid f(x) = 3x^3 - 13x^2 + 6x - 8\}$
2. $\{(x, f(x)) \mid f(x) = x^4 - x^2 - 4x + 4\}$
3. $\{(x, f(x)) \mid f(x) = x^4 + x^3 + 2x - 4\}$
4. $\{(x, f(x)) \mid f(x) = 5x^3 + 11x^2 - 2x - 8\}$
5. $\{(x, P(x)) \mid P(x) = 2x^4 - 3x^3 - 8x^2 - 5x - 3\}$
6. $\{(x, P(x)) \mid P(x) = 3x^4 - 40x^3 + 130x^2 - 120x + 27\}$

Find all rational zeros of each function.

7. $\{(x, f(x)) \mid f(x) = 2x^3 + 3x^2 - 14x - 21\}$
8. $\{(x, f(x)) \mid f(x) = 3x^4 - 11x^3 + 9x^2 + 13x - 10\}$
9. $\{(x, P(x)) \mid P(x) = 4x^4 - 13x^3 - 7x^2 + 41x - 14\}$
10. $\{(x, P(x)) \mid P(x) = 2x^3 - 4x^2 + 3x + 9\}$
11. $\{(x, Q(x)) \mid Q(x) = 2x^3 - 7x^2 + 10x - 6\}$
12. $\{(x, Q(x)) \mid Q(x) = x^3 + 3x^2 - 4x - 12\}$

Find all complex zeros of each function. Hint: First find all rational zeros.

13. $\{(x, P(x)) \mid P(x) = 3x^3 - 5x^2 - 14x - 4\}$
14. $\{(x, P(x)) \mid P(x) = x^3 - 4x^2 - 5x + 14\}$
15. $\{(x, P(x)) \mid P(x) = 2x^4 + 3x^3 + 2x^2 - 1\}$
16. $\{(x, P(x)) \mid P(x) = 8x^4 - 22x^3 + 29x^2 - 66x + 15\}$
17. $\{(x, P(x)) \mid P(x) = 12x^4 + 7x^3 + 7x - 12\}$
18. $\{(x, P(x)) \mid P(x) = 6x^4 - 13x^3 + 2x^2 - 4x + 15\}$

19. Factor the polynomial $2x^3 + 3x^2 - 2x - 3$ over C.
20. Factor the polynomial $x^4 - 6x^3 - 3x^2 - 24x - 28$ over C.
21. Show that $\sqrt{3}$ is irrational. *Hint:* Consider the equation $x^2 - 3 = 0$.
22. Show that $\sqrt{2}$ is irrational.

11.5　　*Irrational Zeros of Polynomial Functions*

The location theorem can often be used to isolate some real zeros of polynomial functions on intervals of the domain. Once we have isolated such zeros, various means exist for obtaining decimal approximations to them, in particular to zeros that are irrational. We shall be concerned with only one such means herein, namely, linear interpolation.

Consider the function

$$P = \{(x,\ P(x)) \mid P(x) = x^3 - 3x^2 - 2x + 5\}.$$

By means of synthetic division and the remainder theorem, we can establish that $(1, 1)$ and $(2, -3)$ are in the function, and hence, by the location theorem, there is at least one zero between 1 and 2. Actually, since $P(x)$ is positive for x large and positive, and $P(x)$ is negative for x large in absolute value and negative, one of the remaining two roots is >2 and the other <1, so that there is exactly one zero between 1 and 2. Since the only rational zero for P would have to be an integer (the leading coefficient is 1), any zeros between 1 and 2 must be irrational. Figure 11.1-a shows the two points $P_1(1, 1)$, and $P_2(2, -3)$, and a straight line segment joining them.

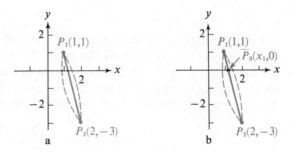

Figure 11.1

The dashed lines show possibilities for the graph of P on the interval $1 \le x \le 2$; but since we are uncertain of the curvature, we cannot be sure on which side of the line segment the graph actually lies. In either case, however, the point where the segment intersects the x-axis clearly is close (in some sense) to the point where the graph of P intersects this axis. In Figure 11.1-b, we show the same segment, this time with an additional detail.

If we can find the value of the x-intercept x_1 of the line segment P_1P_2, then we will have a first approximation to a zero for P. Since the slope of P_1P_3 is the same as the slope for P_1P_2, we have

$$\frac{0-1}{x_1 - 1} = \frac{-3-1}{2-1},$$

from which

$$x_1 = \frac{5}{4}.$$

To find $P(5/4)$, we divide $P(x)$ synthetically by 1.25, as follows.

$$
\begin{array}{r|rrrr}
1.25 & 1 & -3 & -2 & 5 \\
     &   & 1.25 & -2.1875 & -5.234375 \\
\hline
     & 1 & -1.75 & -4.1875 & -0.234375
\end{array}
$$

Thus, $P(5/4) \approx -0.2344$. Figure 11.2-a shows our present situation, from which

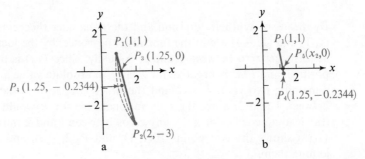

Figure 11.2

it is clear that the graph of P crosses the x-axis to the left of P_3, so that, at least insofar as this point is concerned, the graph of P is concave upward on this interval. We can now repeat the linear-interpolation process, using P_1 and P_4, to obtain another approximation to the zero for P. Figure 11.2-b shows the necessary detail. Again, since the slope of P_1P_5 is the same as the slope of P_1P_4, we have

$$\frac{0-1}{x_2 - 1} = \frac{-0.2344 - 1}{1.25 - 1},$$

from which

$$x_2 - 1 = \frac{0.25}{1.2344} \approx 0.2025,$$

$$x_2 \approx 1.2025.$$

Thus, a second approximation to the desired zero of P is 1.2025. This process can be continued as long as necessary to obtain any desired degree of accuracy. To three decimal places, the zero sought here is 1.202.

Exercise 11.5

Find to one decimal place the indicated real zero(s) of each function.

1. $\{(x, P(x))| P(x) = x^3 - 3x + 1\}$; between 1 and 2
2. $\{(x, P(x))| P(x) = 2x^3 - x^2 + 3x + 1\}$; between 0 and -1
3. $\{(x, P(x))| P(x) = x^3 - 2x - 5\}$; between 2 and 3
4. $\{(x, P(x))| P(x) = x^3 + 2x^2 - 1\}$; between 0 and 1
5. $\{(x, P(x))| P(x) = x^3 + 3x^2 - 6x - 3\}$; the greatest positive
6. $\{(x, P(x))| P(x) = 2x^3 - 5x^2 - x + 5\}$; the least positive
7. $\{(x, P(x))| P(x) = x^3 + x - 1\}$; all
8. $\{(x, P(x))| P(x) = x^4 - 4x^3 - 4x + 12\}$; all

9. Find a decimal numeral for $\sqrt[3]{5}$ to two decimal places. *Hint:* Consider the equation $x^3 - 5 = 0$.

10. Find a decimal numeral for $\sqrt[5]{2}$ to two decimal places.

Chapter Review

[11.1]   *Find the value of the polynomial for the given value of x.*

1. $P(x) = 2x^3 - 3x^2 + 15x - 6$; -3
2. $Q(x) = x^4 - 3x^3 + 2x^2 + 15x - 12$; i

3. Find the quotient and reminder when $x^3 - 2x^2 + 3x - 5$ is divided by $x - i$.
4. Show that $x - i$ is a factor of $2x^3 - 3x^2 + 2x - 3$.

[11.2]   *Find the other zeros of the given polynomial function if one zero is as given.*

5. $\{(x, P(x))| P(x) = x^3 - 2x^2 + 4x - 8\}$; $2i$ is one zero.
6. $\{(x, Q(x))| Q(x) = 2x^3 - 11x^2 + 28x - 24\}$; $2 + 2i$ is one zero.

[11.3]   7. Discuss the nature of the roots of $x^4 + 3x^3 - 2x^2 - x + 2 = 0$ in accordance with Descartes' Rule of Signs.

8. Find an upper bound and a lower bound for the real roots of
$$x^4 + 2x^3 - 7x^2 - 10x + 10 = 0.$$

9. Use synthetic division and Theorem 11.9 to show that $\{(x, P(x))| 2x^3 + x^2 - 4x - 2\}$ has a zero between 1 and 2.

[11.4]   **10.** Find all integral zeros of $\{(x, Q(x)) \,|\, Q(x) = x^4 - 3x^3 - x^2 - 11x - 4\}$.

11. Find all rational zeros of $\{(x, P(x)) \,|\, P(x) = 2x^3 - 11x^2 + 12x + 9\}$.

12. Find all complex zeros of $\{(x, P(x)) \,|\, P(x) = x^3 + 2x^2 + 2x + 4\}$.

[11.5]   **13.** Find to one decimal place the real zero of

$$\{(x, P(x)) \,|\, P(x) = x^3 - 4x^2 + 4\}$$

between 1 and 2.

14. Find to one decimal place the real zero of

$$\{(x, Q(x)) \,|\, Q(x) = x^3 + 3x^2 - 9x + 4\}$$

between 0 and 1.

12 Sequences and Series

12.1 Mathematical Induction

The material in the present section depends on a special property of the set of natural numbers, or positive integers, $N = \{1, 2, 3, \ldots\}$:

a. $1 \in S$

b. If $k \in S$, then $k + 1 \in S$.

In addition, N *contains no elements not implied by properties a and b*. These properties underlie the following theorem, called the **principle of mathematical induction**, which we state without proof.

Theorem 12.1   *If a given sentence involving natural numbers n is true for n = 1, and if its truth for n = k implies its truth for n = k + 1, then it is true for every natural number n.*

Requirements of a proof by mathematical induction

We can exploit Theorem 12.1 to prove a number of assertions. Although the technique we shall use is called **proof by mathematical induction**, the argument we shall employ is deductive, as have been all of the other arguments in this book. Proofs by mathematical induction require two things:

a. A demonstration that the assertion to be proved is true for the natural number 1.
b. A demonstration that the truth of the assertion for a natural number k implies its truth for $k + 1$.

When these two demonstrations have been made, the principle of mathematical induction assures us that the assertion is true for every natural number.

Example

Prove that the sum of the first n natural numbers is $\dfrac{n(n + 1)}{2}$.

(Solution overleaf)

Solution   We wish to show that

$$1 + 2 + 3 + \cdots + n = \frac{n(n+1)}{2}.$$

We must do two things:

1. We must first show that the assertion is true for $n = 1$, that is, that

$$1 = \frac{1(1+1)}{2},$$

which is true.

2. We must next show that the truth of

$$1 + 2 + 3 + \cdots + k = \frac{k(k+1)}{2}$$

implies the truth of

$$1 + 2 + 3 + \cdots + k + (k+1) = \frac{(k+1)[(k+1)+1]}{2}.$$

That is, we must show that the truth of the assertion for $n = k$ implies its truth for $n = k + 1$. Now, assume the truth of

$$1 + 2 + 3 + \cdots + k = \frac{k(k+1)}{2}.$$

Then, by adding $k + 1$ to each member of this equation, we obtain

$$1 + 2 + 3 + \cdots + k + (k+1) = \frac{k(k+1)}{2} + (k+1)$$

$$= (k+1)\left(\frac{k}{2} + 1\right)$$

$$= (k+1)\left(\frac{k+2}{2}\right)$$

$$= \frac{(k+1)[(k+1)+1]}{2}.$$

Thus the second fact necessary for our proof is established. By the principle of mathematical induction, the assertion is true for every natural number n.

Figure 12.1

This method of proof is often compared to lining up a row of dominoes, with the assumption that whenever one domino is toppled, the one following will topple. One then needs only to topple the first domino ($n = 1$) and the whole row behind it will topple, as indicated in Figure 12.1.

Exercise 12.1

By mathematical induction, prove the validity of the formulas in Problems 1–10 for all positive integral values of n.

1. $\dfrac{1}{2} + \dfrac{2}{2} + \dfrac{3}{2} + \cdots + \dfrac{n}{2} = \dfrac{n(n+1)}{4}$

2. $1 + 3 + 5 + \cdots + (2n - 1) = n^2$

3. $2 + 4 + 6 + \cdots + 2n = n(n+1)$

4. $2 + 6 + 10 + \cdots + (4n - 2) = 2n^2$

5. $1^2 + 2^2 + 3^2 + \cdots + n^2 = \dfrac{n(n+1)(2n+1)}{6}$

6. $2 + 2^2 + 2^3 + \cdots + 2^n = 2^{n+1} - 2$

7. $1^3 + 3^3 + 5^3 + \cdots + (2n - 1)^3 = n^2(2n^2 - 1)$

8. $\dfrac{1}{1 \cdot 2} + \dfrac{1}{2 \cdot 3} + \dfrac{1}{3 \cdot 4} + \cdots + \dfrac{1}{n(n+1)} = \dfrac{n}{n+1}$

9. $1 \cdot 2 + 2 \cdot 3 + 3 \cdot 4 + \cdots + n(n+1) = \dfrac{n(n+1)(n+2)}{3}$

10. $1 \cdot 4 + 2 \cdot 9 + 3 \cdot 16 + \cdots + n(n+1)^2 = \dfrac{1}{12} n(n+1)(n+2)(3n+5)$

11. Show that if $2 + 4 + 6 + \cdots + 2n = n(n+1) + 2$ is true for $n = k$, then it is true for $n = k + 1$. Is it true for every $n \in N$?

12. Show that $n^3 + 11n = 6(n^2 + 1)$ is true for $n = 1$, 2, and 3. Is it true for every $n \in N$?

12.2   Sequences

Let us consider a class of functions in which each function has as its domain either the set N of positive integers or a subset of successive members of N.

Definition 12.1   *A **sequence function** is a function having as its domain the set N of positive integers 1, 2, 3, ... A **finite-sequence function** has as its domain the set of positive integers 1, 2, 3, ..., n, for some fixed n.*

For example, the function defined by

$$s(n) = n + 3, \quad n \in \{1, 2, 3, \ldots\}, \tag{1}$$

is a sequence function. The elements in the range of such a function, considered in the order

$$s(1), s(2), s(3), s(4), \ldots,$$

are said to form a **sequence**. Similarly, the elements of a finite-sequence function, considered in order, constitute a **finite sequence**.

For example, the sequence associated with (1) is found by successively substituting the numbers $1, 2, 3, \ldots,$ for n:

$$s(1) = 1 + 3 = 4,$$
$$s(2) = 2 + 3 = 5,$$
$$s(3) = 3 + 3 = 6,$$
$$s(4) = 4 + 3 = 7,$$

and so on. Thus the first four terms of (1) are 4, 5, 6, and 7. The nth term, or general term, is $n + 3$. As another example, the first five terms of the sequence defined by the equation

$$s(n) = \frac{3}{2n - 1}, \quad n \in \{1, 2, 3, \ldots\},$$

are $3/1$, $3/3$, $3/5$, $3/7$, and $3/9$; and the twenty-fifth term is

$$s(25) = \frac{3}{2(25) - 1} = \frac{3}{49}.$$

Given several terms in a sequence, we are often able to construct an expression for a general term of a sequence to which they belong. Such a sequence is not unique. Thus, if the first three terms in a sequence are $2, 4, 6, \ldots,$ we may *surmise* that the general term is $s(n) = 2n$. Note, however, that the sequences for both

$$s(n) = 2n$$

and

$$s(n) = 2n + (n - 1)(n - 2)(n - 3)$$

start with 2, 4, 6, but that the two sequences differ for terms following the third.

Sequence notation

The notation ordinarily used for the terms in a sequence is not function notation as such. It is customary to denote a term in a sequence by means of a subscript. Thus, the sequence $s(1), s(2), s(3), s(4), \ldots$ would appear as $s_1, s_2, s_3, s_4, \ldots$.

Arithmetic progressions

Let us next consider two special kinds of sequences that have many applications. The first kind can be defined as follows:

Definition 12.2   An **arithmetic progression** *is a sequence defined by equations of the form*

$$s_1 = a, \qquad s_{n+1} = s_n + d,$$

where $a, d \in R$, *and* $n \in N$.

Since each term in such a sequence is obtained from the preceding term by adding d, d is called the **common difference**. Thus, $3, 7, 11, 15, \ldots$ is an arithmetic progression, in which $s_1 = 3$ and $d = 4$.

For each arithmetic progression, the general term is established by the following.

Theorem 12.2   *The nth term in the sequence defined by*

$$s_1 = a, \qquad s_{n+1} = s_n + d,$$

where $a, d \in R$, and $n \in N$, is

$$s_n = a + (n - 1)d. \tag{2}$$

Proof We shall use mathematical induction. That (2) is true for the natural number 1 is evident by direct substitution of 1 in (2):

$$s_1 = a + (1 - 1)d = a.$$

If now we assume that (2) is true for the natural number k, then we have

$$s_k = a + (k - 1)d.$$

By the defining equation, we accordingly have

$$s_{k+1} = s_k + d.$$

Replacing s_k in this expression with $a + (k - 1)d$, we obtain

$$\begin{aligned} s_{k+1} &= a + (k - 1)d + d \\ &= a + kd \\ &= a + [(k + 1) - 1]d, \end{aligned}$$

and the principle of mathematical induction assures us that the relationship (2) is valid for all natural numbers.

Geometric progressions   The second kind of sequence we shall consider can be defined as follows:

Definition 12.3   *A **geometric progression** is a sequence defined by equations of the form*

$$s_1 = a, \qquad s_{n+1} = rs_n,$$

where $a, r \in R$, $a, r \neq 0$, and $n \in N$.

Thus, $3, 9, 27, 81, \ldots$ is a geometric progression in which each term except the first is obtained by multiplying the preceding term by 3. Since the effect of multiplying the terms in this way is to produce a fixed ratio between any two successive terms, the multiplier, r, is called the **common ratio**.

The general term for a geometric progression is that established by the following theorem. The proof by induction is left as an exercise.

Theorem 12.3   *The nth term in the sequence defined by*

$$s_1 = a, \qquad s_{n+1} = rs_n,$$

where $a, r \in R, a \neq 0, r \neq 0$, and $n \in N$, is

$$s_n = ar^{n-1}.$$

Exercise 12.2

Find the first four terms in the sequence with the general term as given.

Examples   a. $\quad s_n = \dfrac{n(n+1)}{2}$ b. $\quad s_n = (-1)^n 2^n$

Solutions   a. $\quad s_1 = \dfrac{1(1+1)}{2} = 1$ b. $\quad s_1 = (-1)^1 2^1 = -2$

           $s_2 = \dfrac{2(2+1)}{2} = 3$ $s_2 = (-1)^2 2^2 = 4$

           $s_3 = \dfrac{3(3+1)}{2} = 6$ $s_3 = (-1)^3 2^2 = -8$

           $s_4 = \dfrac{4(4+1)}{2} = 10;$ $s_4 = (-1)^4 2^4 = 16;$

           1, 3, 6, 10 $-2, 4, -8, 16$

1. $\ s_n = n - 5$ **2.** $\ s_n = 2n - 3$ **3.** $\ s_n = \dfrac{n^2 - 2}{2}$

4. $\ s_n = \dfrac{3}{n^2 + 1}$ **5.** $\ s_n = 1 + \dfrac{1}{n}$ **6.** $\ s_n = \dfrac{n}{2n - 1}$

7. $\ s_n = \dfrac{n(n-1)}{2}$ **8.** $\ s_n = \dfrac{5}{n(n+1)}$ **9.** $\ s_n = (-1)^n$

10. $\ s_n = (-1)^{n+1}$ **11.** $\ s_n = \dfrac{(-1)^n(n-2)}{n}$ **12.** $\ s_n = (-1)^{n-1} 3^{n+1}$

Write the next three terms in each of the following arithmetic progressions.

Examples   a. $\ 5, 9, \ldots$ b. $\ x, x - a, \ldots$

Solutions   Find the common difference and then continue the sequence.

           a. $\ d = 9 - 5 = 4;$ b. $\ d = (x - a) - x = -a;$

             13, 17, 21 $x - 2a, x - 3a, x - 4a$

13. $\ 3, 7, \ldots$ **14.** $\ -6, -1, \ldots$ **15.** $\ x, x + 1, \ldots$

16. $\ a, a + 5, \ldots$ **17.** $\ 2x + 1, 2x + 4, \ldots$ **18.** $\ 3a, 5a, \ldots$

Write the next four terms in each of the following geometric progressions.

Examples   a. $\ 3, 6, \ldots$ b. $\ x, 2, \ldots$

Solutions   Find the common ratio, and then continue the sequence.

a. $r = \dfrac{6}{3} = 2;$ b. $r = \dfrac{2}{x}$ $(x \neq 0);$

12, 24, 48, 96 $\dfrac{4}{x}, \dfrac{8}{x^2}, \dfrac{16}{x^3}, \dfrac{32}{x^4}$

19. 2, 8, ...   **20.** 4, 8, ...   **21.** $\dfrac{2}{3}, \dfrac{4}{3}, \ldots$

22. $\dfrac{1}{2}, -\dfrac{3}{2}, \ldots$ **23.** $\dfrac{a}{x}, -1, \ldots$ **24.** $\dfrac{a}{b}, \dfrac{a}{bc}, \ldots$

Example   Find the general term and the fourteenth term of the arithmetic progression $-6, -1, \ldots$.

Solution   Find the common difference.

$$d = -1 - (-6) = 5$$

Use $s_n = a + (n-1)d.$

$$s_n = -6 + (n-1)5 = 5n - 11$$
$$s_{14} = 5(14) - 11 = 59$$

25. Find the general term and the seventh term in the arithmetic progression 7, 11, ...

26. Find the general term and the twelfth term in the arithmetic progression $2, \dfrac{5}{2}, \ldots$.

27. Find the general term and the twentieth term in the arithmetic progression $3, -2, \ldots$.

28. Find the general term and the ninth term in the arithmetic progression $\dfrac{3}{4}, 2, \ldots$.

Example   Find the general term and also the ninth term of the geometric progression $-24, 12, \ldots$.

Solution   Find the common ratio.

$$r = \frac{12}{-24} = -\frac{1}{2}$$

Use $s_n = ar^{n-1}.$

$$s_n = -24\left(-\frac{1}{2}\right)^{n-1}$$

$$s_9 = -24\left(-\frac{1}{2}\right)^8 = -\frac{3}{32}$$

29. Find the general term and the sixth term in the geometric progression 48, 96, ...

30. Find the general term and the eighth term in the geometric progression $-3, \dfrac{3}{2}, \ldots$.

31. Find the general term and the seventh term in the geometric progression $-\dfrac{1}{3}, 1, \ldots$.

32. Find the general term and the ninth term in the geometric progression $-81, 27, \ldots$.

Example   Find the first term in an arithmetic progression in which the third term is 7 and the eleventh term is 55.

Solution   Find the common difference by considering an arithmetic progression with first term 7 and with ninth term 55. Use $s_n = a + (n-1)d$.

$$s_9 = 7 + (9-1)d$$
$$55 = 7 + 8d$$
$$d = 6$$

Use the difference to find the first term in an arithmetic progression in which the third term is 7. Again use $s_n = a + (n-1)d$.

$$s_3 = a + (3-1)6$$
$$7 = a + 12$$
$$a = -5$$

33. If the third term in an arithmetic progression is 7 and the eighth term is 17, find the common difference. What are the first and the twentieth terms?

34. If the fifth term of an arithmetic progression is -16 and the twentieth term is -46, what is the twelfth term?

35. Which term in the arithmetic progression 4, 1, ... is -77?

36. What is the twelfth term in an arithmetic progression in which the second term is x and the third term is y?

37. Find the first term of a geometric progression with fifth term 48 and ratio 2.

38. Find two different values for x so that $-\dfrac{3}{2}, x, -\dfrac{8}{27}$ will be in geometric progression.

39. By mathematical induction, prove Theorem 12.3.

12.3   Series

Associated with any sequence is a *series*.

Definition 12.4   *A series is the indicated sum of the terms in a sequence.*

For example, with the finite sequence

$$4, 7, 10, \ldots, 3n + 1,$$

for a given counting number n, there is associated the finite series

$$S_n = 4 + 7 + 10 + \cdots + (3n + 1);$$

similarly, with the finite sequence

$$x, x^2, x^3, x^4, \ldots, x^n,$$

there is associated the finite series

$$S_n = x + x^2 + x^3 + x^4 + \cdots + x^n.$$

Since the terms in the series are the same as those in the sequence, we can refer to the first term or the second term or the general term of a series in the same manner as we do for a sequence.

Sum of the first n terms of an arithmetic progression

Consider the series S_n of the first n terms of the general arithmetic progression,

$$S_n = a + (a + d) + (a + 2d) + \cdots + [a + (n - 1)d], \tag{1}$$

and then consider the same series written as

$$S_n = s_n + (s_n - d) + (s_n - 2d) + \cdots + [s_n - (n - 1)d], \tag{2}$$

where the terms are displayed in reverse order. Adding (1) and (2) term by term, we have

$$S_n + S_n = (a + s_n) + (a + s_n) + (a + s_n) + \cdots + (a + s_n),$$

where the term $(a + s_n)$ occurs n times. Then

$$2S_n = n(a + s_n),$$

$$S_n = \frac{n}{2}(a + s_n). \tag{3}$$

If (3) is rewritten as

$$S_n = n\left(\frac{a + s_n}{2}\right),$$

we observe that the sum is given by the product of the number of terms in the series and the average of the first and last terms. The validity of (3) can be established by mathematical induction and is left as an exercise.

An alternate form for (3) is obtained by substituting $a + (n - 1)d$ for s_n in (3) to obtain

$$S_n = \frac{n}{2}(a + [a + (n - 1)d],$$

$$S_n = \frac{n}{2}[2a + (n - 1)d], \tag{3a}$$

where the sum is now expressed in terms of a, n, and d.

Sum of the first n terms of a geometric progression

To find an explicit representation for the sum of a given number of terms in a geometric progression in terms of a, r, and n, we employ a device somewhat similar to the one used in finding the sum in an arithmetic progression. Consider the geometric series (4) containing n terms, and the series (5) obtained by multiplying both members of (4) by r:

$$S_n = a + ar + ar^2 + ar^3 + \cdots + ar^{n-2} + ar^{n-1}, \tag{4}$$

$$rS_n = ar + ar^2 + ar^3 + ar^4 + \cdots + ar^{n-1} + ar^n. \tag{5}$$

When we subtract (5) from (4), all terms in the right-hand members except the first term in (4) and the last term in (5) vanish, yielding

$$S_n - rS_n = a - ar^n.$$

Factoring S_n from the left-hand member gives

$$(1 - r)S_n = a - ar^n,$$

$$S_n = \frac{a - ar^n}{1 - r}, \tag{6}$$

if $r \neq 1$, and we have a formula for the sum of the first n terms of a geometric progression. Establishing the validity of Equation (6), which can be accomplished by mathematical induction, is left as an exercise.

An alternative expression for (6) can be obtained by first writing

$$S_n = \frac{a - r(ar^{n-1})}{1 - r},$$

and then, since $s_n = ar^{n-1}$, expressing this as

$$S_n = \frac{a - rs_n}{1 - r}, \tag{7}$$

where the sum is now given in terms of a, s_n, and r.

Sigma notation

A series for which the general term is known can be represented in a very convenient, compact way by means of what is called **sigma**, or **summation**, **notation**. The Greek letter $\sum$ (sigma) is used to denote a sum. For example.

$$S_n = 4 + 7 + 10 + \cdots + (3n + 1)$$

can be written

$$S_n = \sum_{j=1}^{n} (3j + 1),$$

where we understand that S_n is the series having terms obtained by replacing j in the expression $3j + 1$ with the numbers $1, 2, 3, \ldots, n$, successively. Similarly,

$$S = \sum_{j=3}^{6} j^2$$

appears in expanded form as

$$S = 3^2 + 4^2 + 5^2 + 6^2,$$

where the first value for j is 3 and the last is 6.

The variable used in conjunction with summation notation is called the **index of summation**, and the set of integers over which we sum (in this case, {3, 4, 5, 6}) is called the **range of summation**.

Notation for an infinite sum

To indicate that a series has an infinite number of terms, we cannot use the notation S_n for the sum, because there is no value to substitute for n. We therefore adopt notation such as

$$S_\infty = \sum_{j=1}^{\infty} \frac{1}{2^j} \tag{8}$$

to indicate that there is no last term in a series. In expanded form, the infinite series (8) is given by

$$S_\infty = \frac{1}{2} + \frac{1}{4} + \frac{1}{8} + \cdots.$$

The meaning, if any, of such an infinite sum will be discussed in Section 12.4.

Exercise 12.3

Write each series in expanded form.

Examples

a. $\displaystyle\sum_{j=2}^{5}(j^2 + 1)$ b. $\displaystyle\sum_{k=1}^{\infty}(-1)^k 2^{k+1}$

Solutions

a. $j = 2$, $2^2 + 1 = 5$;
   $j = 3$, $3^2 + 1 = 10$;
   $j = 4$, $4^2 + 1 = 17$;
   $j = 5$, $5^2 + 1 = 26$.

   $\displaystyle\sum_{j=2}^{5}(j^2 + 1) = 5 + 10 + 17 + 26$

b. $k = 1$, $(-1)^1 2^{1+1} = (-1)(4) = -4$;
   $k = 2$, $(-1)^2 2^{2+1} = (1)(8) = 8$;
   $k = 3$, $(-1)^3 2^{3+1} = (-1)(16) = -16$.

   $\displaystyle\sum_{k=1}^{\infty}(-1)^k 2^{k+1} = -4 + 8 - 16 + \cdots$

1. $\displaystyle\sum_{j=1}^{4} j^2$ 2. $\displaystyle\sum_{j=1}^{4}(3j - 2)$ 3. $\displaystyle\sum_{j=1}^{3}\frac{(-1)^j}{2^j}$

4. $\displaystyle\sum_{k=3}^{5}\frac{(-1)^{k+1}}{k-2}$ 5. $\displaystyle\sum_{k=0}^{\infty}\frac{1}{2^k}$ 6. $\displaystyle\sum_{k=0}^{\infty}\frac{k}{1+k}$

Write each series in sigma notation.

Examples

a. $5 + 8 + 11 + 14$ b. $x^2 + x^4 + x^6$ c. $\dfrac{3}{5} + \dfrac{5}{7} + \dfrac{7}{9} + \cdots$

(Solutions overleaf)

Solutions   Find an expression for the general term and write in sigma notation.

a. $3j + 2$ b. x^{2j} c. $\dfrac{2j + 1}{2j + 3}$

$\displaystyle\sum_{j=1}^{4} (3j + 2)$ $\displaystyle\sum_{j=1}^{3} x^{2j}$ $\displaystyle\sum_{j=1}^{\infty} \dfrac{2j + 1}{2j + 3}$

7. $x + x^3 + x^5 + x^7$ **8.** $x^3 + x^5 + x^7 + x^9 + x^{11}$

9. $1 + 4 + 9 + 16 + 25$ **10.** $\dfrac{1}{3} + \dfrac{1}{9} + \dfrac{1}{27} + \dfrac{1}{81}$

11. $1 \cdot 2 + 2 \cdot 3 + 3 \cdot 4 + 4 \cdot 5 + \cdots$ **12.** $\dfrac{1}{2} + \dfrac{2}{3} + \dfrac{3}{4} + \dfrac{4}{5} + \cdots$

13. $\dfrac{2}{1} + \dfrac{3}{2} + \dfrac{4}{3} + \dfrac{5}{4} + \cdots$ **14.** $\dfrac{1}{1} + \dfrac{2}{3} + \dfrac{3}{5} + \dfrac{4}{7} + \cdots$

Find each of the following sums.

Example $\displaystyle\sum_{j=1}^{12} (4j + 1)$

Solution   Write the first two or three terms in expanded form:

$$5 + 9 + 13 + \cdots.$$

This is an arithmetic series. The first term is 5 and the common difference is 4. Therefore we can use

$$S_n = \frac{n}{2} [2a + (n - 1)d]$$

to obtain

$$S_{12} = \frac{12}{2} [2(5) + (12 - 1)4] = 324.$$

15. $\displaystyle\sum_{j=1}^{7} (2j + 1)$ **16.** $\displaystyle\sum_{j=1}^{21} (3j - 2)$ **17.** $\displaystyle\sum_{j=3}^{15} (7j - 1)$

18. $\displaystyle\sum_{j=10}^{20} (2j - 3)$ **19.** $\displaystyle\sum_{k=1}^{8} \left(\frac{1}{2}k - 3\right)$ **20.** $\displaystyle\sum_{k=1}^{100} k$

Example $\displaystyle\sum_{j=2}^{5} \left(\frac{1}{3}\right)^{j}$

Solution   Write the first two terms in expanded form:

$$\left(\frac{1}{3}\right)^{2} + \left(\frac{1}{3}\right)^{3} + \cdots.$$

This is a geometric series in which the first term is $\frac{1}{9}$, the ratio is $\frac{1}{3}$, and $n = 4$.

Therefore we can use $S_n = \dfrac{a - ar^n}{1 - r}$ to obtain

$$S_4 = \frac{\dfrac{1}{9} - \dfrac{1}{9}\left(\dfrac{1}{3}\right)^4}{1 - \dfrac{1}{3}} = \frac{40}{243}.$$

21. $\displaystyle\sum_{j=1}^{6} 3^j$ **22.** $\displaystyle\sum_{j=1}^{4} (-2)^j$ **23.** $\displaystyle\sum_{k=3}^{7} \left(\frac{1}{2}\right)^{k-2}$

24. $\displaystyle\sum_{j=3}^{12} 2^{j-5}$ **25.** $\displaystyle\sum_{j=1}^{6} \left(\frac{1}{3}\right)^j$ **26.** $\displaystyle\sum_{j=1}^{4} (3 + 2^j)$

27. Find the sum of all even integers n, for $13 < n < 29$.

28. Find the sum of all integral multiples of 7 between 8 and 110.

29. How many bricks will there be in a pile one brick in thickness if there are 27 bricks in the bottom row, 25 in the second row, and so forth, to the top row, which has one brick?

30. If there are a total of 256 bricks in a pile arranged in the manner of those in Problem 29, how many bricks are there in the third row from the bottom of the pile?

31. Find $\displaystyle\sum_{j=1}^{n} \left(\frac{1}{2}\right)^j$ for $n = 2, 3, 4,$ and 5. What value do you think $\displaystyle\sum_{j=1}^{n} \left(\frac{1}{2}\right)^j$ approximates as n becomes larger and larger?

32. Find p if $\displaystyle\sum_{j=1}^{5} pj = 14$.

33. Fine p and q if $\displaystyle\sum_{j=1}^{4} (pj + q) = 28$ and $\displaystyle\sum_{j=2}^{5} (pj + q) = 44$.

34. Consider $S_n = \displaystyle\sum_{j=1}^{n} f(j)$. Explain why this equation defines a sequence function. What is the variable denoting an element in the domain? The range?

35. By using mathematical induction, prove that for an arithmetic progression, $S_n = \dfrac{n}{2}(a + s_n)$ for all positive integral values of n.

36. Show that the sequence formed by adding the corresponding terms in two arithmetic progressions is an arithmetic progression.

37. Show that the sum of the terms in two series with terms in arithmetic progression can be written as a series with terms in arithmetic progression.

38. By using mathematical induction, prove that, for a geometric progression, $S_n = \dfrac{a - ar^n}{1 - r}$ for all positive integral values of n, provided $r \neq 1$.

12.4   *Limits of Sequences and Series*

A sequence that is strictly increasing but bounded

Consider the sequence function defined by

$$s_n = \frac{n}{n+1}, \quad n \in N. \tag{1}$$

If we write the range of (1) in the form

$$\frac{1}{2}, \frac{2}{3}, \frac{3}{4}, \frac{4}{5}, \dots, \frac{n}{n+1}, \dots,$$

then it is clear that each of the terms is greater than the preceding term; indeed, the difference of consecutive terms is

$$\frac{n+1}{n+2} - \frac{n}{n+1} = \frac{(n^2+2n+1)-(n^2+2n)}{(n+1)(n+2)} = \frac{1}{(n+1)(n+2)} > 0.$$

Such a sequence is said to be **strictly increasing**. On the other hand, it is also clear that, no matter how large a value is assigned to n, we have

$$\frac{n}{n+1} < 1,$$

because the denominator is one larger than the numerator; in fact, we have

$$1 - \frac{n}{n+1} = \frac{(n+1)-n}{n+1} = \frac{1}{n+1} > 0.$$

Thus we have a sequence in which each term is greater than the preceding term and yet no term is equal to or greater than 1.

Limit of a sequence

We note, however—and this is a very basic consideration—that the value of $n/(n+1)$ is as close to 1 as we please if n is large enough. For example, the difference satisfies

$$1 - \frac{n}{n+1} = \frac{1}{n+1}, \quad \text{and we have} \quad \frac{1}{n+1} < \frac{1}{1000}$$

provided $n + 1 > 1000$—that is, $n > 999$. If it is true that the nth term in a sequence differs from the number L by as little as we please for all sufficiently large n, we say that **the sequence approaches the number L as a limit**. The symbolism

$$\lim_{n \to \infty} s_n = L$$

(read "the limit, as n increases without bound, of s_n is L") is used to denote this situation. A thorough discussion of the notion of a limit is included in courses in calculus, and will not be attempted here. A few elementary ideas, however, are in order.

A sequence in which the nth term approaches a number L as $n \to \infty$ is said to be a **convergent sequence**, and the sequence is said to **converge** to L. It is not necessary for convergence that a sequence be strictly increasing. For example,

$$1, \frac{1}{2}, \frac{1}{3}, \frac{1}{4}, \dots, \frac{1}{n}, \dots$$

converges to 0, but each term in the sequence is less than, instead of greater than, the term that precedes it. Again, the sequence

$$-1, \frac{1}{2}, -\frac{1}{3}, \frac{1}{4}, \ldots, \frac{(-1)^n}{n}, \ldots$$

converges to 0 but is neither increasing nor decreasing. We can rephrase the definition of convergence of a sequence as follows:

Definition 12.5   *A sequence $s_1, s_2, \ldots, s_n, \ldots$ **converges** to the number L,*

$$\lim_{n \to \infty} s_n = L,$$

if and only if the absolute value of the difference between the nth term in the sequence and the number L is as small as we please for all sufficiently large n. Thus the sequence converges to the number L if and only if

$$\lim_{n \to \infty} |L - s_n| = 0.$$

For example, the **alternating** (because the signs alternate) **sequence**

$$\frac{-2}{3}, \frac{4}{9}, \frac{-8}{27}, \ldots, \left(\frac{-2}{3}\right)^n, \ldots$$

converges to 0 since the absolute value of the difference between $(-2/3)^n$ and 0, that is, $|0 - (-2/3)^n|$, is as small as we please for n large enough. We express this by writing

$$\lim_{n \to \infty} \left(\frac{-2}{3}\right)^n = 0.$$

On the other hand, the alternating sequence

$$\frac{1}{2}, -\frac{2}{3}, \frac{3}{4}, -\frac{4}{5}, \ldots, (-1)^{n+1} \frac{n}{n+1}, \ldots$$

does not converge. As n increases, the nth term oscillates back and forth from the neighborhood of $+1$ to the neighborhood of -1, and we cannot find a number L such that $\lim_{n \to \infty} |L - s_n| = 0$. Such a sequence is said to **diverge**. A sequence such as

$$1, 2, 3, \ldots, n, \ldots$$

also is said to diverge. An answer to the logical question of what we mean by "enough" when we say "n large enough" requires a more precise definition of limit than we have given here. As remarked earlier, a course in the calculus will treat this in detail.

Sequence of
partial sums
of a series

For an infinite series,

$$S_\infty = \sum_{j=1}^{\infty} s_j,$$

we can consider the infinite sequence of **partial sums**:

$$S_1 = s_1$$
$$S_2 = s_1 + s_2$$

. . . . .

$$S_n = s_1 + s_2 + \cdots + s_n$$

. . . .

Definition 12.6   *An infinite series*

$$S_\infty = \sum_{j=1}^{\infty} s_j$$

converges if and only if $S_1, S_2, \ldots, S_n, \ldots$, the corresponding sequence of partial sums, converges.

If the sequence of partial sums converges to the number L,

$$\lim_{n \to \infty} S_n = L,$$

then L is said to be the **sum** of the infinite series, and we write

$$S_\infty = \sum_{j=1}^{\infty} s_j = L.$$

If the sequence of partial sums diverges, then the series is said to **diverge**.

Sum of an infinite geo-metric pro-gression

We recall from Section 12.3 that the sum of n terms (the nth partial sum) of a geometric progression is given, for $r \neq 1$, by

$$S_n = \frac{a - ar^n}{1 - r}. \tag{2}$$

If $|r| < 1$, that is, if $-1 < r < 1$, then $|r|^n$ becomes smaller and smaller for increasingly large n. For example, if $r = 1/2$, then

$$r^2 = \frac{1}{4}, \quad r^3 = \frac{1}{8}, \quad r^4 = \frac{1}{16},$$

and so forth; and $(1/2)^n$ is as small as we please if n is sufficiently large. Writing (2) as

$$S_n = \frac{a}{1 - r}(1 - r^n), \tag{3}$$

we see that the value of the factor $(1 - r^n)$ is as close as we please to 1 provided $|r| < 1$ and n is taken large enough. Since this argument shows that the sequence of partial sums (3) converges to

$$\frac{a}{1 - r},$$

we have the following result.

Theorem 12.4   *The sum of an infinite geometric progression,*

$$a + ar + ar^2 + \cdots + ar^n + \cdots,$$

with $|r| < 1$, is

$$S_\infty = \lim_{n \to \infty} S_n = \frac{a}{1 - r}.$$

An interesting application of this sum arises in connection with repeating decimals—that is, decimal numerals that, after a finite number of decimal places, have endlessly repeating groups of digits. For example,

$$0.2121\overline{21},$$

$$0.138512512\overline{512}$$

are repeating decimals. The bar denotes that the numerals appearing under it are repeated endlessly. Consider the problem of expressing such a decimal fraction as an arithmetic fraction. We illustrate the process involved with the first example above. The decimal $0.2121\overline{21}$ can be written as

$$0.21 + 0.0021 + 0.000021 + \cdots, \tag{4}$$

which is a geometric progression with ratio $r = 0.01$. Since the ratio is less than 1 in absolute value, we can use Theorem 12.4 to find the sum of the infinite series (4). Thus

$$S_\infty = \frac{a}{1 - r} = \frac{0.21}{1 - 0.01} = \frac{21}{99} = \frac{7}{33},$$

and the given decimal fraction is equivalent to 7/33.

Exercise 12.4

Discuss the limiting behavior of each expression as $n \to \infty$.

Example $\dfrac{n^2 + 3}{n^2}$

Solution   By writing $\dfrac{n^2 + 3}{n^2}$ as $\dfrac{n^2}{n^2} + \dfrac{3}{n^2}$ and then as $1 + \dfrac{3}{n^2}$, we observe that

$$\lim_{n \to \infty} \frac{n^2 + 3}{n^2} = \lim_{n \to \infty} \left(1 + \frac{3}{n^2}\right) = 1 + 0 = 1.$$

1. $\dfrac{1}{n}$ 2. $1 + \dfrac{1}{n^2}$ 3. $\dfrac{n + 1}{n}$ 4. $\dfrac{n + 3}{n^2}$

5. $2n$ 6. $(-1)^n$ 7. $\dfrac{1}{2^n}$ 8. $(-1)^n \dfrac{1}{n}$

State which of the following sequences are convergent.

Example $1, \dfrac{3}{2}, \dfrac{7}{4}, \dfrac{15}{8}, \ldots, \dfrac{2^n - 1}{2^{n-1}}, \ldots$

Solution   Writing the general term as $\dfrac{2^n}{2^{n-1}} - \dfrac{1}{2^{n-1}}$, or $2 - \dfrac{1}{2^{n-1}}$, we observe that

$$\lim_{n \to \infty} \frac{2^n - 1}{2^{n-1}} = \lim_{n \to \infty} \left(2 - \frac{1}{2^{n-1}} \right) = 2 - 0 = 2.$$

The sequence is convergent.

9. $\dfrac{1}{2}, \dfrac{1}{4}, \dfrac{1}{8}, \dfrac{1}{16}, \ldots, \dfrac{1}{2^n}, \ldots$ **10.** $2, \dfrac{3}{2}, \dfrac{4}{3}, \dfrac{5}{4}, \ldots, \dfrac{n+1}{n}, \ldots$

11. $1, 2, 3, 4, 5, \ldots, n, \ldots$ **12.** $2, 4, 6, 8, \ldots, 2n, \ldots$

13. $1, -\dfrac{1}{2}, \dfrac{1}{4}, -\dfrac{1}{8}, \ldots, (-1)^{n-1}\dfrac{1}{2^{n-1}}, \ldots$

14. $1, -1, 1, -1, \ldots, (-1)^{n+1}, \ldots$

Find the sum of each of the following infinite geometric series. If the series has no sum, so state.

Examples   a. $3 + 2 + \cdots$ b. $\dfrac{1}{81} - \dfrac{1}{54} + \cdots$

Solutions   a. $r = \dfrac{2}{3}$; series has a sum since $|r| < 1$. Using Theorem 12.4, we have

$$S_\infty = \frac{a}{1 - r} = \frac{3}{1 - \dfrac{2}{3}} = 9.$$

b. $r = -\dfrac{1}{54} \div \dfrac{1}{81} = -\dfrac{3}{2}$; series does not have a sum since $|r| > 1$.

15. $12 + 6 + \cdots$ **16.** $2 + 1 + \cdots$ **17.** $\dfrac{1}{36} + \dfrac{1}{30} + \cdots$

18. $\dfrac{1}{16} - \dfrac{1}{8} + \cdots$ **19.** $\displaystyle\sum_{j=1}^{\infty} \left(\dfrac{2}{3}\right)^j$ **20.** $\displaystyle\sum_{j=1}^{\infty} \left(-\dfrac{1}{4}\right)^j$

Find an arithmetic fraction equal to each of the given decimal numerals.

Example $0.818\overline{181}$

Solution   Rewrite as a series: $0.81 + 0.0081 + 0.000081 + \cdots$.

Find the common ratio: $r = 0.01$. Use $S_\infty = \dfrac{a}{1 - r}$.

$$S_\infty = \frac{0.81}{1 - 0.01} = \frac{81}{99} = \frac{9}{11}.$$

21. $0.31\overline{3131}$ **22.** $0.454\overline{545}$ **23.** $2.410\overline{410}$

24. $3.027\overline{027}$ **25.** $0.12\overline{8888}$ **26.** $0.8\overline{3333}$

27. A force is applied to a particle moving in a straight line in such a fashion that each second it moves only one half of the distance it moved the preceding second. If the particle moves ten centimeters the first second, approximately how far will it move before coming to rest?

28. The arc length through which the bob on a pendulum moves is nine-tenths of its preceding arc length. Approximately how far will the bob move before coming to rest if the first arc length is 12 inches?

12.5   *The Binomial Theorem*

Factorial Notation

There are situations, as in the binomial expansion on page 305, in which it is necessary to write the product of several consecutive positive integers. To facilitate writing products of this type, we use a special symbol $n!$ (read "n factorial" or "factorial n"), which is defined by

$$n! = n(n - 1)(n - 2) \ldots (3)(2)(1).$$

Thus

$$5! = 5 \cdot 4 \cdot 3 \cdot 2 \cdot 1 \quad \text{(read ``five factorial''),}$$

and

$$8! = 8 \cdot 7 \cdot 6 \cdot 5 \cdot 4 \cdot 3 \cdot 2 \cdot 1 \quad \text{(read ``eight factorial'').}$$

Factorial notation can also be used to represent products of consecutive positive integers, beginning with integers different from 1. For example,

$$8 \cdot 7 \cdot 6 \cdot 5 = \frac{8!}{4!},$$

because

$$\frac{8!}{4!} = \frac{8 \cdot 7 \cdot 6 \cdot 5 \cdot 4 \cdot 3 \cdot 2 \cdot 1}{4 \cdot 3 \cdot 2 \cdot 1} = 8 \cdot 7 \cdot 6 \cdot 5.$$

Since

$$n! = n(n - 1)(n - 2)(n - 3) \cdots 5 \cdot 4 \cdot 3 \cdot 2 \cdot 1$$

and

$$(n - 1)! = (n - 1)(n - 2)(n - 3) \cdots 5 \cdot 4 \cdot 3 \cdot 2 \cdot 1,$$

for $n > 1$ we can write the recursive relationship

$$n! = n(n - 1)!. \tag{1}$$

For example,

$$7! = 7 \cdot 6!,$$

$$27! = 27 \cdot 26!,$$

$$(n + 2)! = (n + 2)(n + 1)!.$$

If (1) is to hold also for $n = 1$, then we must have

$$1! = 1 \cdot (1 - 1)!$$

or

$$1! = 1 \cdot 0!.$$

Therefore, for consistency, we define 0! by

$$\mathbf{0! = 1.}$$

A special case of the use of factorial notation occurs in the form

$$\binom{n}{r} = \frac{n!}{r!(n - r)!}.$$

The symbol $\binom{n}{r}$ is read "the number of combinations of n things taken r at a time"; it denotes the number of r-element subsets of a set containing n members. Combinations will be discussed in Section 13.2. As examples, we have

$$\binom{5}{3} = \frac{5!}{3!(5 - 3)!} = \frac{5!}{3!2!} = \frac{5 \cdot 4 \cdot 3!}{3!2 \cdot 1} = 10,$$

$$\binom{5}{1} = \frac{5!}{1!(5 - 1)!} = \frac{5!}{4!} = \frac{5 \cdot 4!}{4!} = 5,$$

and

$$\binom{5}{0} = \frac{5!}{0!(5 - 0)!} = \frac{5!}{5!} = 1.$$

Binomial expansions

The series obtained by expanding a binomial of the form

$$(a + b)^n$$

is particularly useful in certain branches of mathematics. Starting with familiar examples, where n takes the values 1, 2, 3, 4, and 5 in turn, we can show by direct multiplication that

$$(a + b)^1 = a + b$$

$$(a + b)^2 = a^2 + 2ab + b^2$$

$$(a + b)^3 = a^3 + 3a^2b + 3ab^2 + b^3$$

$$(a + b)^4 = a^4 + 4a^3b + 6a^2b^2 + 4ab^3 + b^4$$

$$(a + b)^5 = a^5 + 5a^4b + 10a^3b^2 + 10a^2b^3 + 5ab^4 + b^5.$$

We observe that in each case:

1. The first term is a^n.

2. The variable factors of the second term are $a^{n-1}b^1$, and the coefficient is n, which can be written in the form

$$\frac{n}{1!}.$$

3. The variable factors of the third term are $a^{n-2}b^2$, and the coefficient can be written in the form

$$\frac{n(n-1)}{2!}.$$

4. The variable factors of the fourth term are $a^{n-3}b^3$, and the coefficient can be written in the form

$$\frac{n(n-1)(n-2)}{3!}.$$

The foregoing expansions suggest the following result, known as the **binomial theorem**. Since its proof, which requires the use of mathematical induction, is quite lengthy, it is omitted.

Theorem 12.5   *For each natural number n,*

$$(a+b)^n = a^n + \frac{n}{1!}a^{n-1}b + \frac{n(n-1)}{2!}a^{n-2}b^2 + \frac{n(n-1)(n-2)}{3!}a^{n-3}b^3$$

$$+ \cdots + \frac{n(n-1)(n-2)\cdots(n-r+2)}{(r-1)!}a^{n-r+1}b^{r-1} + \cdots + b^n, \qquad (2)$$

where r is the number of the term.

For example,

$$(x-2)^4 = x^4 + \frac{4}{1!}x^3(-2)^1 + \frac{4\cdot 3}{2!}x^2(-2)^2 + \frac{4\cdot 3\cdot 2}{3!}x(-2)^3 + \frac{4\cdot 3\cdot 2\cdot 1}{4!}(-2)^4$$

$$= x^4 - 8x^3 + 24x^2 - 32x + 16.$$

In this case, $a = x$ and $b = -2$ in the binomial expansion.

Observe that the coefficients of the terms in the binomial expansion (2) can be represented as follows.

1st term: $\dbinom{n}{0} = \dfrac{n!}{0!n!} = 1,$

2nd term: $\dbinom{n}{1} = \dfrac{n!}{1!(n-1)!} = \dfrac{n\cdot(n-1)!}{1\cdot(n-1)!} = n,$

3rd term: $\dbinom{n}{2} = \dfrac{n!}{2!(n-2)!} = \dfrac{n(n-1)(n-2)!}{2!(n-2)!} = \dfrac{n(n-1)}{2!},$

rth term: $\dbinom{n}{r-1} = \dfrac{n!}{(r-1)!(n-r+1)!}$

$$= \frac{n(n-1)(n-2)\cdots(n-r+2)(n-r+1)!}{(r-1)!(n-r+1)!}$$

$$= \frac{n(n-1)(n-2)\cdots(n-r+2)}{(r-1)!}.$$

Hence, the binomial expansion (2) can be represented by the expression

$$(a+b)^n = \binom{n}{0}a^n + \binom{n}{1}a^{n-1}b + \binom{n}{2}a^{n-2}b^2 + \binom{n}{3}a^{n-3}b^3 + \cdots$$

$$+ \binom{n}{r-1}a^{n-r+1}b^{r-1} + \cdots + \binom{n}{n}b^n. \quad (3)$$

For example,

$$(x-2)^4 = \binom{4}{0}x^4 + \binom{4}{1}x^3(-2)^1 + \binom{4}{2}x^2(-2)^2 + \binom{4}{3}x(-2)^3 + \binom{4}{4}(-2)^4.$$

Using the formula for $\binom{n}{r}$ to find the coefficients, we obtain the same result as above,

$$(x-2)^4 = x^4 - 8x^3 + 24x^2 - 32x + 16.$$

rth term in
a binomial
expansion

Note that the rth term in a binomial expansion is given by

$$\binom{n}{r-1}a^{n-r+1}b^{r-1} = \frac{n!}{(r-1)!(n-r+1)!}\,a^{n-r+1}b^{r-1}$$

$$= \frac{n(n-1)(n-2)\cdots(n-r+2)}{(r-1)!}\,a^{n-r+1}b^{r-1}. \quad (4)$$

For example, by the left-hand member of (4), the seventh term of $(x-2)^{10}$ is

$$\binom{10}{6}x^4(-2)^6 = \frac{10!}{6!4!}\,x^4(-2)^6 = \frac{10\cdot9\cdot8\cdot7\cdot6!}{6!\cdot4\cdot3\cdot2\cdot1}\,x^4(64)$$

$$= 13440\,x^4,$$

while, by the right-hand member of (4), we have

$$\frac{10\cdot9\cdot8\cdot7\cdot6\cdot5}{6\cdot5\cdot4\cdot3\cdot2\cdot1}\,x^4(64) = 13440\,x^4.$$

Exercise 12.5

1. Write $(2n)!$ in expanded form for $n = 4$.
2. Write $2n!$ in expanded form for $n = 4$.
3. Write $n(n-1)!$ in expanded form for $n = 6$.
4. Write $2n(2n-1)!$ in expanded form for $n = 2$.

Write in expanded form and simplify.

Examples

a. $\dfrac{7!}{4!}$ b. $\dfrac{4!6!}{8!}$

Solutions

a. $\dfrac{7\cdot6\cdot5\cdot4!}{4!}$ b. $\dfrac{4\cdot3\cdot2\cdot1\cdot6!}{8\cdot7\cdot6!}$

   210 $\dfrac{3}{7}$

5. $5!$ **6.** $7!$ **7.** $\dfrac{9!}{7!}$ **8.** $\dfrac{12!}{11!}$

9. $\dfrac{5!\,7!}{8!}$ **10.** $\dfrac{12!\,8!}{16!}$ **11.** $\dfrac{8!}{2!\,(8-2)!}$ **12.** $\dfrac{10!}{4!\,(10-4)!}$

Write each product in factorial notation.

Examples   a. $1\cdot 2\cdot 3\cdot 4\cdot 5\cdot 6$ b. $11\cdot 12\cdot 13\cdot 14$ c. 150

Solutions   a. $6!$ b. $\dfrac{14!}{10!}$ c. $\dfrac{150!}{149!}$

13. $1\cdot 2\cdot 3$ **14.** $1\cdot 2\cdot 3\cdot 4\cdot 5$ **15.** $3\cdot 4\cdot 5\cdot 6$

16. 7 **17.** $8\cdot 7\cdot 6$ **18.** $28\cdot 27\cdot 26\cdot 25\cdot 24$

Write each expression in factorial notation and simplify.

Examples   a. $\dbinom{6}{2}$ b. $\dbinom{4}{4}$

Solutions   a. $\dbinom{6}{2} = \dfrac{6!}{2!\,(6-2)!} = \dfrac{6!}{2!\,4!}$ b. $\dbinom{4}{4} = \dfrac{4!}{4!\,(4-4)!}$

$$= \dfrac{6\cdot 5\cdot 4!}{2\cdot 1\cdot 4!} = 15 \qquad\qquad\qquad = \dfrac{4!}{4!\,0!} = 1$$

19. $\dbinom{6}{5}$ **20.** $\dbinom{4}{2}$ **21.** $\dbinom{3}{3}$ **22.** $\dbinom{5}{5}$

23. $\dbinom{7}{0}$ **24.** $\dbinom{2}{0}$ **25.** $\dbinom{5}{2}$ **26.** $\dbinom{5}{3}$

Write each expression in factored form and show the first three factors and the last three factors.

Example $(2n+1)!$

Solution $(2n+1)(2n)(2n-1)\cdot \,\cdots\, \cdot 3\cdot 2\cdot 1$

27. $n!$ **28.** $(n+4)!$ **29.** $(3n)!$

30. $3n!$ **31.** $(n-2)!$ **32.** $(3n-2)!$

Simplify each expression.

Examples   a. $\dfrac{(n-1)!}{(n-3)!}$ b. $\dfrac{(n-1)!\,(2n)!}{2n!\,(2n-2)!}$

(Solutions overleaf)

Solutions   a. $\dfrac{(n-1)(n-2)(n-3)!}{(n-3)!}$ b. $\dfrac{(n-1)!\,(2n)(2n-1)(2n-2)!}{2(n)(n-1)!\,(2n-2)!}$

$(n-1)(n-2)$ $2n-1$

33. $\dfrac{(n+2)!}{n!}$ **34.** $\dfrac{(n+2)!}{(n-1)!}$ **35.** $\dfrac{(n+1)(n+2)!}{(n+3)!}$

36. $\dfrac{(2n+4)!}{(2n+2)!}$ **37.** $\dfrac{(2n)!\,(n-2)!}{4(2n-2)!\,(n)!}$ **38.** $\dfrac{(2n+1)!\,(2n-1)!}{[(2n)!]^2}$

Expand.

Example $(a-3b)^4$

Solution   From the binomial expansion (2) on page 305,

$$(a-3b)^4 = a^4 + \frac{4}{1!}a^3(-3b) + \frac{4\cdot 3}{2!}a^2(-3b)^2 + \frac{4\cdot 3\cdot 2}{3!}a(-3b)^3$$

$$+ \frac{4\cdot 3\cdot 2\cdot 1}{4!}(-3b)^4$$

$$= a^4 - 12a^3b + 54a^2b^2 - 108ab^3 + 81b^4.$$

Alternatively, from the abbreviation of the binomial expansion (3) on page 306, we have

$$(a-3b)^4 = \binom{4}{0}a^4 + \binom{4}{1}a^3(-3b) + \binom{4}{2}a^2(-3b)^2 + \binom{4}{3}a(-3b)^3 + \binom{4}{4}(-3b)^4,$$

which also simplifies to the expression obtained above.

39. $(x+3)^5$ **40.** $(2x+y)^4$ **41.** $(x-3)^4$ **42.** $(2x-1)^5$

43. $\left(2x-\dfrac{y}{2}\right)^3$ **44.** $\left(\dfrac{x}{3}+3\right)^5$ **45.** $\left(\dfrac{x}{2}+2\right)^6$ **46.** $\left(\dfrac{2}{3}-a^2\right)^4$

Write the first four terms in each expansion. Do not simplify the terms.

Example $(x+2y)^{15}$

Solution   From the binomial expansion (2),

$$(x+2y)^{15} = x^{15} + \frac{15}{1!}x^{14}(2y) + \frac{15\cdot 14}{2!}x^{13}(2y)^2 + \frac{15\cdot 14\cdot 13}{3!}x^{12}(2y)^3$$

or alternatively from the form (3),

$$(x+2y)^{15} = \binom{15}{0}x^{15} + \binom{15}{1}x^{14}(2y) + \binom{15}{2}x^{13}(2y)^2 + \binom{15}{3}x^{12}(2y)^3.$$

47. $(x + y)^{20}$ **48.** $(x - y)^{15}$ **49.** $(a - 2b)^{12}$

50. $(2a - b)^{12}$ **51.** $(x - \sqrt{2})^{10}$ **52.** $\left(\dfrac{x}{2} + 2\right)^8$

Find each power to the nearest hundredth.

Example $(0.97)^7$

Solution   Either form of the binomial expansion (2) or (3) can be used. We shall use (2).

$$(0.97)^7 = (1 - 0.03)^7$$

$$= 1^7 + \frac{7}{1!}(1)^6(-0.03)^1 + \frac{7 \cdot 6}{2!}(1)^5(-0.03)^2 + \frac{7 \cdot 6 \cdot 5}{3!}(1)^4(-0.03)^3 + \cdots$$

$$= 1 - 0.21 + 0.0189 - 0.000945 + \cdots$$

$$= 0.807955^+$$

Hence, to the nearest hundredth, $(0.97)^7 = 0.81$.

53. $(1.02)^{10}$ *Hint:* $1.02 = (1 + 0.02)$
54. $(1.01)^{15}$
55. If an amount of money (P) is invested at 4% compounded annually, the amount (A) present at the end of (n) years is given by $A = P(1 + 0.04)^n$. Find the amount A (to the nearest dollar) if \$1000 was invested for 10 years.
56. In Problem 55, find the amount present at the end of 20 years.

Find each specified term.

Example $(x - 2y)^{12}$, the seventh term.

Solution   In Formula (4), page 306, use $n = 12$ and $r = 6$.

$$\frac{12!}{6!\,(12 - 6)!}\,x^6(-2y)^6 = \frac{12 \cdot 11 \cdot 10 \cdot 9 \cdot 8 \cdot 7}{6!}\,x^6(-2y)^6 = 59{,}136x^6y^6$$

57. $(a - b)^{15}$, the sixth term   **58.** $(x + 2)^{12}$, the fifth term
59. $(x - 2y)^{10}$, the fifth term   **60.** $(a^3 - b)^9$, the seventh term

61. Given that the binomial formula holds for $(1 + x)^n$ where n is a negative integer:
  a. Write the first four terms of $(1 + x)^{-1}$.
  b. Find the first four terms of the quotient $1/(1 + x)$, by dividing 1 by $(1 + x)$. Compare the results of (a) and (b).

62. Given that the binomial formula holds as an infinite "sum" for $(1 + x)^n$, where n is a rational number and $|x| < 1$, find to two decimal places.
  a. $\sqrt{1.02}$ b. $\sqrt{0.99}$

Chapter Review

[12.1]   *By mathematical induction, prove each formula for all positive integral values of n.*

   1. $3 + 6 + 9 + \cdots + 3n = \dfrac{3n(n + 1)}{2}$

   2. $\dfrac{1}{2} + \dfrac{1}{4} + \dfrac{1}{8} + \cdots + \dfrac{1}{2^n} = 1 - \dfrac{1}{2^n}$

[12.2]   *Write the next three terms of each of the following arithmetic progressions.*

   3. $7, 10, \ldots$ **4.** $a, a - 2, \ldots$

   Write the next three terms in each of the following geometric progressions.

   5. $-2, 6, \ldots$ **6.** $\dfrac{2}{3}, 1, \ldots$

   7.   Find the general term and the seventh term of the arithmetic progression $-3, 2, \ldots$.

   8.   Find the general term and the fifth term of the geometric progression $-2, \dfrac{2}{3}, \ldots$.

   9.   If the fourth term of an arithmetic progression is 13 and the ninth term is 33, find the seventh term.

   10.   Which term in a geometric progression $-\dfrac{2}{9}, \dfrac{2}{3}, \ldots$ is 54?

[12.3]   **11.**   Write $\sum\limits_{k=2}^{5} k(k - 1)$ in expanded form.

   12.   Write $x^2 + x^3 + x^4 + \cdots$ in sigma notation.

   13.   Find the value for $\sum\limits_{j=3}^{9} (3j - 1)$.

   14.   Find the value for $\sum\limits_{j=1}^{5} \left(\dfrac{1}{3}\right)^j$.

[12.4]   **15.**   Specify the limit of $\dfrac{3n^2 - 1}{n^2}$ as $n \to \infty$.

   16.   Find the value of the geometric series $4 - 2 + 1 - \dfrac{1}{2} + \cdots$

   17.   Find the value of $\sum\limits_{i=1}^{\infty} \left(\dfrac{1}{3}\right)^i$.

   18.   Find a fraction equivalent to $0.44\overline{4}$.

[12.5]   **19.**   Write $n(n - 3)!$ in expanded form for $n = 5$.

   Write each of the following expressions in expanded form and simplify.

   20. $\dfrac{8! \, 3!}{7!}$ **21.** $\dbinom{7}{2}$ **22.** $\dfrac{(n - 1)!}{n!(n + 1)!}$

   23.   Write the first four terms of the binomial expansion of $(x - 2y)^{10}$.

   24.   Find the eighth term in the expansion of $(x - 2y)^{10}$.

13 *Probability*

13.1 *Basic Counting Principles; Permutations*

Associated with each finite set A is a nonnegative integer n, namely the number of elements in A. Hence, we have a function from the set of all finite sets to the set of nonnegative integers. The symbolism $n(A)$ is used to denote elements in the range of this set function n. For example, if

$$A = \{5, 7, 9\}, \quad B = \{1/2, 0, 3, -5, 7\}, \quad C = \emptyset,$$

then

$$n(A) = 3, \quad n(B) = 5, \quad \text{and} \quad n(C) = 0.$$

Counting properties

All the sets with which we shall hereafter be concerned are assumed to be finite sets. We then have the following properties, called **counting properties**, for the function n.

   I $n(A \cup B) = n(A) + n(B), \quad \text{if} \quad A \cap B = \emptyset.$

Thus, if A and B are disjoint sets, then the number of elements in their union is the sum of the number of elements in A and the number of elements in B.

For example, suppose there are five roads from town R to town S, and two railroads from town R to town S. If A is the set of roads and B the set of railroads from R to S, then $n(A) = 5$, $n(B) = 2$, and $n(A \cup B) = 5 + 2 = 7$; thus there are seven ways one can go from town R to town S by driving or riding on a train.

   II $n(A \cup B) = n(A) + n(B) - n(A \cap B), \quad \text{if} \quad A \cap B \neq \emptyset.$

That is, if A and B overlap, then to count the number of elements in $A \cup B$, we might add the number of elements in A to the number of elements in B. But since any elements in the intersection of A and B are counted twice in this process (once in A and once in B), we must subtract the number of such elements from the sum $n(A) + n(B)$ to obtain the number of elements in $A \cup B$, as suggested in Figure 13.1.

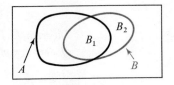

Figure 13.1

311

For example, suppose there are fifteen unrelated girls and seventeen unrelated boys in a mathematics class, and suppose that there are precisely two brother-sister pairs in the class. If A denotes the set of different families represented by the girls and B denotes the set of different families represented by the boys, then the number of different families represented by all of the members of the class is

$$n(A) + n(B) - n(A \cap B) = 15 + 17 - 2 = 30.$$

Actually, Property II is a consequence of Property I.

III $n(A \times B) = n(A) \cdot n(B).$

This asserts that the number of elements in the Cartesian product of sets A and B is the product of the number of elements in A and the number of elements in B.

For example, suppose that there are five roads from town R to town S (set A), and further suppose that there are three roads from town S to town T (set B). Then for each element of A there are three elements of B, and the total possible ways one can drive from R to T via S is

$$n(A \times B) = n(A) \cdot n(B) = 5 \cdot 3 = 15.$$

Permutations   Given the set of digits $A = \{1, 2, 3\}$, how many different three-digit numerals can be constructed from the members of A if no member is used more than once? The answer to this question can be obtained by simply listing the different three-digit numerals, 1 2 3, 1 3 2, 2 1 3, 2 3 1, 3 1 2, 3 2 1, and counting them. Such a procedure would be quite impracticable, however, if the number of members of the given set of digits were very large. Another way to arrive at the same conclusion is by applying the third counting property. If we let A denote the set of possible first digits in the foregoing numerals, then $n(A) = 3$. Since no numeral may be used more than once, and since one numeral has already been used for a first digit, there remain but two possibilities for the second digit. If B denotes the set of possible second digits after the first digit has been chosen, then $n(B) = 2$. By similar reasoning, if C is the set of possible third digits after the first two have been chosen, then $n(C) = 1$. By the third counting property (applied twice), we find that

$$n(A \times B \times C) = [n(A) \cdot n(B)] \cdot n(C) = 3 \cdot 2 \cdot 1 = 6.$$

Each of the three-digit numerals discussed above is called a *permutation* of the elements of the set of numerals $\{1, 2, 3\}$.

Definition 13.1   A *permutation* of a set A is an ordering (*first, second, etc.*) of the members of A.

With Definition 13.1, we can state the following result.

Theorem 13.1   Let $P_{n,n}$ denote the number of distinct permutations of the members of a set A containing n members. Then

$$P_{n,n} = n! \tag{1}$$

The symbol $P_{n,n}$ (or sometimes $_nP_n$ or P_n^n) is read "the number of permutations of n things taken n at a time."

Proof Let A_1 denote the set of possible selections for the first member. Then $A_1 = A$, and $n(A_1) = n(A) = n$. Having made a first selection, let A_2 denote the set of possible second selections. Then $A_2 \subset A$, and $n(A_2) = n(A_1) - 1 = n - 1$. A continuation of this procedure, together with successive application of the third counting principle, leads to

$$P_{n,n} = n(A_1) \cdot n(A_2) \cdots \cdot n(A_n) = n \cdot (n - 1) \cdot (n - 2) \cdots \cdot 1 = n!,$$

as was to be proved.

Example

In how many ways can nine men be assigned positions to form distinct baseball teams?

Solution

Let A denote the set of men, so that $n(A) = 9$. The total number of ways in which 9 men can be assigned 9 positions on a team, or, in other words, the number of possible permutations of the members of a 9-element set, is, by Equation (1),

$$P_{9,9} = 9! = 9 \cdot 8 \cdot 7 \cdots 1 = 362,880.$$

Theorem 13.2   *Let $P_{n,r}$ denote the number of permutations of the members, taken r at a time, of a set A containing n members; that is, let $P_{n,r}$ be the number of distinct orderings of r elements when there is a set A of n elements from which to choose. Then*

$$P_{n,r} = n(n - 1)(n - 2) \cdots (n - r + 1). \tag{2}$$

The proof follows the proof of Theorem 13.1, except that the last subset considered is such that $n(A_r) = n - (r - 1) = n - r + 1$.

Example

In how many ways can a basketball team be formed by choosing players for the five positions from a set of ten players?

Solution

Let A denote the set of players, so that $n(A) = 10$. Then from (2) and the fact that a basketball team consists of 5 players, we have

$$P_{10,5} = 10 \cdot 9 \cdot 8 \cdots (10 - 5 + 1) = 10 \cdot 9 \cdot 8 \cdot 7 \cdot 6 = 30,240.$$

An alternative expression for $P_{n,r}$ can be obtained by observing that

$$P_{n,r} = n(n - 1)(n - 2) \cdots (n - r + 1)$$
$$= \frac{n(n - 1)(n - 2) \cdots (n - r + 1)(n - r)!}{(n - r)!},$$

so that

$$P_{n,r} = \frac{n!}{(n - r)!}. \tag{3}$$

Distinguish-
able permu-
tations

The problem of finding the number of distinguishable permutations of n objects taken n at a time, if some of the objects are identical, requires a little more careful analysis. As an example, consider the number of permutations of the letters of the word *DIVISIBLE*. We can make a distinction between the three *I*'s by assigning subscripts to each so that we have nine distinct letters,

$$D, I_1, V, I_2, S, I_3, B, L, E.$$

The number of permutations of these nine letters is of course 9!. If the letters other than I_1, I_2, and I_3 are retained in the positions they occupy in a permutation of the above nine letters, I_1, I_2, and I_3 can be permuted among themselves 3! ways. Thus, if P is the number of *distinguishable* permutations of the letters

$$D, I, V, I, S, I, B, L, E,$$

then, since for each of these there are 3! ways in which the *I*'s can be permuted without otherwise changing the order of the other letters, it follows that

$$3! \cdot P = 9!,$$

from which

$$P = \frac{9!}{3!}.$$

As another example, consider the letters of the word *MISSISSIPPI*. There would exist 11! distinguishable permutations of the letters in this word if each letter were distinct. Note, however, that the letters *S* and *I* each appear four times and the letter *P* appears twice. Reasoning as we did in the previous example, we see that the number P of distinguishable permutations of the letters in *MISSISSIPPI* is given by

$$4! \, 4! \, 2! \, P = 11!,$$

from which

$$P = \frac{11!}{4! \, 4! \, 2!}.$$

Exercise 13.1

Given the following sets, find $n(A \cap B)$, $n(A \cup B)$, *and* $n(A \times B)$.

Example $A = \{a, b, c\}, \quad B = \{c, d\}$

Solution

$A \cap B = \{c\}$. Therefore $n(A \cap B) = 1$.

$n(A \cup B) = n(A) + n(B) - n(A \cap B) = 3 + 2 - 1 = 4$.

$n(A \times B) = n(A) \cdot n(B) = 3 \cdot 2 = 6$.

1. $A = \{d, e\}, \quad B = \{e, f, g, h\}$
2. $A = \{e\}, \quad B = \{a, b, c, d\}$
3. $A = \{1, 2, 3, 4\}, \quad B = \{3, 4, 5, 6\}$
4. $A = \{1, 2\}, \quad B = \{3, 4, 5\}$
5. $A = \{1, 2\}, \quad B = \{1, 2\}$
6. $A = \emptyset, \quad B = \{2, 3, 4\}$

Example   In how many ways can three members of a class be assigned a grade of A, B, C, or D?

Solution

Sometimes a simple diagram, such as —— , —— , —— , designating a sequence, is a helpful preliminary device. Since each of the students may receive any one of four different grades, the sequence would appear as

$$\underline{4}, \underline{4}, \underline{4}.$$

From counting property III, there are $4 \cdot 4 \cdot 4$, or 64, possible ways the grades may be assigned.

Example

In how many different ways can three members of a class be assigned a grade of A, B, C, or D so that no two members receive the same grade?

Solution

Since the first student may receive any one of four different grades, the second student may then receive any one of three different grades, and the third student may then receive any one of two different grades, the sequence would appear as

$$\underline{4}, \underline{3}, \underline{2}.$$

From counting property III, there are $4 \cdot 3 \cdot 2$, or 24, possible ways the grades may be assigned. In this case, we could have obtained the same result directly from Theorem 13.2, since $P_{4,3} = 4 \cdot 3 \cdot 2 = 24$.

In each problem, a digit or letter may be used more than once unless stated otherwise.

7. How many different two-digit numerals can be formed from the digits 5 and 6?

8. How many different two-digit numerals can be formed from the digits 7, 8, and 9?

9. In how many different ways can four students be seated in a row?

10. In how many different ways can five students be seated in a row?

11. In how many different ways can four questions on a true-false test be answered?

12. In how many different ways can five questions on a true-false test be answered?

13. In how many ways can you write different three-digit numerals from {2, 3, 4, 5}?

14. How many different seven-digit telephone numbers can be formed from the set of digits {1, 2, 3, 4, 5, 6, 7, 8, 9, 0}?

15. In how many ways can you write different three-digit numerals, using {2, 3, 4, 5}, if no digit is to be used more than once in each numeral?

16. How many different seven-digit telephone numbers can be formed from the set of digits {1, 2, 3, 4, 5, 6, 7, 8, 9, 0} if no digit is to be used more than once in any number?

17. How many three-letter arrangements can be formed from {A, N, S, W, E, R}?

18. If no letter is to be used more than once in any arrangement, how many different three-letter arrangements can be formed from {A, N, S, W, E, R}?

19. How many four-digit numerals for positive odd integers can be formed from {1, 2, 3, 4, 5}?

20. How many four-digit numerals for positive even integers can be formed from {1, 2, 3, 4, 5}?

21. How many numerals for positive integers less than 500 can be formed from {3, 4, 5}?

22. How many numerals for positive odd integers less than 500 can be formed from {3, 4, 5}?

23. How many numerals for positive even integers less than 500 can be formed from {3, 4, 5}?

24. How many numerals for positive even integers between 400 and 500, inclusive, can be formed from {3, 4, 5}?

25. How many permutations of the elements of {P, R, I, M, E} end in a vowel?

26. How many permutations of the elements of {P, R, O, D, U, C, T} end in a vowel?

27. Find the number of distinguishable permutations of the letters in the word

LIMIT.

28. Find the number of distinguishable permutations of the letters in the word

COMBINATION.

29. Find the number of distinguishable permutations of the letters in the word

COLORADO.

30. Find the number of distinguishable permutations of the letters in the word

TALLAHASSEE.

31. Show that $P_{5,3} = 5(P_{4,2})$.   32. Show that $P_{5,r} = 5(P_{4,r-1})$.

33. Show that $P_{n,3} = n(P_{n-1,2})$.

34. Show that $P_{n,3} - P_{n,2} = (n-3)(P_{n,2})$.

35. Solve for n: $P_{n,5} = 5(P_{n,4})$.   36. Solve for n: $P_{n,5} = 9(P_{n-1,4})$.

Example

In how many ways can four students be seated around a circular table?

Solution

In any such arrangements (which is called a **circular permutation**), there is no first position. Each person can take four different initial positions without affecting the arrangement. Thus, there are

$$\frac{4!}{4} = 6 \quad \text{arrangements.}$$

(In general, there are $n!/n$, or $(n-1)!$, circular permutations of n things taken n at a time).

37. In how many ways can five students be seated around a circular table?

38. In how many ways can six students be seated around a circular table?

39. In how many ways can six students be seated around a circular table if a certain two must be seated together?

40. In how many ways can three different keys be arranged on a key ring? *Hint:* Arrangements should be considered identical if one can be obtained from the other by turning the ring over. In general, there are only $(1/2)(n-1)!$ distinct arrangements of n keys on a ring $(n \geq 3)$.

An additional counting concept is needed before we turn our attention to probability—namely, finding the number of distinct r-element subsets of an n-element set with no reference to relative order of the elements in the subset. For example, five different cards can be arranged in 5! permutations, but to a poker player they represent the same hand. The set of five cards (with no reference to the arrangement of the cards) is called a *combination*.

Definition 13.2   *A subset of an n-element set A is called a **combination**.*

The counting of combinations is related to the counting of permutations. From Theorem 13.2, we know that the number of distinct permutations of n elements of a set A taken r at a time is given by

$$P_{n,r} = \frac{n!}{(n-r)!}.$$

With this in mind, consider the following result concerning the number $\binom{n}{r}$ that was introduced in Section 12.5.

Theorem 13.3   *The number* $\binom{n}{r}$ *of distinct combinations of the members, taken r at a time, of a set containing n members is given by*

$$\binom{n}{r} = \frac{P_{n,r}}{r!}. \tag{1}$$

Proof   There are, by definition, $\binom{n}{r}$ r-element subsets of the set A, where $n(A) = n$. Also, from Theorem 13.1, each of these subsets has $r!$ permutations of its members. There are therefore $\binom{n}{r}r!$ permutations of n elements of A taken r at a time. Thus

$$P_{n,r} = \binom{n}{r} r!,$$

from which we obtain

$$\binom{n}{r} = \frac{P_{n,r}}{r!},$$

as was to be shown.

Thus, to find the number of r-element subsets of an n-element set A, we count the number of permutations of the elements of A taken r at a time, and then divide by the number of possible permutations of an r-element set. This seems very much like counting a set of people by counting the number of arms and legs and dividing the result by 4, but this approach gives us a very useful expression for the number we seek, $\binom{n}{r}$. Since

$$P_{n,r} = n(n-1)(n-2)\cdots(n-r+1),$$

it follows that

$$\binom{n}{r} = \frac{P_{n,r}}{r!} = \frac{n(n-1)(n-2)\cdots(n-r+1)}{r!} \tag{2}$$

Example   In how many ways can a committee of five be selected from a set of twelve persons?

Solution   What we wish here is the number of 5-element subsets of a 12-element set. From (2), we have

$$\binom{12}{5} = \frac{12\cdot 11\cdot 10\cdot 9\cdot 8}{5\cdot 4\cdot 3\cdot 2\cdot 1} = 792.$$

By Equation (3) on page 313, we have the alternative expression

$$\binom{n}{r} = \frac{P_{n,r}}{r!} = \frac{n!}{r!\,(n-r)!}. \tag{3}$$

Since the numbers $\binom{n}{r}$ are the coefficients in the binomial expansion, and since these coefficients are symmetric, we have the following plausible assertion.

Theorem 13.4 $\displaystyle \binom{n}{r} = \binom{n}{n-r}.$

Proof   From (3), we have both

$$\binom{n}{r} = \frac{n!}{r!\,(n-r)!}$$

and

$$\binom{n}{n-r} = \frac{n!}{(n-r)!\,[n-(n-r)]!} = \frac{n!}{(n-r)!\,r!},$$

and the theorem is proved.

Theorem 13.4 is plausible also since each time a distinct set of r objects is chosen, a distinct set of $n-r$ objects remains unchosen.

Exercise 13.2

Example   How many different amounts of money can be formed from a penny, a nickel, a dime and a quarter?

Solution   We want to find the total number of combinations that can be formed by taking
the coins 1, 2, 3, and 4 at a time. By Equation (3) we have,

$$\binom{4}{1} = \frac{4!}{1!\,3!} = 4, \quad \binom{4}{2} = \frac{4!}{2!\,2!} = 6, \quad \binom{4}{3} = \frac{4!}{3!\,1!} = 4, \quad \binom{4}{4} = \frac{4!}{4!\,0!} = 1,$$

and the total number of combinations is 15. Clearly each combination gives a
different amount.

1. How many different amounts of money can be formed from a penny, a nickel,
   and a dime?

2. How many different amounts of money can be formed from a penny, a nickel,
   a dime, a quarter, and a half-dollar?

3. How many different committees of four persons each can be chosen from a
   group of six persons?

4. How many different committees of four persons each can be chosen from a
   group of ten persons?

5. In how many different ways can a set of five cards be selected from a standard
   bridge deck containing 52 cards?

6. In how many different ways can a set of 13 cards be selected from a standard
   bridge deck of 52 cards?

7. In how many different ways can a hand consisting of five spades, five hearts,
   and three diamonds be selected from a standard bridge deck of 52 cards?

8. In how many different ways can a hand consisting of ten spades, one heart,
   one diamond, and one club be selected from a standard bridge deck of 52
   cards?

9. In how many different ways can a hand consisting of either five spades, five
   hearts, five diamonds, or five clubs be selected from a standard bridge deck?

10. In how many different ways can a hand consisting of three aces and two cards
    that are not aces be selected from a standard bridge deck?

11. A combination of three balls is picked at random from a box containing five
    red, four white, and three blue balls. In how many ways can the set chosen
    contain at least one white ball?

12. In Problem 11, in how many ways can the set chosen contain at least one white
    and one blue ball?

13. A set of five distinct points lies on a circle. How many inscribed triangles can
    be drawn having all their vertices in this set?

14. A set of ten distinct points lies on a circle. How many inscribed quadrilaterals
    can be drawn having all their vertices in this set?

15. A set of ten distinct points lies on a circle. How many inscribed hexagons can
    be drawn having all their vertices in this set?

16. Given $\binom{n}{3} = \binom{50}{47}$, find n.

17. Given $\binom{n}{7} = \binom{n}{5}$, find n.

13.3   *Probability Functions*

Sample spaces, outcomes, and events

When an experiment of some kind is undertaken, associated with the experiment is a set of possible results. For example, when a die is rolled, it will come to a stop with the number of spots on its upper face corresponding to one of the numerals 1, 2, 3, 4, 5, 6. This exhausts all possibilities. Of course, in the absence of chicanery, exactly which one of these random results will occur cannot be precisely specified in advance of the experiment. A question of great practical importance regarding the result of casting a die, and the result of any comparable experiment in which the outcome is uncertain, is "Can we assign some kind of measure to the degree of uncertainty involved?" The answer is "Yes," but before we assign a measure, let us define some necessary terms.

Definition 13.3   *The set of all possible results of an experiment is called a* ***sample space*** *for the experiment.*

Definition 13.4   *Each element of a sample space is called an* ***outcome***, *or* ***sample point***.

Definition 13.5   *Any subset of a sample space is called an* ***event***, *and is commonly denoted by the letter E.*

The reason for the terminology in Definition 13.5 is that, in conducting an experiment, one may be interested in sets of outcomes rather than in individual outcomes. In the tossing of a die, for example, if the sample space is taken as $\{1, 2, 3, 4, 5, 6\}$, then the event that an outcome (a numeral) denotes an even integer is the set $\{2, 4, 6\}$, which is a subset of the sample space. The event that an outcome denotes an odd integer is the set $\{1, 3, 5\}$. These two events are complements of each other, and are examples of **complementary events**, that is, two events whose intersection is $\emptyset$ and whose union is the entire sample space.

Probability

We can define a *probability function P* on a sample space S by means of either a priori or a posteriori considerations, as follows.

A priori considerations involve physical, geometrical, and other inherent properties of the experiment in question. They involve no sampling of outcomes. One way to assign a probability to an event is simply to use the ratio of the number of outcomes in the event to the number of possible outcomes. Thus, when a die is cast and we admit as outcomes the die's stopping with any of its six different faces uppermost, then without making any trial throws we would assign the value 1/6 as the probability of each of the six possible outcomes.

More generally, we have the following definition.

Definition 13.6   *If E is any subset containing $n(E)$ members (outcomes) of a sample space containing $n(S)$ equally likely outcomes, then the* ***probability*** *of the occurrence of E, $P(E)$, is given by*

$$P(E) = \frac{n(E)}{n(S)}.$$ (1)

Example

If two dice are cast, what is the (a priori) probability that the sum of the number of dots showing on the top faces of the dice is less than 6?

Solution

For our sample spaces,, let us consider A the set of possible outcomes for one die and B the set for the other. Then

$$n(A) = 6 \quad \text{and} \quad n(B) = 6.$$

The possible outcomes for both would be $S = A \times B$, the Cartesian product of A and B, and $n(A \times B) = n(S) = 36$. Each outcome here is an ordered pair (a, b), where a is the numeral on the upper face of the first die, and b is that of the second die. The event we seek is $\{(a, b) \,|\, a + b < 6\}$. The lattice in the figure shows the situation schematically. Since

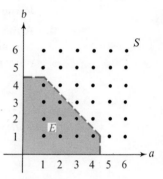

$$E = \{(1, 1), (1, 2), (1, 3), (1, 4), (2, 1),$$
$$(2, 2), (2, 3), (3, 1), (3, 2), (4, 1)\},$$

we have $n(E) = 10$, and

$$P(E) = \frac{n(E)}{n(S)} = \frac{10}{36} = \frac{5}{18}.$$

Care must be taken in interpreting the meaning of "the probability of an event." In the foregoing example, 5/18 does not assure us, for instance, that E will occur 5 times out of 18 casts, or, indeed, that even one cast out of 18 will produce the event described. What it does imply, however, is that if you cast the dice a very great number of times, then you can *expect* the sum on the exposed faces to be less than 6 about 5/18 of the time.

A posteriori considerations involve testing the experiment a certain number of times. Mortality tables give probability functions of this sort. Actually, (1) can still be used in defining such probability functions, provided we interpret the function $n(E)$ as being the number of times the event E occurred in the test and $n(S)$ as being the total number of times the experiment was performed in the test.

Although diagrams are often useful in illustrating an event and a sample space, the number of outcomes $N(E)$ in the event and the number $N(S)$ of possible outcomes in the sample space ordinarily are computed directly.

Example

If three marbles are drawn at random from an urn containing six white marbles and four blue marbles, what is the probability that the three marbles are all white?

Solution

Since the number of ways of drawing three white marbles from six white marbles is $\binom{6}{3}$ and the number of ways of drawing three marbles from the total number of marbles is $\binom{10}{3}$, we have

$$P(E) = \frac{N(E)}{N(S)} = \frac{\binom{6}{3}}{\binom{10}{3}} = \frac{\dfrac{6!}{3!\,3!}}{\dfrac{10!}{3!\,7!}} = \frac{1}{6}.$$

Exercise 13.3

1. A die is cast. List the outcomes in the sample space. List the outcomes in the event E that the number on the upper face of the die is greater than 2. Determine $P(E)$.

2. A coin is tossed. List the outcomes in the sample space. List the outcomes in the event E that a head appears. Determine $P(E)$.

3. Two coins are tossed. List the outcomes in the sample space. List the outcomes in the event E that both coins show the same face. Determine $P(E)$.

4. A die is cast and a coin is tossed. List the outcomes in the sample space. List the outcomes in the event E that the coin shows a head and the die shows a numeral greater than 4. Determine $P(E)$.

In Exercises 5–10, consider an experiment in which two dice are cast. Determine the probability that the specified event occurs.

5. The sum of the numbers of dots shown is 7.

6. The sum of the numbers of dots shown is 8.

7. The sum of the numbers of dots shown is 3.

8. The sum of the numbers of dots shown is 12.

9. At least one of the numbers of dots shown is less than 3.

10. Both dice show the same number of dots.

In Exercises 11–14, consider an experiment in which a single card is drawn at random from a standard pack of 52 cards. Determine the probability that the specified event occurs.

11. The card is the King of hearts. 12. The card is an ace.

13. The card is a black face card. 14. The card is a 5, 6, 7, or 8.

In Exercises 15–18, consider an experiment in which two cards are drawn at random from a standard pack of 52 cards. Determine the probability that the specified event occurs.

15. Both cards are spades. 16. Both cards are red.

17. Both cards are face cards. 18. Both cards are aces.

In Exercises 19–22, consider an experiment in which 2 marbles are drawn at random from an urn containing 8 red, 6 blue, 4 green, and 2 white marbles. Determine the probability that the specified event occurs.

19. Both marbles are white. 20. Both marbles are green.

21. Both marbles are blue. 22. Both marbles are red.

13.4   Probability of the Union of Events

Probability of the complement of an event

If we denote the complement of an event E in a sample space S by E', then $P(E')$ denotes the probability of the occurrence of E'. Since an outcome in S must lie in either E or E', but not both, and since $P(S) = 1$, it follows that

$$P(E) + P(E') = 1.$$

or

$$P(E') = 1 - P(E).$$

We can often use this fact to simplify the computing of a probability.

Example

If two marbles are drawn at random from an urn containing 6 white, 2 red, and 5 green marbles, what is the probability that not both are white, that is, that at least one is not white?

Solution

Rather than consider all of the possible pairs of marbles in which not both are white, we simply compute the probability of the event E that they *are* both white and subtract the result from 1. We have

$$P(E) = \frac{\binom{6}{2}}{\binom{13}{2}} = \frac{15}{78} = \frac{5}{26}.$$

Then

$$P(E') = 1 - \frac{5}{26} = \frac{21}{26},$$

and this is just the probability that not both marbles are white.

The ratio of the probability of an event to the probability of its complement is given a special name.

Definition 13.7   *The **odds** that an experiment with sample space S will result in an event E are given by*

$$\frac{P(E)}{P(E')} = \frac{P(E)}{1 - P(E)}, \quad P(E') \neq 0.$$

Example

Find the odds that the sum of the number determined by a single cast of two dice will be less than 6.

Solution

From the example on page 321, the probability of the event is $5/18$. Hence

$$P(E') = 1 - P(E) = 1 - \frac{5}{18} = \frac{13}{18}.$$

(*Solution continued overleaf*)

Then, by Definition 13.7,

$$\frac{P(E)}{P(E')} = \frac{5/18}{13/18} = \frac{5}{13},$$

and the odds that the sum of the numbers will be less than 6 are $5/13$, or 5 to 13.

Probability of the union of events

By the second counting principle on page 311, if E_1 and E_2 are events in a sample space S, then

$$n(E_1 \cup E_2) = n(E_1) + n(E_2) - n(E_1 \cap E_2).$$

This leads directly to the following theorem.

Theorem 13.5   *If S is a sample space, and E_1 and E_2 are any events in S, then*

$$P(E_1 \text{ or } E_2) = P(E_1 \cup E_2) = P(E_1) + P(E_2) - P(E_1 \cap E_2).$$

Example

If two cards are drawn from a standard deck of playing cards, what is the probability that either both are red or both are Jacks?

Solution

A deck of cards contains 52 cards, and an outcome here consists of two cards. Hence, the number of elements of the sample space is the number of ways (combinations) one can draw two cards from 52, which is $\binom{52}{2}$. Let E_1 be the event that both are red. Since there are 26 red cards in a deck, the number of outcomes (combinations) in the event E_1 is

$$n(E_1) = \binom{26}{2}.$$

Let E_2 be the event that both cards are Jacks. Then, because there are four Jacks,

$$n(E_2) = \binom{4}{2}.$$

Since there is only one red pair of Jacks, $n(E_1 \cap E_2) = 1$. We then have

$$P(E_1 \cup E_2) = P(E_1) + P(E_2) - P(E_1 \cup E_2)$$

$$= \frac{\binom{26}{2}}{\binom{52}{2}} + \frac{\binom{4}{2}}{\binom{52}{2}} - \frac{1}{\binom{52}{2}} = \frac{\binom{26}{2} + \binom{4}{2} - 1}{\binom{52}{2}}$$

$$= \frac{\dfrac{26 \cdot 25}{1 \cdot 2} + \dfrac{4 \cdot 3}{1 \cdot 2} - 1}{\dfrac{52 \cdot 51}{1 \cdot 2}} = \frac{325 + 6 - 1}{1326} = \frac{330}{1326} = \frac{165}{663} = \frac{55}{221}.$$

Probability of the union of disjoint events

Of course, if E_1 and E_2 are *disjoint*, then $E_1 \cap E_2 = \emptyset$, so that $P(E_1 \cap E_2) = 0$, and then the equation in Theorem 13.5 reduces to

$$P(E_1 \cup E_2) = P(E_1) + P(E_2).$$

Disjoint events are said to be **mutually exclusive**.

Example   A card is drawn at random from a standard deck of 52 cards. What is the probability that the card is either a face card (Jack, Queen, or King) or a four?

Solution   Let E_1 be the event that the card is a four. There are four such cards in a deck, so that $n(E_1) = 4$. Let E_2 be the event that the card is a Jack, Queen, or King. There are twelve such cards in a deck. Hence $n(E_2) = 12$. Since the sample space is just the entire deck, $n(S) = 52$, and since E_1 and E_2 are mutually exclusive,

$$P(E_1 \cup E_2) = P(E_1) + P(E_2)$$

$$= \frac{4}{52} + \frac{12}{52} = \frac{16}{52} = \frac{4}{13}.$$

Therefore, the probability is 4/13.

Exercise 13.4

Two dice are cast. Let E be the event that both dice show the same numeral. Let F be the event that the sum of the numbers thrown is greater than eight. Find each of the following probabilities.

1. $P(E)$ 2. $P(F)$ 3. $P(E \cup F)$
4. $P(E')$ 5. $P(F')$ 6. $P(E' \cup F')$

A box contains five red, four white, and three blue marbles. Two marbles are drawn from the box. Let RR be the event that both marbles are red, WW that both marbles are white, BB that both marbles are blue, and RW, RB, BW that a red and a white, a red and a blue, and a blue and a white are drawn, respectively. Find each probability.

7. $P(RR)$ 8. $P(BB)$ 9. $P(WW)$
10. $P(RW)$ 11. $P(RB)$ 12. $P(BW)$
13. What is the probability that neither is white?
14. What is the probability that neither is blue?
15. What is the probability that at least one is red?
16. What is the probability that either one is red or else both are white?
17. What is the probability that by drawing a single card from a standard deck o 52 cards, one will get a 2, 3, or 4?
18. What is the probability that if two cards are drawn from a standard deck of 52 cards they will be of the same suit? Different suits?

*If the probability of the event E that a person will receive k dollars is P(E), then the person's **mathematical expectation** relative to this event is kP(E).*

19. A lottery offers a prize of $50, and 70 tickets are sold. What is the mathematical expectation of a person who buys three tickets? If each ticket costs $1, is the person's expectation greater or less than his outlay?

20. The odds that a certain horse will win the Irish Sweepstakes are 2 to 7. If you hold a ticket on this horse to pay $100,000 if he wins, what is your mathematical expectation?

If E_1, E_2, E_3, etc., are mutually exclusive events, and the return to you is k_1 if E_1 occurs, k_2 if E_2 occurs, etc., then your mathematical expection is $\sum_{i=1}^{n} k_i P(E_i)$.

21. One coin is selected at random from a collection containing a penny, a nickel, and a dime. What is the expectation?

22. One coin is selected at random from a collection containing a dime, a quarter, and half-dollar. What is the expectation?

23. Three $1 bills and four $5 bills are hidden from view. What is the expectation on a single selection?

24. Three $1 bills, four $5 bills, and one $10 bill are hidden from view. What is the expectation on a single draw?

13.5   *Probability of the Intersection of Events*

In some experiments, we may be interested in events that are not dependent on each other, in the sense that the occurrence of one may have no effect on the probability of the occurrence or nonoccurrence of the other.

Independent events

Consider an experiment in which two cards are drawn at random, one after the other, from a deck of ten cards, six of which are red and four blue. We can inquire into the probability that the first card drawn is red and the second blue. The simplest such situation would be one in which the first card is drawn, observed, and returned to the deck, which is then shuffled thoroughly before the second card is drawn. In this case, the sample space would consist of a set of ordered pairs (x, y), where x is the result of the first draw and y the result of the second draw. Since there are ten possibilities in each case, the sample space would consist of $10 \times 10 = 100$ ordered pairs. In the Cartesian graph of Figure 13.2, r_i and b_i are used to designate the drawing of red and blue cards, respectively.

The events E_1 that the first card drawn is a red card and E_2 that the

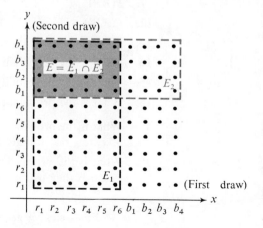

Figure 13.2

second card drawn is a blue card are outlined in the figure. The event E that both E_1 and E_2 occur is the intersection of E_1 and E_2; that is, $E = E_1 \cap E_2$. By inspection,

$$P(E_1) = \frac{60}{100} = \frac{3}{5}, \qquad P(E_2) = \frac{40}{100} = \frac{2}{5},$$

and

$$P(E) = P(E_1 \cap E_2) = \frac{24}{100} = \frac{6}{25}.$$

Moreover, in this example, it is evident that

$$P(E) = P(E_1 \cap E_2) = P(E_1) \cdot P(E_2)$$

Definition 13.8   *If E_1 and E_2 are events in a sample space, and if*

$$P(E_1 \cap E_2) = P(E_1) \cdot P(E_2).$$

*then E_1 and E_2 are **independent events**. If two events are not independent, then they are said to be **dependent**.*

Dependent events

Now consider the same experiment, except that this time the first card is not returned to the deck before the second is taken. Then there will be ten possible first draws, but only nine possible second draws. The sample space will therefore contain 10×9 ordered pairs, such that no ordered pair with first and second components the same remains in the set.

Figure 13.3 shows a graph of the sample space, which is the same as that in the preceding figure except that one diagonal is missing. Sets with graphs that have a missing diagonal are called **deleted Cartesian sets**. Again, the figure shows E_1 and E_2, the events that a red and a blue are obtained on the first and second draw, respectively. The event that both occur is $E = E_1 \cap E_2$, which is also shown in the figure. By inspection,

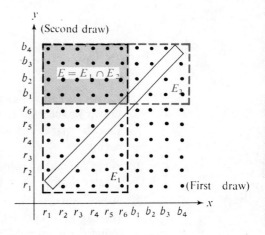

Figure 13.3

$$P(E_1) = \frac{54}{90} = \frac{3}{5}, \quad P(E_2) = \frac{36}{90} = \frac{2}{5}, \quad \text{and} \quad P(E) = P(E_1 \cap E_2) = \frac{24}{90} = \frac{4}{15}.$$

This time,

$$P(E_1 \cap E_2) \neq P(E_1) \cdot P(E_2),$$

so that the events are dependent.

If we write 4/15 as

$$\frac{4}{15} = \frac{3}{5} \cdot \frac{4}{9},$$

then we have

$$P(E_1 \cap E_2) = P(E_1) \cdot \frac{4}{9},$$

where 4/9 can be interpreted as the probability of the occurrence of E_2, *given the occurrence of* E_1. This probability is denoted by the symbol $P(E_2|E_1)$.

Definition 13.9   *If E_1 and E_2 are events in a sample space, and $P(E_1) \neq 0$, then the **conditional probability** $P(E_2|E_1)$ of E_2 given E_1 is*

$$P(E_2|E_1) = \frac{P(E_1 \cap E_2)}{P(E_1)}.$$

If E_1 and E_2 are independent and $P(E_1) \neq 0$, then by Definitions 13.8 and 13.9,

$$P(E_2|E_1) = \frac{P(E_1 \cap E_2)}{P(E_1)} = \frac{P(E_1) \cdot P(E_2)}{P(E_1)} = P(E_2).$$

Exercise 13.5

1. A bag contains four red marbles and ten blue marbles. If two marbles are drawn in succession, and if the first is not replaced, what is the probability that the first is red and the second is blue? Are the two draws independent events?

2. A red die and a green die are cast. What is the probability of obtaining a sum greater than 9, given that the green die shows 4?

3. A bag contains four white and six red marbles. Two marbles are drawn from the bag and replaced, and two more marbles are then drawn from the bag. What is the probability of drawing:
   a. Two red marbles on the first draw and two white ones on the second draw?
   b. A total of two white marbles?
   c. Four white marbles?
   d. Four red marbles?

4. In Problem 3, what is the probability of drawing:
   a. Exactly three white marbles in the two draws?
   b. At least three white marbles in the two draws?
   c. Exactly two red marbles in the two draws?
   d. At least two red marbles in the two draws?

5. A red and a green die are cast. Let E_1 be the event that at least one die shows 3 and E_2 be the event that the sum of the two numbers thrown is 8.
   a. Find $P(E_1)$.   b. Find $P(E_2)$.
   c. Find $P(E_2|E_1)$.   d. Are E_1 and E_2 independent?

6. In Problem 5, let E_1 be the event that neither die shows a result larger than 4 and let E_2 be the event that the dice do not show the same number.
    a. Find $P(E_1)$.
    b. Find $P(E_2)$.
    c. Find $P(E_2 | E_1)$.
    d. Are E_1 and E_2 independent?

7. A coin is tossed three consecutive times. What is the probability that:
    a. The second toss is a head?
    b. The third toss is a head?
    c. Both the second and third tosses are heads?
    d. The first and third tosses are heads?
    e. The first and third tosses are heads but the second is not?

8. In Problem 7, state whether each of the following pairs of events is independent.
    a. a and b
    b. a and c
    c. a and d
    d. a and e
    e. d and e

9. The probability that A will pass a course in college algebra is $5/6$, that B will pass $3/4$, and that C will pass, $2/3$. What is the probability of each of the following events?
    a. At least one of the three will pass.
    b. At least A and C will pass.
    c. A and C will pass but B will not.
    d. At least two of the three will pass.

10. In Problem 9, what is the probability of **c**, given the occurrence of **a**? Are the events **a** and **c** independent?

11. A day is selected at random in some fashion such that any day of the week is an equally likely choice. Let the probability be $1/30$ that a day selected at random will be a rainy day.
    a. What is the probability that a rainy Wednesday will be selected?
    b. What is the probability that a dry Thursday will be selected?
    c. What is the probability that either Monday, Tuesday, or Wednesday will be selected, and that it will not rain that day?
    d. Let E_1 be the selection of Sunday, and let E_2 be the event that it does not rain on the day selected. What is the conditional probability of E_2, given the occurrence of E_1? Are E_1 and E_2 independent?

12. Two identical urns contain marbles. One urn contains all red marbles, while half of the contents of the other urn are white and half are red. If a marble is drawn at random from one of the urns and found to be red, what is the probability that it was drawn from the urn containing only red marbles?

13. Of two dice, one is normal, but the other has three faces showing 4 spots and three faces showing 3 spots. If one of the dice is chosen at random and a 4 is thrown with it, what is the probability that the normal die was chosen?

14. Argue that if E_1 and E_2 are mutually exclusive events with nonzero probabilities, then E_1 and E_2 are not independent.

15. One box contains three red and eight white marbles, and a second box contains five red and two white marbles. If one marble is drawn from each box, what is the probability of drawing:
    a. Two red marbles?
    b. Two white marbles?
    c. One red and one white marble?

16. In Problem 15, what are the odds of drawing:
    a. Two red marbles?
    b. Two white marbles?
    c. One red and one white marble?

Chapter Review

[13.1]

1. How many different two-digit numerals can be formed from {6, 7, 8, 9}?

2. How many different ways can six questions on a true-false test be answered?

3. How many four-digit numerals for positive odd integers can be formed from {3, 4, 5, 6,}?

4. How many distinguishable permutations can be formed using the letters in the word *TENNIS*?

[13.2]

5. How many different committees of 5 persons can be formed from a group of 12 persons?

6. In how many different ways can a hand consisting of 3 spades, 5 hearts, 4 diamonds, and 1 club be selected from a standard bridge deck of 52 cards?

7. A box contains 4 red, 6 white, and 2 blue marbles. In how many ways can you select 3 marbles from the box if at least one of the marbles chosen is white?

8. How many hexagons can be drawn whose vertices are members of a set of 9 fixed points on the circle?

[13.3] *In Questions 9–15, consider an experiment in which two cards are drawn at random from a deck of 52 cards. What is the probability that:*

9. Both cards are clubs?

10. Both cards are tens?

11. Neither card is a face card?

12. Both cards are black?

[13.4] **13.** One card is red and one is black?

14. One card is a face card and the other is not?

15. Both cards are face cards or one card is a ten and the other an ace?

16. What are the odds of randomly drawing one red and one blue marble from a box containing 4 red and 6 blue marbles?

[13.5] *In Questions 17–20, consider an experiment in which two cards are drawn at random from a deck of 52 cards. What is the probability that:*

17. One card is black and the other is the ace of spades?

18. Both cards are red and one (but not the other) is a face card?

19. Both cards are red and at least one is a face card?

20. Neither card is a face card and at least one is red?

Appendix

Appendix.

Mathematical Structure

Here we present the basic substance of the text in compact form. It is a handy guide for following the development of the course and for reviewing material in preparation for tests. Also, you may find it a convenient reference for mathematics courses that you take in the future.

Postulates for Real Numbers

[1.3]
Equality
Postulates

E-1 $a = a$.

E-2   If $a = b$, then $b = a$.

E-3   If $a = b$ and $b = c$, then $a = c$.

E-4   If $a = b$, then a may be replaced by b and b by a in any mathematical statement without altering the truth or falsity of the statement.

[1.3]
Field
Postulates

F-1 $a + b$ is a unique element of R.

F-2 $(a + b) + c = a + (b + c)$.

F-3   For each $a \in R$, there exists an element $0 \in R$ with the property
$$a + 0 = a \quad \text{and} \quad 0 + a = a.$$

F-4   There exists an element $^-a \in R$ with the property
$$a + (^-a) = 0 \quad \text{and} \quad (^-a) + a = 0$$
for all $a \in R$.

F-5 $a + b = b + a$.

F-6 $a \cdot b$ is a unique element of R.

F-7 $(a \cdot b) \cdot c = a \cdot (b \cdot c)$.

F-8 $a \cdot (b + c) = a \cdot b + a \cdot c \quad \text{and} \quad (b + c) \cdot a = b \cdot a + c \cdot a$.

F-9   There exists an element $1 \in R$, $1 \neq 0$, with the property
$$a \cdot 1 = a \quad \text{and} \quad 1 \cdot a = a$$
for all $a \in R$.

F-10 $a \cdot b = b \cdot a$.

F-11   For each element $a \in R$ $(a \neq 0)$, there exists an element $a^{-1} \in R$ with the property
$$a \cdot (a^{-1}) = 1 \quad \text{and} \quad (a^{-1}) \cdot a = 1.$$

[1.5]
Order
Postulates

O-1   If a is a real number, then exactly one of the following statements is true:
$$a \text{ is positive}, \quad a \text{ is zero}, \quad \text{or} \quad -a \text{ is positive}.$$

O-2   If a and b are positive real numbers, then $a + b$ is positive and $a \cdot b$ is positive.

O-3   There is a one-to-one correspondence between the set of real numbers and the set of points on a geometric line.

Properties of Numbers

[1.4]   *Assume all variables represent arbitrary real numbers unless otherwise noted.*

If $a = b$, then
$a + c = b + c$ and $a \cdot c = b \cdot c$.

If $a + b = 0$, then $a = -b$.

If $a \cdot b = 1$ $(a, b \neq 0)$, then $a = \dfrac{1}{b}$.

If $a + c = b + c$, then $a = b$.

If $a \cdot c = b \cdot c$ $(c \neq 0)$, then $a = b$.

For every a, $a \cdot 0 = 0$.

If $a \cdot b = 0$, then
either $a = 0$ or $b = 0$, or both.

$-(-a) = a$.

$(-a) + (-b) = -(a + b)$.

$(-a)(b) = -(ab)$.

$(-a)(-b) = ab$.

$\dfrac{-a}{b} = \dfrac{a}{-b} = -\dfrac{a}{b}$ $(b \neq 0)$.

$\dfrac{-a}{-b} = \dfrac{a}{b}$ $(b \neq 0)$.

$\dfrac{a}{b} = \dfrac{c}{d}$ $(b, d \neq 0)$ if and only if $ad = bc$.

$\dfrac{ac}{bc} = \dfrac{a}{b}$ $(b, c \neq 0)$.

$\dfrac{1}{a} \cdot \dfrac{1}{b} = \dfrac{1}{ab}$ $(a, b \neq 0)$.

$\dfrac{a}{b} \cdot \dfrac{c}{d} = \dfrac{ac}{bd}$ $(b, d \neq 0)$.

$\dfrac{a}{c} + \dfrac{b}{c} = \dfrac{a + b}{c}$ $(c \neq 0)$.

$\dfrac{a}{b} + \dfrac{c}{d} = \dfrac{ad + bc}{bd}$ $(b, d \neq 0)$.

$\dfrac{a}{b} - \dfrac{c}{d} = \dfrac{ad - bc}{bd}$ $(b, d \neq 0)$.

$1 \div \dfrac{a}{b} = \dfrac{b}{a}$ $(a, b \neq 0)$.

$\dfrac{a}{b} \div \dfrac{c}{d} = \dfrac{ad}{bc}$ $(b, c, d \neq 0)$.

If $a < b$ and $b < c$, then $a < c$.

If $a < b$, then $a + c < b + c$.

If $a < b$ and $c > 0$, then $ac < bc$.

If $a < b$ and $c < 0$, then $ac > bc$.

[2.2] $a^m \cdot a^n = a^{m+n}$. $(a^m)^n = a^{mn}$. $(ab)^n = a^n b^n$.

[2.4] $\dfrac{a^m}{a^n} = a^{m-n}$ $(a \neq 0)$. $\left(\dfrac{a}{b}\right)^m = \dfrac{a^m}{b^m}$ $(b \neq 0)$.

[2.5]   If $P(x)$ is a real polynomial and c is any real number, then there exists a unique real polynomial $Q(x)$ and a real number r such that

$$P(x) = (x - c)Q(x) + r.$$

[2.9]   If $r_1 \neq r_2$, then there exist constants $c_1, c_2 \in R$ such that

$$\frac{ax + b}{(x - r_1)(x - r_2)} = \frac{c_1}{x - r_1} + \frac{c_2}{x - r_2}.$$

If $P(x)$ is of degree less than $Q(x)$, and if $Q(x) = (x - r_1)(x - r_2) \cdots (x - r_n)$, where no two factors are identical, then there exist constants $c_1, c_2, \ldots c_n \in R$ such that

$$\frac{P(x)}{Q(x)} = \frac{c_1}{x - r_1} + \frac{c_2}{x - r_2} + \cdots + \frac{c_n}{x - r_n}.$$

If $P(x)$ is of degree less than $Q(x)$, and if $Q(x) = (x - r_1)^n$, then there exist constants $c_1, c_2, \ldots c_n \in R$ such that

$$\frac{P(x)}{Q(x)} = \frac{c_1}{x - r_1} + \frac{c_2}{(x - r_1)^2} + \cdots + \frac{c_n}{(x - r_1)^n}.$$

[3.1] $a^0 = 1 \quad (a \neq 0). \quad a^{-n} = \dfrac{1}{a^n} \quad (a \neq 0).$

[3.2] $(a^{1/n})^n = a \quad$ (n an odd natural number).

$(a^{1/n})^n = |a| \quad$ (n an even natural number).

$(a^{1/n})^m = (a^m)^{1/n} \quad (n \in N, \, m \in J).$

$(a^{1/np})^{mp} = (a^{1/n})^m \quad (m \in J, \, n, \, p \in N).$

$a^{m/n} = (a^{1/n})^m \quad (m \in J, \, n \in N).$

[3.3] $\sqrt[n]{a} = a^{1/n} \quad (a \geq 0).$

$\sqrt[n]{a^n} = a \quad$ (n an odd natural number).

$\sqrt[n]{a^n} = |a| \quad$ (n an even natural number).

$\sqrt[n]{a^m} = (\sqrt[n]{a})^m \quad (n \in N, \, m \in J).$

$\sqrt[n]{a} \, \sqrt[n]{b} = \sqrt[n]{ab} \quad (n \in N).$

$\dfrac{\sqrt[n]{a}}{\sqrt[n]{b}} = \sqrt[n]{\dfrac{a}{b}} \quad (b \neq 0, \, n \in N).$

$\sqrt[cn]{a^{cm}} = \sqrt[n]{a^m} \quad (m \in J, \, n, c \in N).$

[4.1]   If $P(x)$, $Q(x)$, and $R(x)$ are expressions, then for all values of x for which $P(x)$, $Q(x)$, and $R(x)$ are real numbers, the sentence

$$P(x) = Q(x)$$

is equivalent to each of the following:

I $P(x) + R(x) = Q(x) + R(x)$,

II $P(x) \cdot R(x) = Q(x) \cdot R(x)$ for $x \in \{x \mid R(x) \neq 0\}$.

[4.2] $ab = 0 \quad$ if and only if $\quad a = 0 \quad$ or $\quad b = 0 \quad$ or both.

[4.3]   If $ax^2 + bx + c = 0 \quad (a \neq 0)$, then $\quad x = \dfrac{-b \pm \sqrt{b^2 - 4ac}}{2a}.$

[4.4]   The solution set of $U(x) = V(x)$ is a subset of the solution set of $[U(x)]^n = [V(x)]^n$, for each natural number n.

[4.6]   If $P(x)$, $Q(x)$, and $R(x)$ are expressions, then for all values of x for which $P(x)$, $Q(x)$, and $R(x)$ are real numbers, the sentence

$$P(x) < Q(x)$$

is equivalent to each of the following.

I $P(x) + R(x) < Q(x) + R(x)$,

II $P(x) \cdot R(x) < Q(x) \cdot R(x)$ for $x \in \{x \mid R(x) > 0\}$,

III $P(x) \cdot R(x) > Q(x) \cdot R(x)$ for $x \in \{x \mid R(x) < 0\}$.

Similarly, the sentence

$$P(x) \leq Q(x)$$

is equivalent to sentences of the form I-III, with $<$ (or $>$) replaced by $\leq$ (or $\geq$) under the same conditions, $R(x) > 0$ and $R(x) < 0$, as above.

[5.2]   The length of the line segment between two points (x_1, y_1) and (x_2, y_2) is given by

$$d = \sqrt{(x_2 - x_1)^2 + (y_2 - y_1)^2},$$

and the slope is given by

$$m = \frac{y_2 - y_1}{x_2 - x_1} \quad (x_2 \neq x_1).$$

[5.3]   Forms for linear equations:

$$y - y_1 = m(x - x_1) \quad \text{point-slope form,}$$

$$y = mx + b \quad \text{slope-intercept form,}$$

$$\frac{x}{a} + \frac{y}{b} = 1 \quad \text{intercept form.}$$

[6.3] $y = kx$ direct variation,

$y = \dfrac{k}{x}$ inverse variation,

$y = kuvw$ joint variation.

[6.4]   If $P(x) = a_0 x^n + a_1 x^{n-1} + \cdots + a_n$ is a real polynomial equation of degree n, then the graph of

$$\{(x, P(x)) \mid P(x) = a_0 x^n + a_1 x^{n-1} + \cdots + a_n, \ a_0 \neq 0\}$$

is a smooth curve that has at most $n - 1$ turning points.

If $P(x)$ is a real polynomial, then for every real number c there exists a unique real polynomial $Q(x)$ such that

$$P(x) = (x - c)Q(x) + P(c).$$

[6.5]   The graph of the rational function over R defined by $y = P(x)/Q(x)$ has a vertical asymptote at $x = a$ for each value a at which $Q(x)$ vanishes and $P(x)$ does not vanish.

The graph of the rational function over R defined by

$$y = \frac{a_0 x^n + a_1 x^{n-1} + \cdots + a_n}{b_0 x^m + b_1 x^{m-1} + \cdots + b_m},$$

where a_0, $b_0 \neq 0$ and n, m are nonnegative integers, has

I   a horizontal asymptote at $y = 0$ if $n < m$,

II   a horizontal asymptote at $y = a_0/b_0$ if $n = m$,

III   no horizontal asymptotes if $n > m$.

[7.1]   Let $x, y \in Q$, and let $x > y > 0$. Then

$$b^x > b^y \text{ if } b > 1, \quad b^x = b^y \text{ if } b = 1, \quad \text{and} \quad b^x < b^y \text{ if } 0 < b < 1.$$

[7.2] $\log_b (x_1 x_2) = \log_b x_1 + \log_b x_2$.

$\log_b \dfrac{x_2}{x_1} = \log_b x_2 - \log_b x_1$.

$\log_b (x_1)^m = m \log_b x_1$.

[7.6] $\log_a x = \dfrac{\log_b x}{\log_b a}$.

[8.1]   Any ordered pair that satisfies both the equations

$$f(x, y) = 0 \quad \text{and} \quad g(x, y) = 0$$

will also satisfy the equation

$$a \cdot f(x, y) + b \cdot g(x, y) = 0,$$

for all real numbers a and b.

[8.2]   If any equation in the system

$$f(x, y, z) = 0, \quad g(x, y, z) = 0, \quad \text{and} \quad h(x, y, z) = 0$$

is replaced by a linear combination, with nonzero coefficients, of itself and any one of the other equations in the system, then the result is an equivalent system.

[8.5]   Any half-plane is a convex set.

The intersection of two convex sets is a convex set.

[9.1]   If A, B, and C are $m \times n$ matrices with real-number entries, then:

I $(A + B)_{m \times n}$ is a matrix with real-number entries.

II $(A + B) + C = A + (B + C)$.

III   The matrix $\mathbf{0}_{m \times n}$ has the property that for every matrix $A_{m \times n}$,

$$A + \mathbf{0} = A \quad \text{and} \quad \mathbf{0} + A = A.$$

IV   For every matrix $A_{m \times n}$, the matrix $- A_{m \times n}$ has the property that

$$A + (- A) = 0 \quad \text{and} \quad (- A) + A = 0.$$

V $A + B = B + A.$

[9.2]   If A and B are $m \times n$ matrices and $c, d \in R$ then:

I cA is an $m \times n$ matrix   V $1A = A,$

II $c(dA) = (cd)A,$ VI $(- 1)A = - A,$

III $(c + d)A = cA + dA,$ VII $0A = 0,$

IV $c(A + B) = cA + cB,$ VIII $c0 = 0.$

If A, B, and C are $n \times n$ square matrices, then

$$(AB)C = A(BC).$$

If A, B, and C are $n \times n$ square matrices, then

$$A(B + C) = AB + AC \quad \text{and} \quad (B + C)A = BA + CA.$$

For each matrix $A_{n \times n}$, we have

$$A_{n \times n} I_{n \times n} = I_{n \times n} A_{n \times n} = A_{n \times n}.$$

Furthermore, $I_{n \times n}$ is the unique matrix having this property for all matrices $A_{n \times n}$. If A and B are $n \times n$ square matrices, and a is a real number, then

$$a(AB) = (aA)B = A(aB).$$

[9.5]   If each entry in any row, or each entry in any column, of a determinant is 0, then the determinant is equal to 0.

If any two rows (or any two columns) of a determinant are interchanged, the resulting determinant is the negative of the original determinant.

If two rows (or two columns) in a determinant have corresponding entries that are equal, the determinant is equal to 0.

If each of the entries of one row (or column) of a determinant is multiplied by k, the determinant is multiplied by k.

If each entry in a row (or column) of a determinant is written as the sum of two terms, the determinant can be written as the sum of two determinants as follows: If

$$D = \begin{vmatrix} a_{11} & a_{12} & \cdots & a_{1n} \\ \vdots & \vdots & & \vdots \\ b_{i1} + c_{i1} & b_{i2} + c_{i2} & \cdots & b_{in} + c_{in} \\ \vdots & \vdots & & \vdots \\ a_{n1} & a_{n2} & \cdots & a_{nn} \end{vmatrix},$$

then

$$D = \begin{vmatrix} a_{11} & a_{12} & \cdots & a_{1n} \\ \vdots & \vdots & \vdots & \vdots \\ b_{i1} & b_{i2} & \cdots & b_{in} \\ \vdots & \vdots & \vdots & \vdots \\ a_{n1} & a_{n2} & \cdots & a_{nn} \end{vmatrix} + \begin{vmatrix} a_{11} & a_{12} & \cdots & a_{1n} \\ \vdots & \vdots & \vdots & \vdots \\ c_{i1} & c_{i2} & \cdots & c_{in} \\ \vdots & \vdots & \vdots & \vdots \\ a_{n1} & a_{n2} & \cdots & a_{nn} \end{vmatrix},$$

and if

$$D = \begin{vmatrix} a_{11} & \cdots & b_{1j} + c_{1j} & \cdots & a_{1n} \\ a_{21} & \cdots & b_{2j} + c_{2j} & \cdots & a_{2n} \\ \vdots & & \vdots & & \vdots \\ a_{n1} & \cdots & b_{nj} + c_{nj} & \cdots & a_{nn} \end{vmatrix},$$

then

$$D = \begin{vmatrix} a_{11} & \cdots & b_{1j} & \cdots & a_{1n} \\ a_{21} & \cdots & b_{2j} & \cdots & a_{2n} \\ \vdots & & \vdots & & \vdots \\ a_{n1} & \cdots & b_{nj} & \cdots & a_{nn} \end{vmatrix} + \begin{vmatrix} a_{11} & \cdots & c_{1j} & \cdots & a_{1n} \\ a_{21} & \cdots & c_{2j} & \cdots & a_{2n} \\ \vdots & & \vdots & & \vdots \\ a_{n1} & \cdots & c_{nj} & \cdots & a_{nn} \end{vmatrix}.$$

If each entry of one row (or column) of a determinant is multiplied by a real number k and the resulting product is added to the corresponding entry in another row (or column, respectively) in the determinant, the resulting determinant is equal to the original determinant.

[9.6]   If

$$A = \begin{bmatrix} a_{11} & a_{12} & \cdots & a_{1n} \\ a_{21} & a_{22} & \cdots & a_{2n} \\ \vdots & \vdots & & \vdots \\ a_{n1} & a_{n2} & \cdots & a_{nn} \end{bmatrix},$$

and if $\delta(A) \neq 0$, then

$$A^{-1} = \frac{1}{\delta(A)} \begin{bmatrix} A_{11} & A_{21} & \cdots & A_{n1} \\ A_{12} & A_{22} & \cdots & A_{n2} \\ \vdots & \vdots & & \vdots \\ A_{1n} & A_{2n} & \cdots & A_{:n} \end{bmatrix},$$

where A_{ij} is the cofactor of a_{ij} in A. If $\delta(A) = 0$, then A has no inverse.

If A and B are $n \times n$ nonsingular square matrices, then AB has an inverse, namely,

$$(AB)^{-1} = B^{-1}A^{-1}.$$

[9.8]   Cramer's Rule:

$$x = \frac{\delta(A_x)}{\delta(A)}, \quad y = \frac{\delta(A_y)}{\delta(A)}, \quad \text{and} \quad z = \frac{\delta(A_z)}{\delta(A)}, \quad (\delta(A) \neq 0).$$

[10.1]   If $z_1, z_2, z_3 \in C$, then:

F-1 $z_1 + z_2 \in C$.

F-2 $(z_1 + z_2) + z_3 = z_1 + (z_2 + z_3)$.

F-3   There exists an element $z_0 = (0, 0) \in C$ such that

$$z + z_0 = z \quad \text{and} \quad z_0 + z = z$$

for all z in C.

F-4   For each $z \in C$ there exists an element $-z \in C$ such that

$$z + (-z) = z_0 \quad \text{and} \quad (-z) + z = z_0.$$

F-5 $z_1 + z_2 = z_2 + z_1$.

F-6 $z_1 z_2 \in C$.

F-7 $(z_1 z_2) z_3 = z_1 (z_2 z_3)$.

F-8 $z_1(z_2 + z_3) = z_1 z_2 + z_1 z_3$ and $(z_2 + z_3)z_1 = z_2 z_1 + z_3 z_1$.

F-9   There exists an element $z_I = (1, 0) \in C$ such that for all $z \in C$

$$zz_I = z \quad \text{and} \quad z_I z = z.$$

F-10 $z_1 z_2 = z_2 z_1$.

F-11   For each $z \in C$ other than the zero element z_0, there exists an element $z^{-1} \in C$ such that

$$zz^{-1} = z_I \quad \text{and} \quad z^{-1}z = z_I.$$

[10.2]   If $z_1, z_2, z_3 \in C$, then

$$\frac{z_1}{z_2} = \frac{z_1 \, z_3}{z_2 \, z_3} \quad (z_2, z_3 \neq 0).$$

If $(a,b), (c,0) \in C$, then

$$\frac{(a,b)}{(c,0)} = \left(\frac{a}{c}, \frac{b}{c}\right) \quad (c \neq 0).$$

[10.5]   The set V of vectors $\mathbf{v} = (a,b)$, $a,b \in R$, with the set R of real numbers as scalar multipliers, is a vector space over the field R of real numbers. That is, if v_1, v_2, and v_3 are vectors, and c and d are real scalars, then

I $\mathbf{v}_1 + \mathbf{v}_2 \in V$,	VIII $(c + d)\mathbf{v}_1 = c\mathbf{v}_1 + d\mathbf{v}_1$,
II $(\mathbf{v}_1 + \mathbf{v}_2) + \mathbf{v}_3 = \mathbf{v}_1 + (\mathbf{v}_2 + \mathbf{v}_3)$,	IX $c(\mathbf{v}_1 + \mathbf{v}_2) = c\mathbf{v}_1 + c\mathbf{v}_2$,
III $\mathbf{v}_1 + \mathbf{0} = \mathbf{v}_1$ and $\mathbf{0} + \mathbf{v}_1 = \mathbf{v}_1$,	X $1 \cdot \mathbf{v}_1 = \mathbf{v}_1$,
IV $\mathbf{v}_1 + (-\mathbf{v}_1) = \mathbf{0}$,	XI $(-1)\mathbf{v}_1 = -\mathbf{v}_1$,
V $\mathbf{v}_1 + \mathbf{v}_2 = \mathbf{v}_2 + \mathbf{v}_1$,	XII $0 \cdot \mathbf{v}_1 = \mathbf{0}$,
VI $c\mathbf{v}_1 \in V$,	XIII $c \cdot \mathbf{0} = \mathbf{0}$.
VII $c(d\mathbf{v}_1) = (cd)\mathbf{v}_1$,	

[11.1]   If $P(x)$ is a polynomial over the field C of complex numbers and $c \in C$, then there exists a unique polynomial $Q(x)$ and a complex number r, such that

$$P(x) = (x - c)Q(x) + r \quad \text{and} \quad r = P(c).$$

If $P(x)$ is a polynomial over the field C of complex numbers, and $P(r) = 0$, then $(x - r)$ is a factor of $P(x)$.

[11.2]   If $P(z)$ is a polynomial over the field R of real numbers, and $P(z) = 0$ for some $z \in C$, then $P(\bar{z}) = 0$.

Every polynomial function of degree $n \geq 1$ over the field C of complex numbers has at least one complex zero.

If $P(z)$ is a polynomial of degree $n \geq 1$ over the field C of complex numbers, then $P(z)$ can be expressed as a product of a constant and n linear factors of the form $(z - z_k)$.

[11.3]

If $P(x) = a_0 x^n + a_1 x^{n-1} + \cdots + a_n$, $a_j, x \in R$, and if $k \in R$ is between $P(x_1)$ and $P(x_2)$, then there exists at least one $c \in R$ between x_1 and x_2 such that $P(c) = k$.

If $P(x)$ is a polynomial over the field R of real numbers, then the number of positive real zeros of $P(x)$ either is equal to the number of variations in sign occurring in the coefficients of $P(x)$ or else is less than this number by an even natural number. Moreover, the number of negative real solutions of $P(x) = 0$ either is equal to the number of variations in sign occurring in $P(-x)$ or else is less than this number by an even natural number.

Let $P(x)$ be a polynomial over the field R of real numbers.

    I   If $r_1 \geq 0$ and the coefficients of the terms in $Q(x)$ and the term $P(r_1)$ are all of the same sign in the right-hand member of

$$P(x) = (x - r_1)Q(x) + P(r_1),$$

then $P(x) = 0$ can have no solution greater than r_1.

    II   If $r_2 \leq 0$ and the coefficients of the terms in $Q(x)$ and the term $P(r_2)$ alternate in sign (zero suitably denoted by $+0$ and -0) in the right-hand member of

$$P(x) = (x - r_2)Q(x) + P(r_2),$$

then $P(x) = 0$ can have no solution less than r_2.

Let $P(x)$ be a polynomial over the field R of real numbers. If $x_1, x_2 \in R$, with $x_1 < x_2$, and $P(x_1)$ and $P(x_2)$ are opposite in sign, then there exists at least one $c \in R$, $x_1 < c < x_2$, such that $P(c) = 0$.

[11.4]

If a is a composite number, then a is the product of only one set of prime factors; that is, the prime factorization of a is unique except for the ordering of the factors.

If the rational number p/q, in lowest terms, is a solution of

$$P(x) = a_0 x^n + a_1 x^{n-1} + \cdots + a_n = 0,$$

where $a_j \in J$, then p is an integral factor of a_n and q is an integral factor of a_0.

[12.1]

If a given open sentence involving n is true for $n = 1$, and if its truth for $n = k$ implies its truth for $n = k + 1$, then it is true for every natural number n.

[12.2]

The nth term in the sequence defined by

$$s_1 = a, \qquad s_{n+1} = s_n + d,$$

where $a, d \in R$, and $n \in N$, is

$$s_n = a + (n - 1)d.$$

The *n*th term in the sequence defined by

$$s_1 = a, \qquad s_{n+1} = rs_n,$$

where $a, r \in R$, $a \neq 0$, $r \neq 0$, and $n \in N$, is

$$s_n = ar^{n-1}.$$

[12.3]   The sum of the first *n* terms of an arithmetic progression is

$$S_n = \frac{n}{2}(a + s_n), \quad \text{or} \quad S_n = \frac{n}{2}[2a + (n-1)d].$$

The sum of the first *n* terms of a geometric progression is

$$S_n = \frac{a - ar^n}{1 - r}, \quad \text{or} \quad S_n = \frac{a - rs_n}{1 - r} \quad (r \neq 1).$$

[12.4]   The sum of an infinite geometric progression, $a + ar + ar^2 + \cdots + ar^n + \ldots$, with $|r| < 1$, is

$$S_\infty = \lim_{n \to \infty} S_n = \frac{a}{1 - r}.$$

[12.5]   For each natural number *n*,

$$(a + b)^n = a^n + \frac{n}{1!}a^{n-1}b + \frac{n(n-1)}{2!}a^{n-2}b^2 + \frac{n(n-1)(n-2)}{3!}a^{n-3}b^3$$

$$+ \cdots + \frac{n(n-1)(n-2)\cdots(n-r+2)}{(r-1)!}a^{n-r+1}b^{r-1} + \cdots + b^n,$$

where *r* is the number of the term. Alternatively,

$$(a + b)^n = \binom{n}{0}a^n + \binom{n}{1}a^{n-1}b + \binom{n}{2}a^{n-2}b^2 + \binom{n}{3}a^{n-3}b^3 + \cdots$$

$$+ \binom{n}{r-1}a^{n-r+1}b^{r-1} + \cdots + \binom{n}{n}b^n.$$

The *r*th term in a binomial expansion is given by

$$\binom{n}{r-1}a^{n-r+1}b^{r-1} = \frac{n!}{(r-1)!(n-r+1)!}a^{n-r+1}b^{r-1}$$

$$= \frac{n(n-1)(n-2)(n-r+2)}{(r-1)!}a^{n-r+1}b^{r-1}.$$

[13.1]   Counting properties:

$$n(A \cup B) = n(A) + n(B), \quad \text{if } A \cap B = \emptyset,$$

$$n(A \cup B) = n(A) + n(B) - n(A \cap B), \quad \text{if } A \cap B \neq \emptyset,$$

$$n(A \times B) = n(A) \cdot n(B).$$

Let $P_{n,n}$ denote the number of distinct permutations of a set *A*, where $n(A) = n$. Then

$$P_{n,n} = n!.$$

Let $P_{n,r}$ denote the number of permutations of the members, taken r at a time, of a set containing n members; that is, let $P_{n,r}$ be the number of distinct orderings of r elements when there is a set of n elements from which to choose. Then

$$P_{n,r} = n(n-1)(n-2)\cdots[n-(r-1)]$$
$$= n(n-1)(n-2)\cdots(n-r+1)$$
$$= \frac{n!}{(n-r)!}.$$

[13.2]   Let $\binom{n}{r}$ denote the number of distinct combinations of the members, taken r at a time, of a set containing n members. Then

$$\binom{n}{r} = \frac{P_{n,r}}{r!} = \frac{n!}{r!(n-r)!};$$
$$\binom{n}{r} = \binom{n}{n-r}.$$

[13.4] $P(E') = 1 - P(E).$

If S is a sample space, and E_1 and E_2 are any events in S, then

$$P(E_1 \text{ or } E_2) = P(E_1 \cup E_2) = P(E_1) + P(E_2) - P(E_1 \cap E_2),$$
$$P(E_1 \text{ or } E_2) = P(E_1 \cup E_2) = P(E_1) + P(E_2) \quad \text{if } E_1 \cap E_2 = \varnothing.$$

Mathematical Systems

Three operations appear over and over in the algebraic systems of this text, operations called "addition," "multiplication," and "multiplication by a scalar." Operations having the same properties as these occur also in many other systems in pure and applied mathematics.

As a consequence, theorems and the relevant skills developed for one of the systems carry over to any other system whose operations have properties that include the basic properties of the analogous operations in the first system.

Postulates F-1 to F-11 listed on page 334 are the basic properties for addition and multiplication. Some important systems for which various ones of these postulates hold appear in the following table.

System	Binary Operations	Postulates
Group	one	1, 2, 3, 4
Abelian group	one	1, 2, 3, 4, 5
Ring	two	1, 2, 3, 4, 5, 6, 7, 8
Ring with identity	two	1, 2, 3, 4, 5, 6, 7, 8, 9
Commutative ring with identity	two	1, 2, 3, 4, 5, 6, 7, 8, 9, 10
Integral domain	two	1, 2, 3, 4, 5, 6, 7, 8, 9, 10, 11'
Field	two	1, 2, 3, 4, 5, 6, 7, 8, 9, 10, 11

Postulate 11′, which applies to an integral domain, is a corollary of Postulate 11. It is stated as follows:

$$\text{If } a \cdot c = b \cdot c \text{ (or } c \cdot a = c \cdot b \text{) and } c \neq 0, \text{ then } a = b.$$

The above listing of systems and the postulates which hold in the respective system shows that every field is an integral domain, every integral domain is a ring, every ring is an Abelian group, and every Albelian group is a group.

Some of the above systems also satisfy one or more of the order postulates listed on Page 334. A system satisfying Postulates O-1 and O-2 is said to be *ordered*. If it also satisfies O-3 it is said to be *completely ordered*. The **set of complex numbers** constitutes a field but does not satisfy the order postulates; it is not an ordered field. The **set of rational numbers** is a field satisfying Postulates O-1 and O-2 but not O-3; it is an ordered field. The **set of real numbers** is also an ordered field, and it satisfies Postulate O-3; it is a complete ordered field, and is the only complete ordered field in the sense that any other complete ordered field would have to be "just like" (isomorphic with) the complete field of real numbers.

In a mathematical system called a **vector space**, which was briefly mentioned in Chapters 9 and 10, the vectors are the members of an Abelian group and the scalars are the members of a field. The elements of an Abelian group satisfy Postulates F-1 through F-5; the members of a vector space also satisfy postulates I-V (page 215), or VI-X (page 265), for the multiplication of a vector by a scalar. The most familiar vector spaces are the set of ordered pairs $\{(a_1, a_2)\}$ and the set of ordered triples $\{(a_1, a_2, a_3)\}$ of real numbers, with real numbers as scalars. Less familiar are the set of ordered n-triples $\{(a_1, a_2, \ldots, a_n)\}$ and the set of $n \times m$ matrices $\{A_{n \times m}\}$ of real or complex numbers, where n and m are fixed, with real or complex numbers as scalars.

There are still other algebraic systems, with different sets of postulates, which we have not considered in this text. For example, in Boolean algebra, which is important in logic and in the theory of switching circuits, not only does multiplication distribute over addition,

$$A \times (B + C) = (A \times B) + (A \times C),$$

but also addition distributes over multiplication,

$$A + (B \times C) = (A + B) \times (A + C).$$

Table I Common Logarithms

x	0	1	2	3	4	5	6	7	8	9
1.0	.0000	.0043	.0086	.0128	.0170	.0212	.0253	.0294	.0334	.0374
1.1	.0414	.0453	.0492	.0531	.0569	.0607	.0645	.0682	.0719	.0755
1.2	.0792	.0828	.0864	.0899	.0934	.0969	.1004	.1038	.1072	.1106
1.3	.1139	.1173	.1206	.1239	.1271	.1303	.1335	.1367	.1399	.1430
1.4	.1461	.1492	.1523	.1553	.1584	.1614	.1644	.1673	.1703	.1732
1.5	.1761	.1790	.1818	.1847	.1875	.1903	.1931	.1959	.1987	.2014
1.6	.2041	.2068	.2095	.2122	.2148	.2175	.2201	.2227	.2253	.2279
1.7	.2304	.2330	.2355	.2380	.2405	.2430	.2455	.2480	.2504	.2529
1.8	.2553	.2577	.2601	.2625	.2648	.2672	.2695	.2718	.2742	.2765
1.9	.2788	.2810	.2833	.2856	.2878	.2900	.2923	.2945	.2967	.2989
2.0	.3010	.3032	.3054	.3075	.3096	.3118	.3139	.3160	.3181	.3201
2.1	.3222	.3243	.3263	.3284	.3304	.3324	.3345	.3365	.3385	.3404
2.2	.3424	.3444	.3464	.3483	.3502	.3522	.3541	.3560	.3579	.3598
2.3	.3617	.3636	.3655	.3674	.3692	.3711	.3729	.3747	.3766	.3784
2.4	.3802	.3820	.3838	.3856	.3874	.3892	.3909	.3927	.3945	.3962
2.5	.3979	.3997	.4014	.4031	.4048	.4065	.4082	.4099	.4116	.4133
2.6	.4150	.4166	.4183	.4200	.4216	.4232	.4249	.4265	.4281	.4298
2.7	.4314	.4330	.4346	.4362	.4378	.4393	.4409	.4425	.4440	.4456
2.8	.4472	.4487	.4502	.4518	.4533	.4548	.4564	.4579	.4594	.4609
2.9	.4624	.4639	.4654	.4669	.4683	.4698	.4713	.4728	.4742	.4757
3.0	.4771	.4786	.4800	.4814	.4829	.4843	.4857	.4871	.4886	.4900
3.1	.4914	.4928	.4942	.4955	.4969	.4983	.4997	.5011	.5024	.5038
3.2	.5051	.5065	.5079	.5092	.5105	.5119	.5132	.5145	.5159	.5172
3.3	.5185	.5198	.5211	.5224	.5237	.5250	.5263	.5276	.5289	.5302
3.4	.5315	.5328	.5340	.5353	.5366	.5378	.5391	.5403	.5416	.5428
3.5	.5441	.5453	.5465	.5478	.5490	.5502	.5514	.5527	.5539	.5551
3.6	.5563	.5575	.5587	.5599	.5611	.5623	.5635	.5647	.5658	.5670
3.7	.5682	.5694	.5705	.5717	.5729	.5740	.5752	.5763	.5775	.5786
3.8	.5798	.5809	.5821	.5832	.5843	.5855	.5866	.5877	.5888	.5899
3.9	.5911	.5922	.5933	.5944	.5955	.5966	.5977	.5988	.5999	.6010
4.0	.6021	.6031	.6042	.6053	.6064	.6075	.6085	.6096	.6107	.6117
4.1	.6128	.6138	.6149	.6160	.6170	.6180	.6191	.6201	.6212	.6222
4.2	.6232	.6243	.6253	.6263	.6274	.6284	.6294	.6304	.6314	.6325
4.3	.6335	.6345	.6355	.6365	.6375	.6385	.6395	.6405	.6415	.6425
4.4	.6435	.6444	.6454	.6464	.6474	.6484	.6493	.6503	.6513	.6522
4.5	.6532	.6542	.6551	.6561	.6571	.6580	.6590	.6599	.6609	.6618
4.6	.6628	.6637	.6646	.6656	.6665	.6675	.6684	.6693	.6702	.6712
4.7	.6721	.6730	.6739	.6749	.6758	.6767	.6776	.6785	.6794	.6803
4.8	.6812	.6821	.6830	.6839	.6848	.6857	.6866	.6875	.6884	.6893
4.9	.6902	.6911	.6920	.6928	.6937	.6946	.6955	.6964	.6972	.6981
5.0	.6990	.6998	.7007	.7016	.7024	.7033	.7042	.7050	.7059	.7067
5.1	.7076	.7084	.7093	.7101	.7110	.7118	.7126	.7135	.7143	.7152
5.2	.7160	.7168	.7177	.7185	.7193	.7202	.7210	.7218	.7226	.7235
5.3	.7243	.7251	.7259	.7267	.7275	.7284	.7292	.7300	.7308	.7316
5.4	.7324	.7332	.7340	.7348	.7356	.7364	.7372	.7380	.7388	.7396
x	0	1	2	3	4	5	6	7	8	9

Table I (Continued)

x	0	1	2	3	4	5	6	7	8	9
5.5	.7404	.7412	.7419	.7427	.7435	.7443	.7451	.7459	.7466	.7474
5.6	.7482	.7490	.7497	.7505	.7513	.7520	.7528	.7536	.7543	.7551
5.7	.7559	.7566	.7574	.7582	.7589	.7597	.7604	.7612	.7619	.7627
5.8	.7634	.7642	.7649	.7657	.7664	.7672	.7679	.7686	.7694	.7701
5.9	.7709	.7716	.7723	.7731	.7738	.7745	.7752	.7760	.7767	.7774
6.0	.7782	.7789	.7796	.7803	.7810	.7818	.7825	.7832	.7839	.7846
6.1	.7853	.7860	.7868	.7875	.7882	.7889	.7896	.7903	.7910	.7917
6.2	.7924	.7931	.7938	.7945	.7952	.7959	.7966	.7973	.7980	.7987
6.3	.7993	.8000	.8007	.8014	.8021	.8028	.8035	.8041	.8048	.8055
6.4	.8062	.8069	.8075	.8082	.8089	.8096	.8102	.8109	.8116	.8122
6.5	.8129	.8136	.8142	.8149	.8156	.8162	.8169	.8176	.8182	.8189
6.6	.8195	.8202	.8209	.8215	.8222	.8228	.8235	.8241	.8248	.8254
6.7	.8261	.8267	.8274	.8280	.8287	.8293	.8299	.8306	.8312	.8319
6.8	.8325	.8331	.8338	.8344	.8351	.8357	.8363	.8370	.8376	.8382
6.9	.8388	.8395	.8401	.8407	.8414	.8420	.8426	.8432	.8439	.8445
7.0	.8451	.8457	.8463	.8470	.8476	.8482	.8488	.8494	.8500	.8506
7.1	.8513	.8519	.8525	.8531	.8537	.8543	.8549	.8555	.8561	.8567
7.2	.8573	.8579	.8585	.8591	.8597	.8603	.8609	.8615	.8621	.8627
7.3	.8633	.8639	.8645	.8651	.8657	.8663	.8669	.8675	.8681	.8686
7.4	.8692	.8698	.8704	.8710	.8716	.8722	.8727	.8733	.8739	.8745
7.5	.8751	.8756	.8762	.8768	.8774	.8779	.8785	.8791	.8797	.8802
7.6	.8808	.8814	.8820	.8825	.8831	.8837	.8842	.8848	.8854	.8859
7.7	.8865	.8871	.8876	.8882	.8887	.8893	.8899	.8904	.8910	.8915
7.8	.8921	.8927	.8932	.8938	.8943	.8949	.8954	.8960	.8965	.8971
7.9	.8976	.8982	.8987	.8993	.8998	.9004	.9009	.9015	.9020	.9025
8.0	.9031	.9036	.9042	.9047	.9053	.9058	.9063	.9069	.9074	.9079
8.1	.9085	.9090	.9096	.9101	.9106	.9112	.9117	.9122	.9128	.9133
8.2	.9138	.9143	.9149	.9154	.9159	.9165	.9170	.9175	.9180	.9186
8.3	.9191	.9196	.9201	.9206	.9212	.9217	.9222	.9227	.9232	.9238
8.4	.9243	.9248	.9253	.9258	.9263	.9269	.9274	.9279	.9284	.9289
8.5	.9294	.9299	.9304	.9309	.9315	.9320	.9325	.9330	.9335	.9340
8.6	.9345	.9350	.9355	.9360	.9365	.9370	.9375	.9380	.9385	.9390
8.7	.9395	.9400	.9405	.9410	.9415	.9420	.9425	.9430	.9435	.9440
8.8	.9445	.9450	.9455	.9460	.9465	.9469	.9474	.9479	.9484	.9489
8.9	.9494	.9499	.9504	.9509	.9513	.9518	.9523	.9528	.9533	.9538
9.0	.9542	.9547	.9552	.9557	.9562	.9566	.9571	.9576	.9581	.9586
9.1	.9590	.9595	.9600	.9605	.9609	.9614	.9619	.9624	.9628	.9633
9.2	.9638	.9643	.9647	.9652	.9657	.9661	.9666	.9671	.9675	.9680
9.3	.9685	.9689	.9694	.9699	.9703	.9708	.9713	.9717	.9722	.9727
9.4	.9731	.9736	.9741	.9745	.9750	.9754	.9759	.9763	.9768	.9773
9.5	.9777	.9782	.9786	.9791	.9795	.9800	.9805	.9809	.9814	.9818
9.6	.9823	.9827	.9832	.9836	.9841	.9845	.9850	.9854	.9859	.9863
9.7	.9868	.9872	.9877	.9881	.9886	.9890	.9894	.9899	.9903	.9908
9.8	.9912	.9917	.9921	.9926	.9930	.9934	.9939	.9943	.9948	.9952
9.9	.9956	.9961	.9965	.9969	.9974	.9978	.9983	.9987	.9991	.9996
x	0	1	2	3	4	5	6	7	8	9

Table II Exponential Functions

x	e^x	e^{-x}	x	e^x	e^{-x}
0.00	1.0000	1.0000	1.5	4.4817	0.2231
0.01	1.0101	0.9901	1.6	4.9530	0.2019
0.02	1.0202	0.9802	1.7	5.4739	0.1827
0.03	1.0305	0.9705	1.8	6.0496	0.1653
0.04	1.0408	0.9608	1.9	6.6859	0.1496
0.05	1.0513	0.9512	2.0	7.3891	0.1353
0.06	1.0618	0.9418	2.1	8.1662	0.1225
0.07	1.0725	0.9324	2.2	9.0250	0.1108
0.08	1.0833	0.9331	2.3	9.9742	0.1003
0.09	1.0942	0.9139	2.4	11.023	0.0907
0.10	1.1052	0.9048	2.5	12.182	0.0821
0.11	1.1163	0.8958	2.6	13.464	0.0743
0.12	1.1275	0.8869	2.7	14.880	0.0672
0.13	1.1388	0.8781	2.8	16.445	0.0608
0.14	1.1503	0.8694	2.9	18.174	0.0550
0.15	1.1618	0.8607	3.0	20.086	0.0498
0.16	1.1735	0.8521	3.1	22.198	0.0450
0.17	1.1853	0.8437	3.2	24.533	0.0408
0.18	1.1972	0.8353	3.3	27.113	0.0369
0.19	1.2092	0.8270	3.4	29.964	0.0334
0.20	1.2214	0.8187	3.5	33.115	0.0302
0.21	1.2337	0.8106	3.6	36.598	0.0273
0.22	1.2461	0.8025	3.7	40.447	0.0247
0.23	1.2586	0.7945	3.8	44.701	0.0224
0.24	1.2712	0.7866	3.9	49.402	0.0202
0.25	1.2840	0.7788	4.0	54.598	0.0183
0.30	1.3499	0.7408	4.1	60.340	0.0166
0.35	1.4191	0.7047	4.2	66.686	0.0150
0.40	1.4918	0.6703	4.3	73.700	0.0136
0.45	1.5683	0.6376	4.4	81.451	0.0123
0.50	1.6487	0.6065	4.5	90.017	0.0111
0.55	1.7333	0.5769	4.6	99.484	0.0101
0.60	1.8221	0.5488	4.7	109.95	0.0091
0.65	1.9155	0.5220	4.8	121.51	0.0082
0.70	2.0138	0.4966	4.9	134.29	0.0074
0.75	2.1170	0.4724	5.0	148.41	0.0067
0.80	2.2255	0.4493	5.5	244.69	0.0041
0.85	2.3396	0.4274	6.0	403.43	0.0025
0.90	2.4596	0.4066	6.5	665.14	0.0015
0.95	2.5857	0.3867	7.0	1096.6	0.0009
1.0	2.7183	0.3679	7.5	1808.0	0.0006
1.1	3.0042	0.3329	8.0	2981.0	0.0003
1.2	3.3201	0.3012	8.5	4914.8	0.0002
1.3	3.6693	0.2725	9.0	8103.1	0.0001
1.4	4.0552	0.2466	10.0	22026	0.00005

Table III *Natural Logarithms of Numbers*

n	$\log_e n$	n	$\log_e n$	n	$\log_e n$
	*	4.5	1.5041	9.0	2.1972
0.1	7.6974	4.6	1.5261	9.1	2.2083
0.2	8.3906	4.7	1.5476	9.2	2.2192
0.3	8.7960	4.8	1.5686	9.3	2.2300
0.4	9.0837	4.9	1.5892	9.4	2.2407
0.5	9.3069	5.0	1.6094	9.5	2.2513
0.6	9.4892	5.1	1.6292	9.6	2.2618
0.7	9.6433	5.2	1.6487	9.7	2.2721
0.8	9.7769	5.3	1.6677	9.8	2.2824
0.9	9.8946	5.4	1.6864	9.9	2.2925
1.0	0.0000	5.5	1.7047	10	2.3026
1.1	0.0953	5.6	1.7228	11	2.3979
1.2	0.1823	5.7	1.7405	12	2.4849
1.3	0.2624	5.8	1.7579	13	2.5649
1.4	0.3365	5.9	1.7750	14	2.6391
1.5	0.4055	6.0	1.7918	15	2.7081
1.6	0.4700	6.1	1.8083	16	2.7726
1.7	0.5306	6.2	1.8245	17	2.8332
1.8	0.5878	6.3	1.8405	18	2.8904
1.9	0.6419	6.4	1.8563	19	2.9444
2.0	0.6931	6.5	1.8718	20	2.9957
2.1	0.7419	6.6	1.8871	25	3.2189
2.2	0.7885	6.7	1.9021	30	3.4012
2.3	0.8329	6.8	1.9169	35	3.5553
2.4	0.8755	6.9	1.9315	40	3.6889
2.5	0.9163	7.0	1.9459	45	3.8067
2.6	0.9555	7.1	1.9601	50	3.9120
2.7	0.9933	7.2	1.9741	55	4.0073
2.8	1.0296	7.3	1.9879	60	4.0943
2.9	1.0647	7.4	2.0015	65	4.1744
3.0	1.0986	7.5	2.0149	70	4.2485
3.1	1.1314	7.6	2.0281	75	4.3175
3.2	1.1632	7.7	2.0412	80	4.3820
3.3	1.1939	7.8	2.0541	85	4.4427
3.4	1.2238	7.9	2.0669	90	4.4998
3.5	1.2528	8.0	2.0794	100	4.6052
3.6	1.2809	8.1	2.0919	110	4.7005
3.7	1.3083	8.2	2.1041	120	4.7875
3.8	1.3350	8.3	2.1163	130	4.8676
3.9	1.3610	8.4	2.1282	140	4.9416
4.0	1.3863	8.5	2.1401	150	5.0106
4.1	1.4110	8.6	2.1518	160	5.0752
4.2	1.4351	8.7	2.1633	170	5.1358
4.3	1.4586	8.8	2.1748	180	5.1930
4.4	1.4816	8.9	2.1861	190	5.2470

* Subtract 10 for $n < 1$. Thus $\log_e 0.1 = 7.6974 - 10 = -2.3026$.

Table IV   Squares, Square Roots, and Prime Factors

No.	Sq.	Sq. Root	Prime Factors	No.	Sq.	Sq. Root	Prime Factors
1	1	1.000		51	2,601	7.141	$3 \cdot 17$
2	4	1.414	2	52	2,704	7.211	$2^2 \cdot 13$
3	9	1.732	3	53	2,809	7.280	53
4	16	2.000	2^2	54	2,916	7.348	$2 \cdot 3^3$
5	25	2.236	5	55	3,025	7.416	$5 \cdot 11$
6	36	2.449	$2 \cdot 3$	56	3,136	7.483	$2^3 \cdot 7$
7	49	2.646	7	57	3,249	7.550	$3 \cdot 19$
8	64	2.828	2^3	58	3,364	7.616	$2 \cdot 29$
9	81	3.000	3^2	59	3,481	7.681	59
10	100	3.162	$2 \cdot 5$	60	3,600	7.746	$2^2 \cdot 3 \cdot 5$
11	121	3.317	11	61	3,721	7.810	61
12	144	3.464	$2^2 \cdot 3$	62	3,844	7.874	$2 \cdot 31$
13	169	3.606	13	63	3,969	7.937	$3^2 \cdot 7$
14	196	3.742	$2 \cdot 7$	64	4,096	8.000	2^6
15	225	3.873	$3 \cdot 5$	65	4,225	8.062	$5 \cdot 13$
16	256	4.000	2^4	66	4,356	8.124	$2 \cdot 3 \cdot 11$
17	289	4.123	17	67	4,489	8.185	67
18	324	4.243	$2 \cdot 3^2$	68	4,624	8.246	$2^2 \cdot 17$
19	361	4.359	19	69	4,761	8.307	$3 \cdot 23$
20	400	4.472	$2^2 \cdot 5$	70	4,900	8.367	$2 \cdot 5 \cdot 7$
21	441	4.583	$3 \cdot 7$	71	5,041	8.426	71
22	484	4.690	$2 \cdot 11$	72	5,184	8.485	$2^3 \cdot 3^2$
23	529	4.796	23	73	5,329	8.544	73
24	576	4.899	$2^3 \cdot 3$	74	5,476	8.602	$2 \cdot 37$
25	625	5.000	5^2	75	5,625	8.660	$3 \cdot 5^2$
26	676	5.099	$2 \cdot 13$	76	5,776	8.718	$2^2 \cdot 19$
27	729	5.196	3^3	77	5,929	8.775	$7 \cdot 11$
28	784	5.292	$2^2 \cdot 7$	78	6,084	8.832	$2 \cdot 3 \cdot 13$
29	841	5.385	29	79	6,241	8.888	79
30	900	5.477	$2 \cdot 3 \cdot 5$	80	6,400	8.944	$2^4 \cdot 5$
31	961	5.568	31	81	6,561	9.000	3^4
32	1,024	5.657	2^5	82	6,724	9.055	$2 \cdot 41$
33	1,089	5.745	$3 \cdot 11$	83	6,889	9.110	83
34	1,156	5.831	$2 \cdot 17$	84	7,056	9.165	$2^2 \cdot 3 \cdot 7$
35	1,225	5.916	$5 \cdot 7$	85	7,225	9.220	$5 \cdot 17$
36	1,296	6.000	$2^2 \cdot 3^2$	86	7,396	9.274	$2 \cdot 43$
37	1,369	6.083	37	87	7,569	9.327	$3 \cdot 29$
38	1,444	6.164	$2 \cdot 19$	88	7,744	9.381	$2^3 \cdot 11$
39	1,521	6.245	$3 \cdot 13$	89	7,921	9.434	89
40	1,600	6.325	$2^3 \cdot 5$	90	8,100	9.487	$2 \cdot 3^2 \cdot 5$
41	1,681	6.403	41	91	8,281	9.539	$7 \cdot 13$
42	1,764	6.481	$2 \cdot 3 \cdot 7$	92	8,464	9.592	$2^2 \cdot 23$
43	1,849	6.557	43	93	8,649	9.644	$3 \cdot 31$
44	1,936	6.633	$2^2 \cdot 11$	94	8,836	9.695	$2 \cdot 47$
45	2,025	6.708	$3^2 \cdot 5$	95	9,025	9.747	$5 \cdot 19$
46	2,116	6.782	$2 \cdot 23$	96	9,216	9.798	$2^5 \cdot 3$
47	2,209	6.856	47	97	9,409	9.849	97
48	2,304	6.928	$2^4 \cdot 3$	98	9,604	9.899	$2 \cdot 7^2$
49	2,401	7.000	7^2	99	9,801	9.950	$3^2 \cdot 11$
50	2,500	7.071	$2 \cdot 5^2$	100	10,000	10.000	$2^2 \cdot 5^2$

Odd-Numbered Answers

Exercise 1.1 (page 3)   **1.** {3, 4, 5, 6}   **3.** { }
   5. {Sunday, Monday, Tuesday, Wednesday, Thursday, Friday, Saturday}
   7. {natural numbers less than 3} = {1, 2}   **9.** {2} ≠ {−2}   **11.** ∅ ≠ {0}   **13.** 3 ∈ {2, 3, 4}
15. {2} ∉ {2, 3, 4}   **17.** 5 ⊄ {4, 5, 6}   **19.** ∅ ⊂ {4, 5, 6}
21. **a.** {5, 6, 7}   **b.** {5, 6}, {5, 7}, {6, 7}   **c.** {5}, {6}, {7}   **d.** ∅
23. **a.** $A \subset U$ **b.** $C \not\subset A$ **c.** $A \not\subset B$ **d.** $C \subset B$
25. {x | x is an odd natural number}   **27.** {x | $2^x = 5$}   **29.** {x | x ∉ A}

Exercise 1.2 (page 6)   **1.** $A' = \{1, 3, 5, 7, 9\} = C$ **3.** $C' = \{2, 4, 6, 8, 10\} = A$
   5. $A \cup B = \{1, 2, 3, 4, 5, 6, 8, 10\}$ **7.** $A \cap C = \emptyset$
   9. $A' \cup C' = U$ or {1, 2, 3, 4, 5, 6, 7, 8, 9, 10}   **11.** $A' \cup C = C$ or {1, 3, 5, 7, 9}
13.
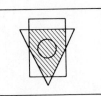
15.
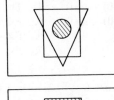
17.
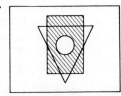
19.
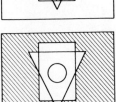
21.
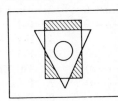
23.
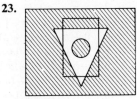
25. $(A')' = A$ **27.** $A \cap U = A$ **29.** $A \cup U = U$ **31.** $A \cup \emptyset = A$ **33.** $\emptyset' \cap \emptyset = \emptyset$
35. $\emptyset \cup \emptyset = \emptyset$
37. **a.** $A \cup B = \emptyset$ if $A = B = \emptyset$ **b.** $A \cup \emptyset = \emptyset$ if $A = \emptyset$ **c.** $A \cap U = U$ if $A = U$
   d. $A \cup B = A$ if $B \subset A$ **e.** $A \cup \emptyset = U$ if $A = U$ **f.** $A' \cap U = U$ if $A = \emptyset$
   g. $A \cap B = A$ if $A \subset B$ **h.** $A' \cup \emptyset = \emptyset$ if $A = U$ **i.** $A \cup B = A \cap B$ if $A = B$

Exercise 1.3 (page 11)   **1.** False   **3.** True   **5.** True   **7.** False   **9.** True   **11.** True
13. {1, 2, 3, 4, 5}   **15.** {1, 2, 3, 4, 5, 6}   **17.** {−9, −8, −7, −6}
19. {x | x ∈ N}   **21.** {x | x ∈ R}   **23.** {x | x ∈ R, x between −4 and 3}
25. **a.** {4}   **b.** {4, −2, 0}   **c.** $\left\{4, -2, \frac{2}{5}, 0, -\frac{3}{4}\right\}$ **d.** $\left\{4, -2, \frac{2}{5}, 0, -\frac{3}{4}, \sqrt{2}, \sqrt{7}\right\}$

27. Transitive law for equality, E-3   **29.** Symmetric law for equality, E-2

31. Transitive law for equality, E-3   **33.** Closure law for multiplication, F-6

35. Associative law for multiplication, F-7   **37.** Additive-inverse law, F-4

39. Commutative law for multiplication, F-10   **41.** Commutative law for addition, F-5

43. Distributive law, F-8   **45.** Commutative law for addition, F-5

47. Commutative law for multiplication, F-10   **49.** Distributive law, F-8

51. Closed   **53.** Closed   **55.** Closed   **57.** Not closed

Exercise 1.4 (page 15)   **1.** 1.1   **3.** 1.8-V   **5.** 1.11-III   **7.** 1.7   **9.** 1.9

11. 1.11-VII   **13.** 1.8-III   **15.** 1.11-IV   **17.** 1.2   **19.** 1.5 or 1.2

For Problems 21–43, the proofs given are not unique. They are presented here in a precise statement-reason format.

21.
1. $a, b, c \in R$ and $a = b$ 1. Hyp.
2. $ac, bc \in R$ 2. F-6
3. $ac = ac$ and $ca = ca$ 3. E-1
4. $ac = bc$ and $ca = cb$ 4. E-4

23.
1. Let $a + b = a$ and $a + b' = a$ 1. Hyp.
2. $a + b = a + b'$ 2. E-3
3. $b = b'$ 3. Th. 1.4
4. $a + 0 = a$ 4. F-3
5. $a + 0 = a + b$ 5. E-4
6. $0 = b$ 6. Th. 1.4
7. $b' = 0$ 7. E-4

25.
1. $a + c = b + c$ 1. Hyp.
2. $(a + c) + (^-c) = (b + c) + (^-c)$ 2. Th. 1.1
3. $a + (c + (^-c)) = b + (c + (^-c))$ 3. F-2
4. $a + 0 = b + 0$ 4. F-4
5. $a = b$ 5. F-3

27.
1. $0 + 0 = 0$ 1. F-3
2. $a(0 + 0) = a \cdot 0$ 2. Th. 1.2
3. $a \cdot 0 + a \cdot 0 = a \cdot 0$ 3. F-8
4. $-(a \cdot 0) + [a \cdot 0 + a \cdot 0] = -(a \cdot 0) + a \cdot 0$ 4. Th. 1.1
5. $[-(a \cdot 0) + a \cdot 0] + a \cdot 0 = -(a \cdot 0) + a \cdot 0$ 5. F-2
6. $0 + a \cdot 0 = 0$ 6. F-4
7. $a \cdot 0 = 0$ 7. F-3

29.
1. $(a + b) + [-(a + b)] = 0$ 1. F-4
2. $-a + (a + b) + [-(a + b)] = -a + 0$ 2. Th. 1.1
3. $(-a + a) + b + [-(a + b)] = -a + 0$ 3. F-2
4. $0 + b + [-(a + b)] = -a + 0$ 4. F-4
5. $b + [-(a + b)] = -a$ 5. F-3
6. $-b + (b + [-(a + b)]) = -b + (-a)$ 6. Th. 1.1
7. $(-b + b) + [-(a + b)] = -b + (-a)$ 7. F-2
8. $0 + [-(a + b)] = -b + (-a)$ 8. F-4
9. $-(a + b) = -b + (-a)$ 9. F-3
10. $-(a + b) = (-a) + (-b)$ 10. F-5

31.
1. $b + (-b) = 0$ 1. F-4
2. $(-a)[b + (-b)] = (-a) \cdot 0$ 2. Th. 1.2
3. $(-a) \cdot b + (-a)(-b) = (-a) \cdot 0$ 3. F-8
4. $(-a) \cdot b + (-a)(-b) = 0$ 4. Th. 1.6
5. $-(ab) + ab = 0$ 5. F-4, F-5
6. $-(ab) + (-a)(-b) = 0$ 6. E-4 (Prob. 30)
7. $-(ab) + (-a)(-b) = -(ab) + ab$ 7. E-3
8. $(-a)(-b) = ab$ 8. Th. 1.4

33. 1. $\dfrac{-a}{-b} = -a \cdot \dfrac{1}{-b}$ 1. Def. 1.13

 2. $= -\left(a \cdot \dfrac{1}{-b}\right)$ 2. Th. 1.8-III

 3. $= -\left(\dfrac{a}{-b}\right)$ 3. Def. 1.13

 4. $= -\left(-\dfrac{a}{b}\right)$ 4. Th. 1.8-V

 5. $\dfrac{-a}{-b} = \dfrac{a}{b}$ 5. Th. 1.8-I; E-3

35. 1. $abc = abc$ 1. E-1
 2. $(ac)b = a(bc)$ 2. Def. 1.11, F-10, F-7
 3. $\dfrac{ac}{bc} = \dfrac{a}{b}$ 3. Th. 1.9

37. 1. $\dfrac{a}{b} \cdot \dfrac{c}{d} = \left(a \cdot \dfrac{1}{b}\right)\left(c \cdot \dfrac{1}{d}\right)$ 1. Def. 1.13

 2. $= ac \cdot \dfrac{1}{b} \cdot \dfrac{1}{d}$ 2. F-10

 3. $= ac \cdot \dfrac{1}{bd}$ 3. Th. 1.11-I

 4. $\dfrac{a}{b} \cdot \dfrac{c}{d} = \dfrac{ac}{bd}$ 4. Def. 1.13; E-3

39. 1. $\dfrac{a}{b} + \dfrac{c}{d} = \dfrac{ad}{bd} + \dfrac{bc}{bd}$ 1. Th. 1.10

 2. $\dfrac{a}{b} + \dfrac{c}{d} = \dfrac{ad+bc}{bd}$ 2. Th. 1.11-III

41. 1. $\dfrac{1}{\frac{a}{b}} = \dfrac{1 \cdot b}{\frac{a}{b} \cdot b}$ 1. Th. 1.10

 2. $= \dfrac{b}{a \cdot \frac{1}{b} \cdot b}$ 2. Def. 1.1.3

 3. $= \dfrac{b}{a}$ 3. F-11

43. **(a)** 1. $\dfrac{a}{b} = q$ 1. Hyp.

 2. $b \cdot \dfrac{a}{b} = bq$ 2. Th. 1.2

 3. $\left(b \cdot \dfrac{1}{b}\right)a = bq$ 3. Def. 1.13

 4. $a = bq$ 4. F-11

 (b) 1. $a = bq$ 1. Hyp.

 2. $\dfrac{1}{b} \cdot a = \dfrac{1}{b} \cdot bq$ 2. Th. 1.2

 3. $\dfrac{a}{b} = \left(\dfrac{1}{b} \cdot b\right)q$ 3. Def. 1.13

 4. $\dfrac{a}{b} = q$ 4. F-11

Exercise 1.5 (page 19) **1.** I   **3.** III   **5.** II   **7.** IV   **9.** $7 > 3$ **11.** $-4 < -3$
13. $-1 \le x \le 1$ **15.** $x > 0$ **17.** $x \ge 0$ **19.** 3   **21.** 7   **23.** -2
25. $3x$ if $x \ge 0$, $-3x$ if $x < 0$ **27.** $x + 1$ if $x \ge -1$, $-(x+1)$ if $x < -1$
29. $y - 3$ if $y \ge 3$, $-(y-3)$ if $y < 3$ **31.** $|-2| < |-5|$ **33.** $-7 < |-1|$
35. $|-3| > 0$ **37.** $2 < 5$ **39.** $7 < 8$ **41.** $|x| < 3$

Chapter 1 Review (page 20) **1.** $\in$ **2.** $\subset$ **3.** $\subset$ **4.** $\in$
5. $\{1, 2, 3, 4\}$, $\{1, 2, 3\}$, $\{1, 3, 4\}$, $\{1, 3\}$ **6.** $\{x \mid x = 5k,\ k \in J\}$ **7.** $\{2, 4, 5, 6, 7, 8, 9\}$
8. $\{6, 8\}$ **9.** $\emptyset$ **10.** $\{5, 7, 9\}$ **11.** $U = \{1, 2, 3, 4, 5, 6, 7, 8, 9\}$ **12.** $\{1, 2, 4\}$
13. $\{3, 21\}$ **14.** $\{-8, -1, 0, 3, 21\}$ **15.** $\left\{-8, \dfrac{-15}{7}, -1, 0, 3, \dfrac{13}{2}, 21\right\}$ **16.** $\{-\sqrt{5}, \sqrt{50}\}$
17. Commutative law for addition   **18.** Commutative law for multiplication

19. Multiplicative inverse.   **20.** Distributive property   **21.** Additive inverse
22. Distributive property   **23.** Theorem 1.1   **24.** Theorem 1.8-IV   **25.** Theorem 1.11-II
26. Theorem 1.10   **27.** Theorem 1.8-VI   **28.** Theorem 1.11-IV   **29.** Theorem 1.11-VI
30. Theorem 1.11-VII   **31.** IV   **32.** II   **33.** $-6 > -9$ **34.** $4 \le y$ **35.** $y \le 0$
36. $y + 2 \ge 0$ **37.** 7   **38.** $x - 5$ if $x - 5 \ge 0$, $5 - x$ if $x - 5 < 0$ **39.** $x < 4$ **40.** $|x| \ge 6$

Exercise 2.1 (page 25)   **1.** $-5z^2 + 8z - 2$ **3.** $-y^3 - 6y^2 + y + 3$ **5.** $x^2 + 2y^2$
7. 9   **9.** $2x + 1,\ 4x - 5$ **11.** $3x^2 - 4x + 2,\ -x^2 + 4$
13. $x^3 + 5x^2 - 4x + 4,\ x^3 + 3x^2 - 2$
15. $x^5 + 3x^4 - 3x^3 - 2x^2 + 4x + 1,\ -x^5 + 3x^4 + 3x^3 - 2x^2 - 4x + 1$ **17.** $3x^2 - 5x + 5$
19. $3x^2 - 4x + 4$ **21.** $x^2 - 2x + 6$ **23.** $-1, -21, 1, -2, 20$ **25.** $1, 1, 0, 1, 0$
27. $3, 2h^4 - h^2 + 3,\ 2h^4 - h^2 + 3$
29. $x^2 + 2hx + h^2 + 2x + 2h - 3,\ x^2 - 2hx + h^2 + 2x - 2h - 3,\ x^4 + 2x^2 - 3$ **31.** $1, 1$
33. $6, 39$ **35.** n, n

Exercise 2.2 (page 28)   **1.** $-6x^3y^4$ **3.** $-8x^4y^5$ **5.** x^{3n} **7.** 9   **9.** $a^2bc - ab^2c + 2abc^2$
11. $x^2 + 7x + 10$ **13.** $x^2 - 4xy + 4y^2$ **15.** $10x^2 + 17x + 3$ **17.** $18a^2 - 8b^2$
19. $-2ac + bc + 6ad - 3bd$ **21.** $x^3 + 6x^2 + 7x - 4$ **23.** $6x^3 + 7x^2 - 6x + 1$ **25.** $a^4 - ab^3$
27. $4a + 4$ **29.** $20x^2 + 4x$ **31.** $5y^2 - 12y + 5$ **33.** $6z^2 - 24z + 32$
35. $P(x - 1) = x^2 - 5x + 11;\ \ P(2 - x) = x^2 - x + 5$ **37.** $[P(3)]^2 = 36;\ \ P(3^2) = -78$
39. $(a + b)^2 - (a^2 + b^2) = 2ab$; for $a^2 + b^2 > (a + b)^2$ the product $a \cdot b$ has to be less than 0 ($a > 0$ and $b < 0$ or $a < 0$ and $b > 0$); for $a^2 + b^2 < (a + b)^2$ the product $a \cdot b$ has to be greater than 0 ($a > 0$ and $b > 0$ or $a < 0$ and $b < 0$).

Exercise 2.3 (page 31)   **1.** $(3x^3y)(3x^2 - x + 2)$ **3.** $(x - 4)(x + 1)$ **5.** $(x - 6)(x - 2)$
7. $(y + 2)(y + 2)$ **9.** $(x + 5)(x - 5)$ **11.** $(2x + 3)(x + 1)$ **13.** $(3a + 1)(2a + 1)$
15. $2(3z + 1)(z + 1)$ **17.** $x^2y^2(x + 1)(x - 1)$ **19.** $(x^n + 1)(x^n - 1)$ **21.** $x^n(x^2 - x + 2)$
23. $(y^2 + 2)(y^2 + 1)$ **25.** $(2a^2 + 1)(a + 1)(a - 1)$ **27.** $(-x + y)(3x - y)[x^2 + (y - 2x)^2]$
29. $(x + y)(x + a)$ **31.** $(a - 2b)(a^2 + 2b^2)$ **33.** $(y - 3x)(y^2 + 3xy + 9x^2)$
35. $(2x - y)(x^2 - xy + y^2)$ **37.** $(a^n - 2)(a^n + 2)$ **39.** $(x^n - y^n)(x^n + y^n)(x^{2n} + y^{2n})$
41. $(3x^{2n} - 1)(x^{2n} - 3)$ **43.** $2(y^n - 30)(y^n + 24)$
45. $ac - ad + bd - bc = a(c - d) - b(c - d) = (a - b)(c - d)$;
also $ac - ad + bd - bc = bd - ad - bc + ac = (b - a)d - (b - a)c = (b - a)(d - c)$
47. $(x^2 + xy + y^2)(x^2 - xy + y^2)$

Exercise 2.4 (page 35)   **1.** $4ay^2$ $(a, y \ne 0)$ **3.** $2a^2b^2$ $(a, b \ne 0)$ **5.** x^n $(x \ne 0)$
7. x^ny^n $(x, y \ne 0)$ **9.** $4a^2 + 2a + 2$ **11.** $x^2 - 4x - 3$ $(x \ne 0)$

13. $3x^2 - 2x + \dfrac{3}{5x}$ $(x \ne 0)$ **15.** $2x - 5$ **17.** $2y^2 - y + \dfrac{13}{2},\ \ R = -\dfrac{3}{2}$ $\left(y \ne -\dfrac{1}{2}\right)$

19. $x^2 - 2x + 3 + \dfrac{6}{2x - 1}$ **21.** $2y^3 - y^2 - \dfrac{1}{2}y - \dfrac{5}{4} + \dfrac{-13y/4 + 9/4}{2y^2 + y + 1}$ **23.** $x^2 - 2x + 3$

Exercise 2.5 (page 38)   **1.** $2x - 5$ **3.** $z - 6$ **5.** $3z^2 + 26z + 99 + \dfrac{400}{z - 4}$

7. $x^3 - x^2 - \dfrac{1}{x - 2}$ $(x \ne 2)$ **9.** $2x^2 - 2x + 3 + \dfrac{-8}{x + 1}$ $(x \ne -1)$

11. $2x^3 + 10x^2 + 50x + 249 + \dfrac{1251}{x - 5}$ $(x \ne 5)$ **13.** $x^2 + 2x - 3 + \dfrac{4}{x + 2}$ $(x \ne -2)$

15. $x^5 + x^4 + 2x^3 + 2x^2 + 2x + 1 + \dfrac{1}{x-1}$ $(x \neq 1)$ **17.** $x^4 + x^3 + x^2 + x + 1$ $(x \neq 1)$

19. Calculations produce only $+1$ for successive sums, hence they always produce 0 when combined with -1 of $x^n - 1$.

Exercise 2.6 (page 40)   **1.** $\dfrac{a}{b}$ $(a, b, c \neq 0)$ **3.** 2 $(x + y \neq 0)$ **5.** -1 $(b - a \neq 0)$

7. $-(x + 1)$ $(x \neq 1)$ **9.** $x^2 - 2x - \dfrac{3}{2}$ $(x \neq 0)$ **11.** $y + 7$ $(y \neq 2)$ **13.** $-(2y + 5)$ $\left(y \neq \dfrac{1}{2}\right)$

15. $\dfrac{x + 3}{x - 1}$ $(x \neq 1, -3)$ **17.** $\dfrac{y - 1}{y + 1}$ $(y \neq 1, -1)$ **19.** $\dfrac{n - 5}{n + 4}$ $(n \neq 3, -4)$ **21.** $\dfrac{3x + 9}{x - 8}$ $(x \neq 3, 8)$

23. $\dfrac{x^2 + xy + y^2}{x + y}$ $(x \neq y, -y)$ **25.** $\dfrac{y^2 + 4}{y^2 + 3}$ $(y \neq 2, -2)$ **27.** $\dfrac{9}{12}$ **29.** $\dfrac{ab^2}{a^2b}$ $(a, b \neq 0)$

31. $\dfrac{3(y - 3)}{y^2 - y - 6}$ $(y \neq -2, 3)$ **33.** $\dfrac{3(a^2 - 3a + 9)}{a^3 + 27}$ $(a \neq -3)$ **35.** No, $x = 1, 2$

37. For $N > 0$, $a - b < 0$ or $a < b$;   for $N < 0$, $a - b > 0$ or $a > b$

Exercise 2.7 (page 44)   **1.** $\dfrac{2x - 1}{2y}$ **3.** 1 **5.** $\dfrac{2a + 9}{9}$ **7.** $\dfrac{5}{2a + 2b}$ **9.** $\dfrac{3}{3 - x}$

11. $\dfrac{-1}{(a + 2)(a + 3)}$ **13.** $\dfrac{-3x^2 - 3y^2}{(2x - y)(x - 2y)}$ **15.** $\dfrac{2y - 1}{(y - 2)(y + 1)(y + 1)}$ **17.** $\dfrac{n^2 - n}{(n + 5)(n + 3)(n - 2)}$

19. $\dfrac{y(y - 11)}{(y - 4)(y + 4)(y - 1)}$ **21.** $\dfrac{x^3 - 2x^2 + 2x - 2}{(x - 1)^2}$ **23.** $\dfrac{4x^3 - 4x^2 + 4x + 4}{(2x + 1)(2x - 1)}$ **25.** $\dfrac{4}{(y + 1)(y + 3)}$

27. $\dfrac{y^2 + xz}{(y - x)(y - z)}$ **29.** 0

Exercise 2.8 (page 47)   **1.** $\dfrac{-b^2}{a}$ **3.** $\dfrac{1}{ax^2y}$ **5.** $\dfrac{2x}{x + 5}$ **7.** $(5ab - 4)(4ab + 3)$ **9.** $x - y$

11. $\dfrac{x^2 - 1}{x^2}$ **13.** $\dfrac{x + 5}{x(x + 1)}$ **15.** $\dfrac{2y^2 - 6y + 3}{3}$ **17.** $\dfrac{7}{10a + 2}$ **19.** $\dfrac{4a^2 - 3a}{4a + 1}$ **21.** $\dfrac{-1}{y - 3}$

23. $\dfrac{a - 2b}{a + 2b}$ **25.** $\dfrac{a + 6}{a - 1}$ **27.** $\dfrac{x - y - z}{x + y + z}$ **29.** $\dfrac{b^2}{3(5b - a)}$

Exercise 2.9 (page 51)   **1.** $\dfrac{2}{x} + \dfrac{-2}{x + 1}$ **3.** $\dfrac{2}{x} + \dfrac{4}{x + 1}$ **5.** $\dfrac{-1/3}{y + 2} + \dfrac{1/3}{y - 1}$ **7.** $\dfrac{14/5}{z + 4} + \dfrac{1/5}{z - 1}$

9. $\dfrac{2}{x} + \dfrac{-3}{x + 2} + \dfrac{1}{x - 1}$ **11.** $\dfrac{3}{x + 2} + \dfrac{7/2}{x + 1} + \dfrac{-7/2}{x - 1}$ **13.** $\dfrac{1/4}{x} + \dfrac{1/12}{x + 4} + \dfrac{-1/3}{x + 1}$ **15.** $\dfrac{-1}{x} + \dfrac{1}{x^2} + \dfrac{1}{x + 1}$

17. $\dfrac{-2}{x} + \dfrac{2}{x - 2} + \dfrac{-3}{(x - 2)^2}$ **19.** $\dfrac{-1}{y^2} + \dfrac{-1/2}{y + 1} + \dfrac{1/2}{y - 1}$ **21.** $2x + \dfrac{1}{x + 1} + \dfrac{1}{x - 1}$

23. $x + 1 + \dfrac{1}{x - 1} + \dfrac{2x}{(x - 1)^2}$

Chapter 2 Review (page 52)   **1.** $y^2 - 1$ **2.** $16x - 3$ **3.** 17 **4.** 24 **5.** $6x^2 + 10x - 4$

6. $20z^2 - 4z - 16$ **7.** $(z - 4)(z - 3)$ **8.** $u(u^2 + 1)(u + 1)(u - 1)$ **9.** $(2y^n + 1)(2y^n - 1)$

10. $(2t - 3)(4t^2 + 6t + 9)$ **11.** $8xy$ **12.** $3z^2 - 2z + 9$ **13.** $2n + 1$ **14.** $r + 1$

15. $2x^2 - 3x + 1$ **16.** $3y^2 - 2y + 3 + \dfrac{1}{y + 3}$ **17.** $2(2rs - s + 3)$ $(r, s \neq 0)$

18. $2(x - 1)$ $(x \neq -1)$ **19.** $\dfrac{x - 2}{x + 2}$ $(x \neq 2, x \neq -2)$ **20.** $\dfrac{4(n + 1)}{4 - n}$ $(n \neq 4, n \neq -4)$

21. $\dfrac{3x + 4}{8}$ **22.** $\dfrac{-1}{2x - 1}$ **23.** $\dfrac{10 - 3x}{2(x - 2)}$ **24.** $\dfrac{y + 4}{[y - (-1)](y - 2)}$ **25.** $\dfrac{5xy}{7ab}$ **26.** 1 **27.** $\dfrac{1}{12rt}$

28. $\dfrac{x}{a}$ **29.** $\dfrac{3(a-b)}{a+b}$ **30.** -1 **31.** $\dfrac{7}{2x^2+5x-3}=\dfrac{2}{2x-1}+\dfrac{-1}{x+3}$

32. $\dfrac{2x^2+3x-1}{x^3-x}=\dfrac{1}{x}+\dfrac{-1}{x+1}+\dfrac{2}{x-1}$

Exercise 3.1 (page 57)   **1.** $\dfrac{1}{5}$ **3.** $\dfrac{1}{9}$ **5.** 27 **7.** $\dfrac{1}{10}$ **9.** $\dfrac{5}{3}$ **11.** $\dfrac{9}{5}$ **13.** $\dfrac{82}{9}$ **15.** $\dfrac{3}{16}$

17. x^2y^3 **19.** x^6y^4 **21.** $\dfrac{9x^2}{y^6}$ **23.** $\dfrac{4x^4}{y}$ **25.** x^4 **27.** $\dfrac{1}{x^6}$ **29.** $\dfrac{y}{x}$ **31.** $4x^5y^2$

33. $\dfrac{y^2-x}{xy^2}$ **35.** $\dfrac{y^2+x^2}{xy}$ **37.** $\dfrac{x}{x-y}$ **39.** $\dfrac{x^2+y^2}{xy}$ **41.** $y+x$ **43.** $\dfrac{xy}{y-x}$ **45.** x^{2n-1}

47. 1 **49.** y^4 **51.** x^6 **53.** $\dfrac{y}{x^{n-1}}$ **55.** $\dfrac{1}{x^{2n-2}}$ **57.** 2.54×10^3 **59.** 6.42×10^5

61. 1.4×10^{-3} **63.** 2.30×10^{-5}

Exercise 3.2 (page 62)   **1.** 4 **3.** 9 **5.** $\dfrac{1}{4}$ **7.** $\dfrac{1}{8}$ **9.** $\dfrac{1}{4}$ **11.** $\dfrac{8}{27}$ **13.** x^2 **15.** $y^{5/4}$

17. $x^{1/3}$ **19.** $y^{1/6}$ **21.** $\dfrac{1}{y^2}$ **23.** $\left(\dfrac{x}{y}\right)^{5/4}$ **25.** $x^n\cdot y^4$ **27.** $x^{3n/2}$ **29.** $x^{5n/2}\cdot y^{3m/2-1}$

31. $x^{n-1}y^{m+1}$ **33.** $x-x^{1/2}$ **35.** $y^{5/3}-y$ **37.** $x-y$ **39.** $(x+y)-(x+y)^{3/2}$
41. $x(x^{1/2}+1)$ **43.** $x^{-1/2}(x^{-1}+1)$ **45.** $(x+1)^{-1/2}(x)$ **47.** 5 **49.** $2|x|$
51. $\dfrac{2}{|x|(x+1)^{1/2}}$ $(x>-1)$

Exercise 3.3 (page 65)   **1.** $\sqrt[3]{2}$ **3.** $3\sqrt[3]{x}$ **5.** $\sqrt{2y}$ **7.** $\sqrt[3]{x^2}$ **9.** $x\sqrt[3]{y^2}$ **11.** $\sqrt[4]{x^3y}$

13. $\sqrt{x-y}$ **15.** $\dfrac{1}{\sqrt[3]{(x^2+y)^2}}$ **17.** $y^{2/3}$ **19.** $(2xy^2)^{1/4}$ **21.** $3(x^3y)^{1/4}$ **23.** $(a+b^2)^{1/3}$

25. 9 **27.** -3 **29.** x^2y **31.** $\dfrac{2}{3}x^3y^5$ **33.** $2x^2\sqrt{x}$ **35.** $xy\sqrt[4]{3xy}$ **37.** $3y\sqrt{2x}$

39. 2 **41.** $b\sqrt[3]{2ab^2}$ **43.** ab^2 **45.** $\dfrac{2\sqrt{3}}{3}$ **47.** $\dfrac{\sqrt{2xy}}{2y}$ **49.** $\dfrac{\sqrt[3]{2x}}{x}$ **51.** $\dfrac{\sqrt[3]{3y^2}}{y}$

53. $\dfrac{1}{\sqrt{3}}$ **55.** $\dfrac{x}{\sqrt{xy}}$ **57.** $\sqrt[3]{9}$ **59.** $2\sqrt{x}$ **61.** $\sqrt{2x}$ **63.** $(x-1)^2$

Exercise 3.4 (page 68)   **1.** $8\sqrt{2}$ **3.** $-\sqrt{5}$ **5.** $8\sqrt{3}$ **7.** $-4\sqrt{2}$ **9.** $5\sqrt[3]{2}$ **11.** $7\sqrt[4]{3}$
13. $2\sqrt{2}-2$ **15.** $2\sqrt{3}+\sqrt{6}$ **17.** $1-\sqrt{5}$ **19.** 1 **21.** $x-\sqrt{3}x-6$ **23.** $2-x$

25. $2(1-\sqrt{3})$ **27.** $\dfrac{x(\sqrt{x}+3)}{x-9}$ **29.** $\dfrac{37-8\sqrt{10}}{27}$ **31.** $\dfrac{-1}{2(1+\sqrt{2})}$ **33.** $\dfrac{1-x-a}{\sqrt{x+a}+x+a}$

35. $-3\sqrt{2}+3$ **37.** $2|x|$ **39.** $3x\sqrt{x-1}$ $(x\geq1)$

41.   Part A: By Definition 3.5, $\sqrt[n]{a^n}=(a^n)^{1/n}$; if n is an odd natural number, a^n is positive or negative as a is positive or negative, and $(a^n)^{1/n}$, by Definition 3.3, is the positive or negative value of a, accordingly.
Part B: By Definition 3.5, $\sqrt[n]{a^n}=(a^n)^{1/n}$; if n is an even natural number, a^n is positive and, by Definition 3.3, $(a^n)^{1/n}$ is the positive root for all values of a.

Exercise 3.5 (page 72)   **1.** $0.\overline{285714}$ **3.** 0.9375 **5.** 0.2 **7.** -4.35 **9.** 3.79 **11.** 0.58
13. -9.66 **15.** 0.09 **17.** 1.12 **19.** $3,162,\ 3.163$ **21.** $1.731010010001\cdots$

Chapter 3 Review (page 72)   **1.** $x^6 y^{12}$ **2.** $\dfrac{x^3}{y^4}$ **3.** x^2 **4.** $\dfrac{x^2}{y^3}$ **5.** $\dfrac{x^2 y^2 + 1}{xy}$ **6.** $\dfrac{y - x^2}{x}$

7. x^{2n-1} **8.** $x^{2n-2} y^4$ **9.** 4.73×10^4 **10.** 4.5×10^{-5} **11.** $x^{11/6}$ **12.** $\dfrac{y^{1/2}}{x^{1/3}}$

13. $\dfrac{1}{x^{n/2} y^n}$ **14.** xy **15.** $y - y^{3/2}$ **16.** $y^2 - y$ **17.** $x^{-1/4}(1 + x^{3/4})$

18. $x^{-2/3}(x^{4/3} - 1)$ **19.** $3xy^2 \sqrt{xy}$ **20.** $4x\sqrt{y}$ **21.** $y\sqrt{2x}$ **22.** $3x$ **23.** $\dfrac{\sqrt[3]{xy^2}}{y}$

24. $\dfrac{1}{2\sqrt{3y}}$ **25.** $\sqrt{3}$ **26.** $\sqrt{2xy}$ **27.** $4\sqrt{2}$ **28.** $10\sqrt[3]{3}$ **29.** 3 **30.** $3x + 2\sqrt{x} - 8$

31. $2 + \sqrt{3}$ **32.** $\dfrac{2x - 3\sqrt{x} + 1}{x - 1}$ **33.** $\dfrac{-7}{2\sqrt{2} - 6}$ **34.** $\dfrac{y - 9}{y + 4\sqrt{y} + 3}$ **35.** 9.53 **36.** 4.16

Exercise 4.1 (page 77)   **1.** $\{-10\}$ **3.** $\{-2\}$ **5.** $\{0\}$ **7.** $\left\{\dfrac{1}{3}\right\}$ **9.** $\{-7\}$ **11.** $\{15\}$

13. $\emptyset$ **15.** $\left\{-\dfrac{14}{5}\right\}$ **17.** $k = v - gt$ **19.** $c = \dfrac{2A - bh}{h}$ $(h \neq 0)$ **21.** $n = \dfrac{l - a + d}{d}$ $(d \neq 0)$

23. $y' = \dfrac{3x + 1}{x^2 - 2y^3}$ $(x^2 \neq 2y^3)$ **25.** $x_1 = \dfrac{x_4}{x_2 - 2x_3}$ $(x_2 \neq 2x_3)$ **27.** $y = 6(x - x_1) + y_1$ $(x \neq x_1)$

29. $-\dfrac{3}{5}$

Exercise 4.2 (page 81)   **1.** $\{0, -2\}$ **3.** $\{2, -7\}$ **5.** $\left\{-\dfrac{1}{2}, 3\right\}$ **7.** $\left\{\dfrac{4}{3}, -\dfrac{1}{2}\right\}$ **9.** $\left\{\dfrac{1}{2}, 1\right\}$

11. $\{1, -\tfrac{10}{3}\}$ **13.** $\{2, -2\}$ **15.** $\{\sqrt{5}, -\sqrt{5}\}$ **17.** $\{6 + \sqrt{5}, 6 - \sqrt{5}\}$ **19.** $\{2, -6\}$

21. $\{-4, -5\}$ **23.** $\left\{\dfrac{1}{2}, -2\right\}$ **25.** $(x - 2)^2 + (y - 2)^2 = 5^2$ **27.** $[x - (-3)]^2 + (y - 1)^2 = 2^2$

29. $\left(x - \dfrac{1}{2}\right)^2 + [y - (-1)]^2 = 2^2$ **31.** $y = (x - 1)^2 + 4$ **33.** $y = [x - (-4)]^2 + (-12)$

35. $y = \left(x - \dfrac{3}{2}\right)^2 + \dfrac{11}{4}$ **37.** $x^2 - 5x + 6 = 0$ **39.** $6x^2 + x - 2 = 0$

Exercise 4.3 (page 84)   **1.** $\{1, 2\}$ **3.** $\left\{2, \dfrac{3}{2}\right\}$ **5.** $\left\{3, -\dfrac{3}{2}\right\}$ **7.** $\{\sqrt{5}\}$ **9.** $\left\{\sqrt{3}, -\dfrac{\sqrt{3}}{2}\right\}$

11. $\{2k, -k\}$ **13.** $\left\{\dfrac{1 + \sqrt{1 - 4ac}}{2a}, \dfrac{1 - \sqrt{1 - 4ac}}{2a}\right\}$ **15.** 4 **17.** Let $r_1 = \dfrac{-b + \sqrt{b^2 - 4ac}}{2a}$ and

$r_2 = \dfrac{-b - \sqrt{b^2 - 4ac}}{2a}$ and then simplify $r_1 + r_2$ and $r_1 \cdot r_2$ **19.** $-\dfrac{3}{2}, -3$ **21.** $-\dfrac{5}{3}$

Exercise 4.4 (page 87)   **1.** $\{64\}$ **3.** $\{-7\}$ **5.** $\{4\}$ **7.** $\{-25\}$ **9.** $\{16\}$ **11.** $\{1\}$ **13.** $\{5\}$

15. $\{16\}$ **17.** $\{4\}$ **19.** $\{1, 3\}$ **21.** $A = \pi r^2$ **23.** $y = \dfrac{1}{x^3}$ **25.** $y = \pm\sqrt{a^2 - x^2}$

Exercise 4.5 (page 88)   **1.** $\{25\}$ **3.** $\left\{\dfrac{\sqrt{2}}{2}, -\dfrac{\sqrt{2}}{2}\right\}$ **5.** $\{2, -7, -3, -2\}$ **7.** $\{64, -8\}$

9. $\left\{\dfrac{1}{4}, -\dfrac{1}{3}\right\}$ **11.** $\{626\}$ **13.** $\{2, -1, -4\}$ **15.** $\{4, 1\}$ **17.** $\{13\}$ **19.** $\{4, 5, -2, -3\}$

Exercise 4.6 (page 91)   **1.** $\{x \mid x > 1\}$ **3.** $\{x \mid x > 3\}$ **5.** $\left\{x \mid x \le \dfrac{13}{2}\right\}$ **7.** $\{x \mid x < -24\}$

9. $\{x \mid -2 \le x < 4\}$ **11.** $\left\{x \mid \dfrac{2}{3} < x < 2\right\}$ **13.** $\{x \mid -2 < x < 5\}$ **15.** $\{x \mid x \ge 13\}$

17. $\left\{x \mid x \le \dfrac{1}{3} \text{ or } x > 11\right\}$ **19.** $\{x \mid x < 0 \text{ or } x > 2\}$ **21.** $k \le -2$ **23.** $k < 0$

Exercise 4.7 (page 95)   **1.** $\{x \mid x < -1 \text{ or } x > 2\}$ **3.** $\{x \mid 0 \le x \le 2\}$ **5.** $\{x \mid x < -1 \text{ or } x > 4\}$

7. $\{x \mid -\sqrt{5} < x < \sqrt{5}\}$ **9.** $\{x \mid x \in R\}$ **11.** $\left\{x \mid x < 0 \text{ or } x \ge \dfrac{1}{2}\right\}$ **13.** $\left\{x \mid -\dfrac{8}{3} < x < -2\right\}$

15. $\{x \mid x < 0 \text{ or } 2 < x \le 4\}$ **17.** $\{x \mid -3 < x < 0 \text{ or } x > 2\}$

19. Since $a, b > 0$ and $(a+b)^2 = a^2 + 2ab + b^2$, we have $a^2 > 0, b^2 > 0, 2ab > 0$ and $(a+b)^2 > 0$; $(a+b)^2 = (a^2 + b^2) + 2ab$; hence $(a+b)^2 > a^2 + b^2$.

21. For all $x \in R$, $(x - 1) \in R$ and $(x-1)^2 \ge 0$, since the square of all positive or negative real numbers is either positive or zero. Then, since $(x-1)^2 = x^2 - 2x + 1$, it follows that $x^2 - 2x + 1 \ge 0^2$; adding $2x$ to both sides produces $x^2 + 1 \ge 2x$.

Exercise 4.8 (page 99)   **1.** $\{6, -6\}$ **3.** $\{-3, 5\}$ **5.** $\left\{\dfrac{1}{3}, 1\right\}$ **7.** $\left\{-\dfrac{3}{2}, -\dfrac{7}{2}\right\}$ **9.** $\left\{\dfrac{1}{2}, \dfrac{7}{2}\right\}$

11. $\left\{\dfrac{22}{3}, -\dfrac{26}{3}\right\}$ **13.** $\{x \mid -2 < x < 2\}$ **15.** $\{x \mid -7 \le x \le 1\}$ **17.** $\emptyset$

19. $\{x \mid x < -1 \text{ or } x > 7\}$ **21.** $\{x \mid x \le -3 \text{ or } x \ge 2\}$ **23.** $\left\{x \mid -\dfrac{13}{2} < x < \dfrac{11}{2}\right\}$

25. $|x - 2| < 1$ **27.** $|x + 8| \le 1$ **29.** $|4x - 5| \le 19$

31. (I) If $a \ge 0$, by Def. 1.15, $|-a| = |-(+a)| = |-a| = a$; also $|a| = a$; hence $|-a| = |a|$.
(II) If $a < 0$, by Def. 1.15, $|-a| = |-(-a)| = |a| = a$; also $|a| = |+(-a)| = |-a| = a$; hence $|-a| = |a|$.

33. $|a^2| = a^2$ (Def. 1.15); $|a|^2 = |a| \cdot |a| = a \cdot a = a^2$; hence $|a^2| = |a|^2 = a^2$.

Exercise 4.9 (page 101)   9 nickels and 7 dimes   **3.** 15 lbs of 32% silver alloy must be melted.

5. \$1400 at 3% and \$600 at 4%   **7.** \$12,000 at 5% and \$36,000 at 3%

9. Rate of automobile—60 mph; rate of plane—180 mph   **11.** 240 miles

13. 7 and 8 are the two numbers.   **15.** 6 and 7; -7 and -6 **17.** $\sqrt{13,000}$

19. $\dfrac{1}{2}$ second to reach 24 ft on way up; $3\dfrac{1}{2}$ seconds to return to original position

21. 12 days for the boy alone and 6 days for his father alone   **23.** 10 ft   **25.** $14°$ to $68°$

Chapter 4 Review (page 105) $\left\{-\dfrac{7}{2}\right\}$ **2.** $\{4\}$ **3.** $\left\{-\dfrac{26}{21}\right\}$ **4.** $\left\{\dfrac{5}{2}\right\}$ **5.** $y = \dfrac{1}{4}x$

6. $x = 4y$ **7.** $\left\{\dfrac{1}{2}, 1\right\}$ **8.** $\left\{-\dfrac{1}{2}, 3\right\}$ **9.** $\{\sqrt{7}, -\sqrt{7}\}$ **10.** $\{-4 + \sqrt{3}, -4 - \sqrt{3}\}$

11. $\left\{-\dfrac{3}{2} + \dfrac{\sqrt{13}}{2}, -\dfrac{3}{2} - \dfrac{\sqrt{13}}{2}\right\}$ **12.** $\left\{\dfrac{1}{2}, -1\right\}$ **13.** $(x-4)^2 + [y-(-3)^2] = (\sqrt{37})^2$

14. $y = \left(x - \dfrac{b}{2}\right)^2 + \dfrac{8 - b^2}{4}$ **15.** $\left\{\dfrac{1 + \sqrt{17}}{4}, \dfrac{1 - \sqrt{17}}{4}\right\}$ **16.** $\left\{\dfrac{-k + \sqrt{k^2 + 16}}{2}, \dfrac{-k - \sqrt{k^2 + 16}}{2}\right\}$

17. $k = 9$ **18.** $k = 4$, or $k = -4$ **19.** $\{4, 1\}$ **20.** $\{8\}$ **21.** $\{2, -2, 1, -1\}$

22. $\left\{\dfrac{1}{5}, -\dfrac{1}{4}\right\}$ **23.** $x \le 27$ **24.** $x > \dfrac{3}{2}$ **25.**

26.   **27.** $-5 < x < 1$ **28.** $x > \dfrac{7}{2}$ or $x < 3$ **29.** $\{3, -4\}$

30. $\left\{1, -\dfrac{1}{3}\right\}$ **31.**   **32.** $|x - 4| \le 7$

33. 2113 and 2263 votes   **34.** $6''$ **35.** $5\dfrac{1}{3}$ hours   **36.** 6 mph going; 4 mph returning

Exercise 5.1 (page 111)   **1.** $\left\{(0, 6), (1, 4), (2, 2), (-3, 12), \left(\dfrac{2}{3}, \dfrac{14}{3}\right)\right\}$

3. $\left\{(0, 0), (1, -3), (2, 3), \left(-3, -\dfrac{9}{7}\right), \left(\dfrac{2}{3}, -\dfrac{9}{7}\right)\right\}$

5. $\left\{(0, \sqrt{11}), (1, \sqrt{14}), (2, \sqrt{17}), (-3, \sqrt{2}), \left(\dfrac{2}{3}, \sqrt{13}\right)\right\}$

7. **(a)** Domain: $\{2, 5, 7\}$; Range: $\{3, 7, 8\}$ **(b)** A function
9. **(a)** Domain: $\{2, 3\}$; Range: $\{-1, 4, 6\}$ **(b)** Not a function
11. **(a)** Domain: $\{5, 6, 7\}$; Range: $\{5, 6, 7\}$ **(b)** A function
13. Domain: $\{x \mid x \in R\}$ **15.** Domain: $\{x \mid x \in R\}$ **17.** Domain: $\{x \mid x \in R, x \ne 2\}$
19. Domain: $\{x \mid x \in R, x \ge 0\}$ **21.** Domain: $\{x \mid x \in R, -2 \le x \le 2\}$
23. Domain: $\{x \mid x \in R, x \ne 0, 1\}$ **25.** Yes   **27.** Yes   **29.** No   **31.** No   **33.** 2
35. -1 **37.** 9   **39.** a^2 **41.** 1   **43.** $a - 2$ **45.** ± 1 **47.** ± 3
49. **a.** 2   **b.** 0   **c.** 2   **d.** x **51.** **a.** even   **b.** odd

Exercise 5.2 (page 117)

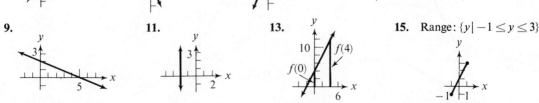

17. Distance, 5; slope, $\dfrac{4}{3}$ **19.** Distance, 13; slope $\dfrac{12}{5}$ **21.** Distance, $\sqrt{2}$; slope, 1

23. Distance, $3\sqrt{5}$; slope, $\frac{1}{2}$ **25.** Distance, 5; slope 0   **27.** Distance, 10; slope, undefined

29. 7, $\sqrt{68}$, $\sqrt{89}$ **31.** 10, 21, 17

33. From Problem 30, the lengths of the sides of the triangle are 15, $3\sqrt{5}$, and $6\sqrt{5}$, respectively; since $15^2 = 225$, $(3\sqrt{5})^2 = 45$, and $(6\sqrt{5})^2 = 180$, it follows that $15^2 = (3\sqrt{5})^2 + (6\sqrt{5})^2$, and the converse of the Pythogorean theorem applies. Hence, the triangle is a right triangle.   **35.** $x - 2y = 8$

37. This follows from the fact that the midpoint of the segment

PR is $\left(\dfrac{x_1 + x_2}{2}, y_1\right)$ and the midpoint of QR is $\left(x_2, \dfrac{y_1 + y_2}{2}\right)$.

Then, since $\triangle PMO$ is similar to $\triangle PQR$, $PO = \dfrac{1}{2} PQ$. Hence,

$\left(\dfrac{x_1 + x_2}{2}, \dfrac{y_1 + y_2}{2}\right)$ is the midpoint of PQ.

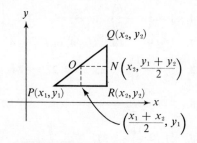

Exercise 5.3 (page 120)   **1.** $4x - y - 7 = 0$ **3.** $x + y - 10 = 0$ **5.** $3x - y = 0$

7. $x + 2y + 2 = 0$ **9.** $3x + 4y + 14 = 0$ **11.** $y - 2 = 0$

13. $y = -x + 3$; slope, -1; intercept, 3   **15.** $y = -\dfrac{3}{2}x + \dfrac{1}{2}$; slope, $-\dfrac{3}{2}$; intercept, $\dfrac{1}{2}$

17. $y = \dfrac{1}{3}x - \dfrac{2}{3}$; slope, $\dfrac{1}{3}$; intercept, $-\dfrac{2}{3}$ **19.** $3x + 2y - 6 = 0$ **21.** $5x + 2y + 10 = 0$

23. $6x - 2y + 3 = 0$ **25.** $3x - y - 5 = 0$ **27.** $2x - 3y + 19 = 0$ **29.** $x - 2 = 0$

31. Substituting $\dfrac{y_2 - y_1}{x_2 - x_1}$ for m in the formula $y - y_1 = m(x - x_1)$ gives the desired result.

33. $F(x) = \dfrac{-x + 11}{3}$

Exercise 5.4 (page 124)

1.

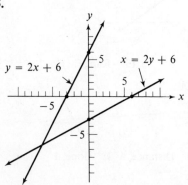

3.

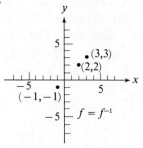

5.

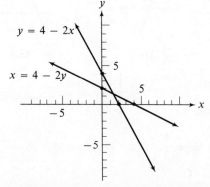

7.

9.

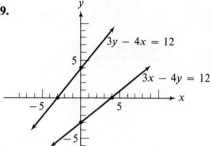

11.

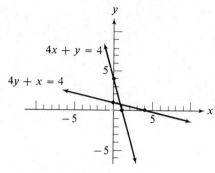

13. $F(x) = x$, $F^{-1}(x) = x$; $F[F^{-1}(x)] = F(x) = x$, $F^{-1}[F(x)] = F^{-1}(x) = x$.

15. $F(x) = 4 - 2x$, $F^{-1}(x) = \dfrac{4-x}{2}$; $F[F^{-1}(x)] = F\left[\dfrac{4-x}{2}\right] = 4 - 2\left(\dfrac{4-x}{2}\right) = 4 - 4 + x = x$,

$F^{-1}[F(x)] = F^{-1}[4 - 2x] = \dfrac{4 - (4 - 2x)}{2} = \dfrac{4 - 4 + 2x}{2} = \dfrac{2x}{2} = x$.

17. $F(x) = \dfrac{3x - 12}{4}$, $F^{-1}(x) = \dfrac{4x + 12}{3}$; $F[F^{-1}(x)] = F\left[\dfrac{4x + 12}{3}\right] = \dfrac{(4)\left(\dfrac{3x - 12}{4}\right) + 12}{3} =$

$\dfrac{3x - 12 + 12}{3} = \dfrac{3x}{3} = x$, $F^{-1}[F(x)] = F^{-1}\left[\dfrac{3x - 12}{4}\right] = \dfrac{(3)\left(\dfrac{4x + 12}{3}\right) - 12}{4} = \dfrac{4x + 12 - 12}{4} = \dfrac{4x}{4} = x$.

Exercise 5.5 (page 127)

1.

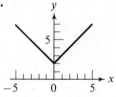

3.

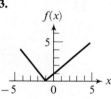

5.

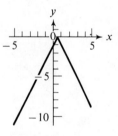

7.

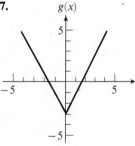

9.

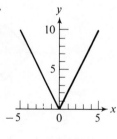

11.

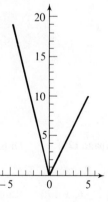

13.

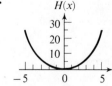

15.

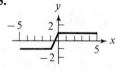

17.

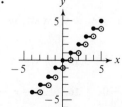

19. 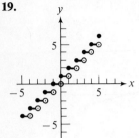   **21.** 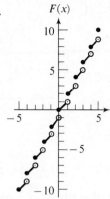   **23.**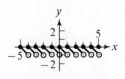

25. Cost in cents: $C = -c[-x], \quad x > 0$

Exercise 5.6 (page 129)

1.   **3.**   **5.** 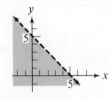   **7.**

9.   **11.**   **13.**   **15.**

17.   **19.** 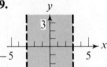   **21.**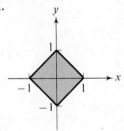

Chapter 5 Review (page 129)   **1.** Domain: $\{4, 2, 3\}$;   Range: $\{-1, -4, -5\}$

2. Domain: $\{1\}$;   Range: $\{2, 3, 6\}$ **3.** Domain: $\{x \mid x \neq -4, x \in R\}$;   Range: $\{y \mid y \neq 0, y \in R\}$

4. Domain: $\{x \mid x \geq 6, x \in R\}$;   Range: $\{y \mid y \geq 0, y \in R\}$ **5.** 13   **6.** -5 **7.** 1

8. $x^2 + 2xh + h^2 + 4$ **9.** $\sqrt{41}, m = -\dfrac{4}{5}$ **10.** $\sqrt{5}, m = -\dfrac{1}{2}$ **11.** $4x - y - 15 = 0$

12. $x - 2y + 12 = 0$ **13.** $m = -4$, y-intercept $= 6$ **14.** $m = \dfrac{3}{2}$, y-intercept $= -8$

15. $2x - 3y + 6 = 0$ **16.** $12x - y - 4 = 0$ **17.** $4x - 3y + 11 = 0$ **18.** $x + 6y + 21 = 0$

19.

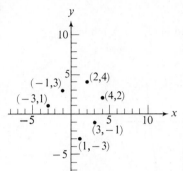

20.

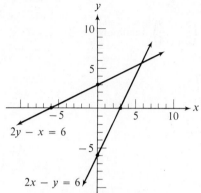

21.

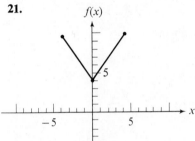

22.

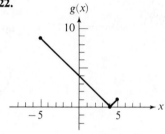

23.

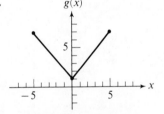

24.

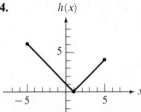

25.

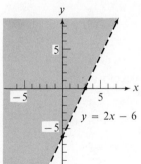

26.

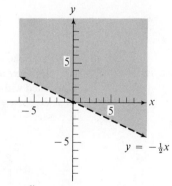

27.

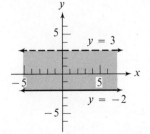

28.

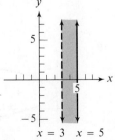

Exercise 6.1 (page 135)

1. y, 4; x, 1 and 4; $\left(\dfrac{5}{2}, -\dfrac{9}{4}\right)$, minimum   **3.** $y, -7$; x, -1 and 7; $(3, -16)$, minimum

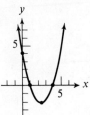
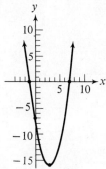

5. $y, -4$; x, 1 and 4; $\left(\dfrac{5}{2}, \dfrac{9}{4}\right)$, maximum   **7.** y, 2; no x; $(0, 2)$, minimum

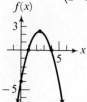

9.   **11.** 4; 4

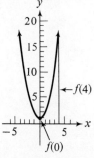

13. Varying k has the effect of translating the graph along the y-axis.

15. **a.** Parabola
b. No; there are two values of y associated with each value of x, except $x = 0$.
c. No

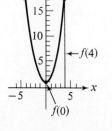

17.   **19.**

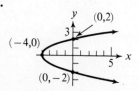
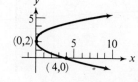

21.
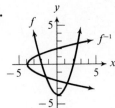

23.
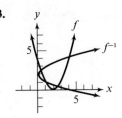

25.
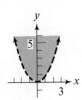

27.

29.

31.
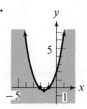

33.
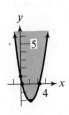

35.
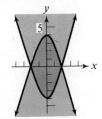

Exercise 6.2 (page 142)   **1. a.** $y = \pm\sqrt{4 - x^2}$ **b.** $y = \sqrt{4 - x^2}$; $y = -\sqrt{4 - x^2}$

  c. Domain $= \{x \mid -2 \le x \le 2\}$ **3. a.** $y = \pm 3\sqrt{4 - x^2}$ **b.** $y = 3\sqrt{4 - x^2}$; $y = -3\sqrt{4 - x^2}$

  c. Domain $= \{x \mid -2 \le x \le 2\}$ **5. a.** $y = \pm\dfrac{1}{2}\sqrt{16 - x^2}$ **b.** $y = \dfrac{1}{2}\sqrt{16 - x^2}$; $y = -\dfrac{1}{2}\sqrt{16 - x^2}$

  c. Domain $= \{x \mid -4 \le x \le 4\}$ **7. a.** $y = \pm\dfrac{1}{\sqrt{3}}\sqrt{24 - 2x^2}$

  b. $y = \dfrac{1}{\sqrt{3}}\sqrt{24 - 2x^2}$; $y = \dfrac{-1}{\sqrt{3}}\sqrt{24 - 2x^2}$ **c.** Domain $= \{x \mid -\sqrt{12} \le x \le \sqrt{12}\}$

9. a. $y = \pm\sqrt{x^2 - 1}$ **b.** $y = \sqrt{x^2 - 1}$; $y = -\sqrt{x^2 - 1}$ **c.** Domain $= \{x \mid x \ge 1 \text{ or } \le -1\}$

11. a. $y = \pm\sqrt{9 + x^2}$ **b.** $y = \sqrt{9 + x^2}$; $y = -\sqrt{9 + x^2}$ **c.** Domain $= \{x \mid x \in R\}$

13.
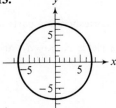

15.
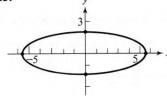

17.
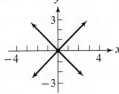

19.
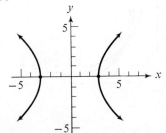

21.
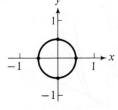

23.
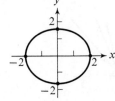

25. Pair of straight lines intersecting at the origin   **27.** A line, the *y*-axis

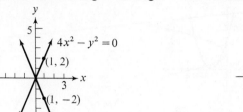

29. Since $x^2 + y^2 \geq 0$ for all $x, y \in R$, and $-1 < 0$, it follows that $x^2 + y^2 \neq -1$ for any $x, y \in R$.

31. Let $P(x, y)$ be a point at distance 4 from $(2, -3)$; then by the distance formula, $\sqrt{(x-2)^2 + (y+3)^2} = 4$; squaring and simplifying yields $(x-2)^2 + (y+3)^2 = 16$.

33. Given $Ax^2 + By^2 = C$ with intercepts $(a, 0)$ and $(0, b)$. If $y = 0$, $x = a$ and $x^2 = \dfrac{C}{A}$; hence $a^2 = \dfrac{C}{A}$. If $x = 0$, $y = b$ and $y^2 = \dfrac{C}{B}$; hence $b^2 = \dfrac{C}{B}$. Because $Ax^2 + By^2 = C$, $\dfrac{x^2}{C/A} + \dfrac{y^2}{C/B} = 1$ and $\dfrac{x^2}{a^2} + \dfrac{y^2}{b^2} = 1$.

35. As $|x|$ increases, so do x^2 and Ax^2 of the fraction $\dfrac{C}{Ax^2}$; the value of this fraction approaches zero, and hence, $\sqrt{1 - \dfrac{C}{Ax^2}}$ approaches 1, and $y = \pm\sqrt{\dfrac{A}{B}}\,|x|\left(\sqrt{1 - \dfrac{C}{Ax^2}}\right)$ approaches $y = \pm\sqrt{\dfrac{A}{B}}\,|x|$.

37.

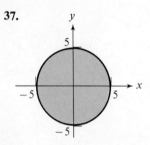

Exercise 6.3 (page 146)   **1.** 1, 16   **3.** 4, 400 ft   **5.** 4, 160 lbs/sq ft   **7.** 187.5, 1,687.5 lbs

9. 16   **11.** 160 lbs/sq ft   **13.** 1,687.5 lbs

15. Let c_1 and d_1 be the circumference and diameter of one circle and c_2 and d_2 be corresponding parts in a second circle; then $c_2 = \pi d_2$ and $\pi = \dfrac{c_2}{d_2}$; similarly $c_1 = \pi d_1$, and substituting for π yields $c_1 = \dfrac{c_2}{d_2}d_1$; and since $c_2 \neq 0$, multiplying both sides by $\dfrac{1}{c_2}$ yields $\dfrac{c_1}{c_2} = \dfrac{d_1}{d_2}$.

17. $k = 3$

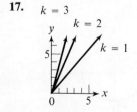

19. "Increases" the curvature   **21.**

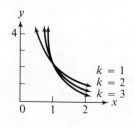

Exercise 6.4 (page 151)   **1.**   3; 17; 47   **3.**   56; 12; 326   **5.** -685; 95; 719

7.

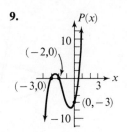

9.

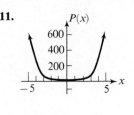

11.

13.

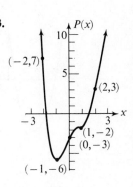

Exercise 6.5 (page 156)   **1.** $x = 3$ **3.** $x = 3$; $x = -2$ **5.** $x = -1$; $x = -4$ **7.** $x = -1$

9.

11.

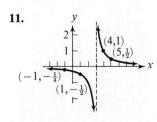

13.

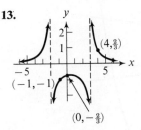

15.

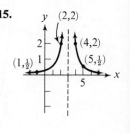

17. $x = 2$; $x = -2$; $y = 0$ **19.** $x = 4$; $y = x + 4$ **21.** $x = 4$; $x = -1$; $y = 1$

23.

25.

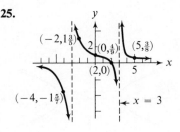

27.

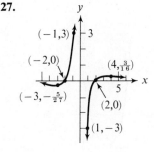

29.

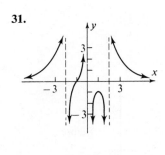

31.

Chapter 6 Review (page 158)   **1.** x-intercepts 3, -2; axis of symmetry $x = \dfrac{1}{2}$; minimum pt. $\left(\dfrac{1}{2}, \dfrac{-25}{4}\right)$

2. x-intercepts 5, 2; axis of symmetry $x = \dfrac{7}{2}$; maximum pt. $\left(\dfrac{7}{2}, \dfrac{9}{4}\right)$

3.

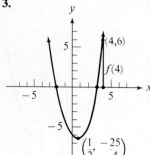

4.

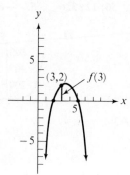

5.

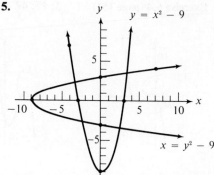

6.

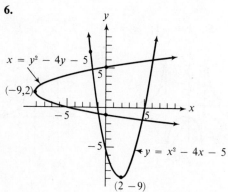

7.

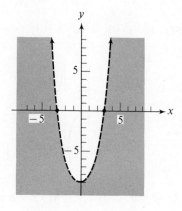

8.

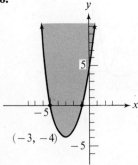

9.

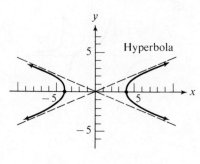

10.

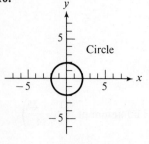

11.

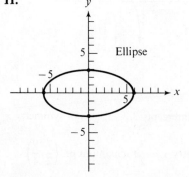

12.

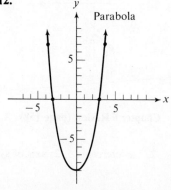

13.

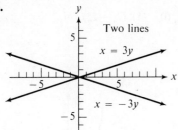

14.

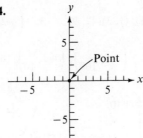

15. $y = 125$ **16.** $r = \dfrac{128}{243}$ **17.** 64 posts   **18.** 432 rpm   **19.** 1   **20.** 53

21. $(-4, -10), (-3, 0), (-2, 0), (-1, -4), (0, -6), (1, 0), (2, 20), (3, 60), (4, 126)$

22.

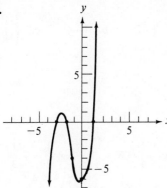

23.

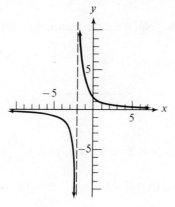

24.

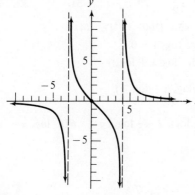

25.

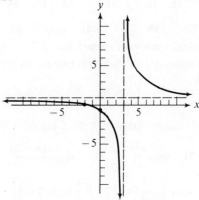

26.

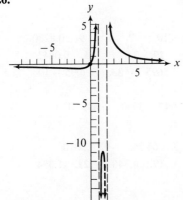

Exercise 7.1 (page 161)   **1.** $(0, 1), (1, 3), (2, 9)$ **3.** $\left(-2, -\dfrac{1}{25}\right), (0, -1), (2, -25)$

5. $(-3, 8), (0, 1), \left(3, \dfrac{1}{8}\right)$ **7.** $\left(-2, \dfrac{1}{100}\right), \left(-1, \dfrac{1}{10}\right), (0, 1)$

9.   **11.**   **13.**   **15.**

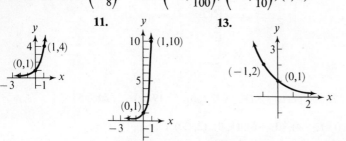

17.   **19.**   No; a constant function   **21.**   **a.** $\{-2\}$ **b.** $\{-4\}$ **c.** $\left\{\dfrac{3}{4}\right\}$

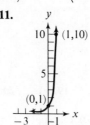

Exercise 7.2 (page 163)   **1.** $\log_4 16 = 2$ **3.** $\log_3 27 = 3$ **5.** $\log_{1/2} \dfrac{1}{4} = 2$ **7.** $\log_8 \dfrac{1}{2} = -\dfrac{1}{3}$

9. $\log_{10} 100 = 2$ **11.** $\log_{10}(0.1) = -1$ **13.** $2^6 = 64$ **15.** $3^2 = 9$ **17.** $\left(\dfrac{1}{3}\right)^{-2} = 9$

19. $10^3 = 1000$ **21.** $10^{-2} = 0.01$ **23.**   2   **25.**   3   **27.** $\dfrac{1}{2}$ **29.** -1 **31.**   1   **33.**   2

35. -1 **37.** $\{2\}$ **39.** $\{2\}$ **41.** $\{64\}$ **43.** $\{-3\}$ **45.** $\{100\}$ **47.** $\{4\}$

49.   By definition, $\log_b 1$ is a number such that $b^{\log_b 1} = 1$. Therefore $\log_b 1 = 0$.

51. $\operatorname{Log}_b b^x = x$ implies $b^x = b^x$, which is true for all $x \in R$.   **53.** $\log_b x + \log_b y$

55. $\log_b x - \log_b y$ **57.** $5 \log_b x$ **59.** $\dfrac{1}{3} \log_b x$ **61.** $\dfrac{1}{2}(\log_b x - \log_b z)$

63. $\dfrac{1}{3}(\log_{10} x + 2 \log_{10} y - \log_{10} z)$ **65.** $\log_{10} 2 + \log_{10} \pi + \dfrac{1}{2} \log_{10} l - \dfrac{1}{2} \log_{10} g$ **67.** $\log_b xy$

69. $\log_b x^2 y^3$ **71.** $\log_b \dfrac{x^3 y}{z^2}$ **73.** $\log_{10} \dfrac{x(x-2)}{z^2}$

75. $\dfrac{1}{4} \log_{10} 8 + \dfrac{1}{4} \log_{10} 2 = \dfrac{1}{4} \log_{10} 2^3 + \dfrac{1}{4} \log_{10} 2 = \left(\dfrac{3}{4} + \dfrac{1}{4}\right) \log_{10} 2 = \log_{10} 2$

77. $\log_{10}[\log_3(\log_5 125)] = \log_{10}[\log_3 (3)] = \log_{10}[1] = 0$

Exercise 7.3 (page 169)   **1.**   2   **3.** -3 or $7 - 10$ **5.** -4 or $6 - 10$ **7.**   4   **9.**   0.8280
11. $9.9101 - 10$ **13.** $8.9031 - 10$ **15.**   2.3945   **17.**   4.10   **19.**   3.67   **21.**   0.0642
23.   5480   **25.**   0.000718   **27.**   0.6246   **29.**   3.1824   **31.**   4.5695   **33.** $9.7095 - 10$
35.   3.225   **37.**   10.52   **39.**   0.05075   **41.**   0.7495
43. $\log_{10} 3.751$;   0.751 is closer to a tabulated value.   **45.**   9.1   **47.**   5000   **49.**   113
51.   **a.** -1.2679 **b.** -3.5813

Exercise 7.4 (page 174)   **1.**   4.014   **3.**   2.299   **5.**   0.000461   **7.**   64.34   **9.**   2.010
11. 3.435×10^{-10} **13.**   0.04582   **15.**   0.2777   **17.**   9.871   **19.**   4.746   **21.**   1.394
23.   3.484   **25.**   2.664   **27.**   1.11 sec.

Exercise 7.5 (page 178)   **1.** $\left\{\dfrac{\log_{10} 7}{\log_{10} 2}\right\}$ **3.** $\left\{\dfrac{\log_{10} 8}{\log_{10} 3} - 1\right\}$ **5.** $\left\{\dfrac{1}{2}\left(\dfrac{\log_{10} 3}{\log_{10} 7} + 1\right)\right\}$

7. $\left\{\sqrt{\dfrac{\log_{10} 15}{\log_{10} 4}}, -\sqrt{\dfrac{\log_{10} 15}{\log_{10} 4}}\right\}$ **9.** $\left\{\dfrac{-1}{\log_{10} 3}\right\}$ **11.** $\left\{1 - \dfrac{\log_{10} 15}{\log_{10} 3}\right\}$ **13.** $n = \dfrac{\log_{10} y}{\log_{10} x}$

15. $t = \dfrac{\log_{10} y}{k \log_{10} e}$ **17.** $\{500\}$ **19.** $\{4\}$ **21.** $\{3\}$

23. 1.343   **25.** 5%   **27.** 2.5%   **29.** 20 yrs   **31.** 12 yrs.   **33.** $7400, $7430

35. 7   **37.** 7.7   **39.** 6.2   **41.** 1.0×10^{-3} **43.** 2.5×10^{-6} **45.** 6.3×10^{-8}

Exercise 7.6 (page 183)   **1.** 3.32   **3.** 3.41   **5.** 1.08   **7.** 0.79   **9.** 1.0986   **11.** 2.8332
13. 5.7900   **15.** 6.1093   **17.** 1.6487   **19.** 29.964   **21.** 1.260   **23.** 0.8607

25. 0.0821   **27.** 0.7600   **29.** $\dfrac{1}{3}$ **31.** 2.10   **33.** 2.86

35. **a.** $\log_9 7 = \dfrac{\log_3 7}{\log_3 9} = \dfrac{\log_3 7}{\log_3 3^2} = \dfrac{\log_3 7}{2 \log_3 3} = \dfrac{1}{2} \log_3 7$

  b. $\log_{a^2} b = \dfrac{\log_a b}{\log_a a^2} = \dfrac{\log_a b}{2 \log_a a} = \dfrac{1}{2} \log_a b$

37. $(\log_{10} 4 - \log_{10} 2) \log_2 10 = \log_{10}\left(\dfrac{4}{2}\right) \cdot \log_2 10 = \log_{10} 2 \cdot \dfrac{\log_{10} 10}{\log_{10} 2} = \log_{10} 10 = 1$

39. 12.048 gr   **41.** 2.2 cm

Chapter 7 Review (page 184)   **1.**   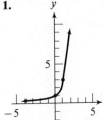   **2.** $x = -3$ **3.** $\log_3 81 = 4$

4. $\log_3 \dfrac{1}{9} = -2$ **5.** $2^3 = 8$ **6.** $10^{-4} = 0.0001$ **7.** $y = 4$ **8.** 1,000

9. $\dfrac{1}{3} \log_{10} x + \dfrac{2}{3} \log_{10} y$ **10.** $\log_{10} 2 + 3 \log_{10} R - \dfrac{1}{2} \log_{10} P - \dfrac{1}{2} \log_{10} Q$ **11.** $\log_b \dfrac{x^2}{\sqrt[3]{y}}$

12. $\log_{10} \dfrac{\sqrt[3]{x^2 y}}{z^3}$ **13.** 1.6232   **14.** 7.4969 − 10   **15.** 2.8340   **16.** 8.6172 − 10

17. 67.4   **18.** 0.0466   **19.** 2.655   **20.** 0.6643   **21.** 363.2   **22.** 49.65   **23.** 1450

24. 0.7711   **25.** $\dfrac{\log_{10} 7}{\log_{10} 5}$ **26.** $\dfrac{\log_{10} 80}{\log_{10} 3} - 1$ **27.** 5   **28.** 4   **29.** 2.18

30. 6.9   **31.** 4.8676   **32.** 121.51   **33.** 0.7866   **34.** 4.059

Exercise 8.1 (page 190)   **1.** $\{(3, 2)\}$ **3.** $\{(2, 1)\}$ **5.** $\{(-5, 4)\}$ **7.** $\left\{\left(0, \dfrac{3}{2}\right)\right\}$ **9.** $\left\{\left(\dfrac{2}{3}, -1\right)\right\}$

11. $\{(1, 2)\}$ **13.** $\left\{\left(-\dfrac{19}{5}, -\dfrac{18}{5}\right)\right\}$ **15.** $a = 1, b = -1$ **17.** $y = -\dfrac{10}{3} x + 2$

19. $C = \dfrac{5}{9}(F - 32)$ **21.** 32 lbs   **23.** $7200 at 4%; $8200 at 5%   **25.** 32 and 64 mph

27. a. From $\dfrac{a_1}{a_2}=\dfrac{b_1}{b_2}$, it follows that $\dfrac{a_1}{b_1}=\dfrac{a_2}{b_2}$. Hence the slopes are equal; the two lines are the same or they are parallel. The equations are either consistent or inconsistent, respectively. Assume that the equations are consistent and the two lines are the same. Then, the y-intercepts $\dfrac{-c_1}{b_1}=\dfrac{-c_2}{b_2}$, from which $\dfrac{b_1}{b_2}=\dfrac{c_1}{c_2}$; but this contradicts the hypothesis $\dfrac{b_1}{b_2}\ne\dfrac{c_1}{c_2}$; hence the equations are inconsistent.   **b.** from (a) above, $\dfrac{a_1}{b_1}=\dfrac{a_2}{b_2}$, and it follows that $\dfrac{a_1}{a_2}=\dfrac{b_1}{b_2}$. Now, either $\dfrac{b_1}{b_2}=\dfrac{c_1}{c_2}$ or $\dfrac{b_1}{b_2}\ne\dfrac{c_1}{c_2}$. Assume $\dfrac{b_1}{b_2}=\dfrac{c_1}{c_2}$; then $\dfrac{c_1}{b_1}=\dfrac{c_2}{b_2}$; the graphs would be the same straight line, and the equations would be consistent. This contradicts the hypothesis that states that the equations are inconsistent; hence $\dfrac{b_1}{b_2}\ne\dfrac{c_1}{c^2}$.

Exercise 8.2 (page 195)   1. $\{(1, 2, -1)\}$ **3.** $\{(2, -2, 0)\}$ **5.** $\{(2, 2, 1)\}$ **7.** $\{(0, 1, 2)\}$
9. Dependent   **11.** $\{(4, -2, 2)\}$ **13.** 3, 6, 6   **15.** 60 nickels, 20 dimes, 5 quarters
17. Thirty-six 1's, thirty-four 5's, twenty-four 10's   **19.** $x^2+y^2+2x+2y-23=0$
21. $a=\dfrac{1}{4},\ b=\dfrac{1}{2},\ c=0$ **23.** $\left\{\left(\dfrac{5}{3},\dfrac{1}{3},0\right),(2, 2, -1)\right\}$
25. Since the solution set for (a), (b), (c) is $\{(1, 1, 1)\}$ and this is not a solution of (d) since $2-1+2\ne4$, the solution set for (a), (b), (c), (d) is $\emptyset$.

Exercise 8.3 (page 200)   1. $\{(-1, -4), (5, 20)\}$ **3.** $\{(2, 3), (3, 2)\}$ **5.** $\{(4, -3), (-3, 4)\}$
7. $\{(-1, 3), (-1, -3), (1, 3), (1, -3)\}$ **9.** $\{(-3, \sqrt{2}), (-3, -\sqrt{2}), (3, \sqrt{2}), (3, -\sqrt{2})\}$
11. $\{(\sqrt{3}, 4), (\sqrt{3}, -4), (-\sqrt{3}, 4), (-\sqrt{3}, -4)\}$ **13.** $\{(1, -2), (-1, 2), (2, -1), (-2, 1)\}$
15. $\{(3, 1), (-3, -1), (-2\sqrt{7}, \sqrt{7}), (2\sqrt{7}, -\sqrt{7})\}$ **17.** $\{(0, 2)\}$ **19.** $\left\{(1, 0), \left(\dfrac{3}{2},\dfrac{1}{2}\right)\right\}$

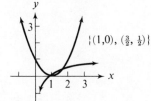

21. $\{(1, 0)\}$

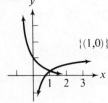

23. $\{(1, 0)\}$ (see graph Problem 21, where $y_1=10^{-x}$ and $y_2=\log_{10} x$.

25. a. 1   **b.** 2   **c.** 4   **27.** $\dfrac{7}{2},\dfrac{5}{2}$ **29.** 6 lbs/sq in.; 5 cu in.   **31.** $\{(2, 2), (-2, -2)\}$

Exercise 8.4 (page 203)

1. 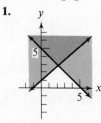   **3.**   **5.**   **7.**

9.
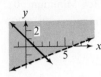

11.
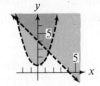

13.
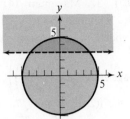

15.
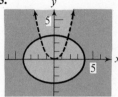

17.

19.
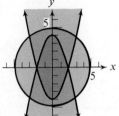

21.
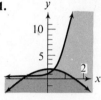

23.
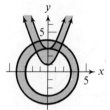

Exercise 8.5 (page 206)

1.
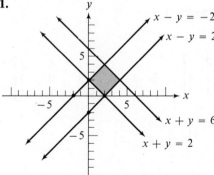

3.
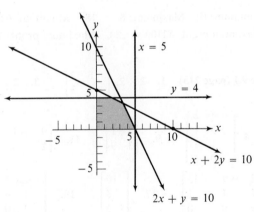

5.
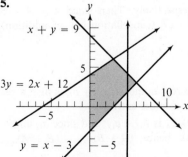

7. No   **9.** Yes

Exercise 8.6 (page 207)   **1.** Maximum: 14;   Minimum: 2   **3.** Maximum: 80/3;   Minimum: 0
5. Maximum: 17;   Minimum: -16 **7.** Maximum profit: \$830   **9.** Maximum profit: \$790
11. 200 standard models and 400 deluxe models

Chapter 8 Review (page 208)   **1.** $\left\{\left(\dfrac{1}{2}, \dfrac{7}{2}\right)\right\}$ **2.** $\{(1, 2)\}$ **3.** $\left\{\left(\dfrac{37}{13}, -\dfrac{10}{13}\right)\right\}$ **4.** $a = 2,\ b = 5$
5. $\{(2, 0, -1)\}$ **6.** $\{(2, 1, -1)\}$ **7.** $\{(2, -1, 3)\}$ **8.** $a = 2,\ b = -3,\ c = 4$
9. $\{(1, 2), (4, -13)\}$ **10.** $\left\{\left(\dfrac{6}{7}, -\dfrac{37}{7}\right), (2, -3)\right\}$ **11.** $\{(2, 3), (2, -3), (-2, 3), (-2, -3)\}$
12. $\left\{\left(\dfrac{7}{8}, \dfrac{17}{8}\right)\right\}$ **13.**   **14.**

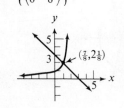   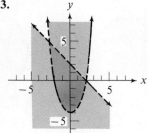   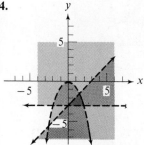

15.   **16.**

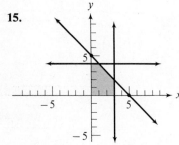   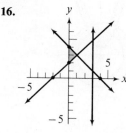

17. Minimum: 0;   Maximum: 8   **18.** Minimum: 0;   Maximum: 11
19. Maximum profit: \$1100   **20.** Maximum profit: \$210

Exercise 9.1 (page 213)   **1.** 2×2, $\begin{bmatrix} 6 & 2 \\ -1 & 3 \end{bmatrix}$ **3.** 2×3, $\begin{bmatrix} 2 & 1 \\ -7 & 4 \\ 3 & 0 \end{bmatrix}$ **5.** 3×3, $\begin{bmatrix} 2 & 4 & -2 \\ 3 & 0 & 3 \\ -1 & 1 & 1 \end{bmatrix}$

7. 2×4, $\begin{bmatrix} 4 & 2 \\ -3 & 1 \\ -1 & 1 \\ 0 & 6 \end{bmatrix}$ **9.** $\begin{bmatrix} 3 & 1 \\ 3 & 9 \end{bmatrix}$ **11.** $\begin{bmatrix} 9 & -1 & -1 \\ 2 & 3 & 6 \end{bmatrix}$ **13.** $\begin{bmatrix} -2 & -3 \\ 4 & 0 \end{bmatrix}$

15. $\begin{bmatrix} 2 & -9 & -13 \\ 7 & -4 & 1 \\ 9 & 0 & 0 \end{bmatrix}$ **17.** $\begin{bmatrix} 10 \\ 3 \\ -3 \end{bmatrix}$ **19.** $\begin{bmatrix} 2 & 3 & 4 \\ -1 & 6 & 2 \\ 1 & 0 & 3 \end{bmatrix}$

21. Let X, $X + A$, and B have each entry x_{ij}, $x_{ij} + a_{ij}$, and b_{ij}, respectively. Then since $X + A = B$, $x_{ij} + a_{ij} = b_{ij}$ and since x_{ij}, a_{ij}, and b_{ij} are real numbers, $x_{ij} = b_{ij} - a_{ij}$ (cancellation law of addition); thus, $X = B - A$.

23. $\begin{bmatrix} 2 & 2 \\ -1 & 1 \end{bmatrix}$ **25.** $\begin{bmatrix} 1 & 1 \\ -4 & 3 \end{bmatrix}$

27. The i, j entry of $(A + B)_{m \times n} = a_{ij} + b_{ij}$; and since a_{ij} and b_{ij} are real numbers, $a_{ij} + b_{ij}$ is also a real number (closure law of addition); thus, $(A + B)_{m \times n}$ is a matrix with real-number entries.

29. The i, j entry of $(A + B)_{m \times n}$ is $a_{ij} + b_{ij}$, and similarly of $(B + A)_{m \times n}$ is $b_{ij} + a_{ij}$; and since a_{ij} and b_{ij} are real numbers, $a_{ij} + b_{ij} = b_{ij} + a_{ij}$ (commutative law of addition); hence, $(A + B)_{m \times n} = (B + A)_{m \times n}$, by Def. 9.1.

Exercise 9.2 (page 218) **1.** $\begin{bmatrix} 0 & -5 & 5 \\ -15 & 5 & -10 \end{bmatrix}$ **3.** $[-1]$ **5.** $\begin{bmatrix} 1 & -13 \\ 4 & -7 \end{bmatrix}$ **7.** $\begin{bmatrix} 30 & -39 \\ 29 & 14 \end{bmatrix}$

9. $\begin{bmatrix} -5 & -1 \\ 8 & -1 \end{bmatrix}$ **11.** $\begin{bmatrix} -1 & 0 & -2 \\ 1 & 2 & 8 \\ 0 & 1 & 3 \end{bmatrix}$ **13.** $\begin{bmatrix} 1 & 0 & 0 \\ 0 & 1 & 0 \\ 0 & 0 & 1 \end{bmatrix}$ **15.** $\begin{bmatrix} 1 & 0 \\ -1 & 2 \end{bmatrix}$ **17.** $\begin{bmatrix} 1 & -2 \\ 1 & 2 \end{bmatrix}$

19. $\begin{bmatrix} -2 & 3 \\ 2 & -4 \end{bmatrix}$ **21.** $\begin{bmatrix} -1 & 1 \\ -1 & -1 \end{bmatrix}$ **23.** $\begin{bmatrix} 1 & -1 \\ 1 & 0 \end{bmatrix}$

25. Since $A + B = \begin{bmatrix} 0 & 2 \\ -1 & 4 \end{bmatrix}$, $A - B = \begin{bmatrix} -2 & 2 \\ 1 & -1 \end{bmatrix}$, $A^2 = \begin{bmatrix} 1 & 0 \\ 0 & 1 \end{bmatrix}$, $B^2 = \begin{bmatrix} 1 & 0 \\ -3 & 4 \end{bmatrix}$

and $AB = \begin{bmatrix} -3 & 4 \\ -1 & 2 \end{bmatrix}$, then **(a)** $(A + B)(A + B) = \begin{bmatrix} 0 & 2 \\ -1 & 4 \end{bmatrix} \cdot \begin{bmatrix} 0 & 2 \\ -1 & 4 \end{bmatrix} = \begin{bmatrix} 2 & 8 \\ -4 & 14 \end{bmatrix}$

and $A^2 + 2AB + B^2 = \begin{bmatrix} 1 & 0 \\ 0 & 1 \end{bmatrix} + 2\begin{bmatrix} -3 & 4 \\ -1 & 2 \end{bmatrix} + \begin{bmatrix} 1 & 0 \\ -3 & 4 \end{bmatrix} = \begin{bmatrix} -4 & 8 \\ -5 & 9 \end{bmatrix}$;

hence $(A + B)(A + B) \neq A^2 + 2AB + B^2$. **(b)** $(A + B)(A - B) = \begin{bmatrix} 0 & 2 \\ -1 & 4 \end{bmatrix} \cdot \begin{bmatrix} -2 & 2 \\ 1 & -1 \end{bmatrix} = \begin{bmatrix} 2 & -2 \\ 6 & -6 \end{bmatrix}$

and $A^2 - B^2 = \begin{bmatrix} 1 & 0 \\ 0 & 1 \end{bmatrix} - \begin{bmatrix} 1 & 0 \\ -3 & 4 \end{bmatrix} = \begin{bmatrix} 0 & 0 \\ 3 & -3 \end{bmatrix}$; hence $(A + B)(A - B) \neq A^2 - B^2$.

For Problems 27 and 29, let $A = \begin{bmatrix} a_{11} & a_{12} \\ a_{21} & a_{22} \end{bmatrix}$, $B = \begin{bmatrix} b_{11} & b_{12} \\ b_{21} & b_{22} \end{bmatrix}$, and $C = \begin{bmatrix} c_{11} & c_{12} \\ c_{21} & c_{22} \end{bmatrix}$.

27. Since $AB = \begin{bmatrix} a_{11}b_{11} + a_{12}b_{21} & a_{11}b_{12} + a_{12}b_{22} \\ a_{21}b_{11} + a_{22}b_{21} & a_{21}b_{12} + a_{22}b_{22} \end{bmatrix}$,

$(AB)C = \begin{bmatrix} (a_{11}b_{11} + a_{12}b_{21})c_{11} + (a_{11}b_{12} + a_{12}b_{22})c_{21} & (a_{11}b_{11} + a_{12}b_{21})c_{12} + (a_{11}b_{12} + a_{12}b_{22})c_{22} \\ (a_{21}b_{11} + a_{22}b_{21})c_{11} + (a_{21}b_{12} + a_{22}b_{22})c_{21} & (a_{21}b_{11} + a_{22}b_{21})c_{12} + (a_{21}b_{12} + a_{22}b_{22})c_{22} \end{bmatrix}$.

The element in the first row, first column is given by $a_{11}b_{11}c_{11} + a_{12}b_{21}c_{11} + a_{11}b_{12}c_{21} + a_{12}b_{22}c_{21} =$

$(a_{11}b_{11}c_{11} + a_{11}b_{12}c_{21}) + (a_{12}b_{21}c_{11} + a_{12}b_{22}c_{21}) = a_{11}(b_{11}c_{11} + b_{12}c_{21}) + a_{12}(b_{21}c_{11} + b_{22}c_{21})$.

When the remaining elements are treated in a similar manner, $(AB)C =$

$\begin{bmatrix} a_{11}(b_{11}c_{11} + b_{12}c_{21}) + a_{12}(b_{21}c_{11} + b_{22}c_{21}) & a_{11}(b_{11}c_{12} + b_{12}c_{22}) + a_{12}(b_{21}c_{12} + b_{22}c_{22}) \\ a_{21}(b_{11}c_{11} + b_{12}c_{21}) + a_{22}(b_{21}c_{11} + b_{22}c_{21}) & a_{21}(b_{11}c_{12} + b_{12}c_{22}) + a_{22}(b_{21}c_{12} + b_{22}c_{22}) \end{bmatrix} =$

$\begin{bmatrix} a_{11} & a_{12} \\ a_{21} & a_{22} \end{bmatrix} \cdot \begin{bmatrix} b_{11}c_{11} + b_{12}c_{21} & b_{11}c_{12} + b_{12}c_{22} \\ b_{21}c_{11} + b_{22}c_{21} & b_{21}c_{12} + b_{22}c_{22} \end{bmatrix} = \begin{bmatrix} a_{11} & a_{12} \\ a_{21} & a_{22} \end{bmatrix} \cdot \left(\begin{bmatrix} b_{11} & b_{12} \\ b_{21} & b_{22} \end{bmatrix} \cdot \begin{bmatrix} c_{11} & c_{12} \\ c_{21} & c_{22} \end{bmatrix} \right) = A(BC)$.

29. $B + C = \begin{bmatrix} b_{11} + c_{11} & b_{12} + c_{12} \\ b_{21} + c_{21} & b_{22} + c_{22} \end{bmatrix}$,

hence $(B + C)A = \begin{bmatrix} (b_{11} + c_{11})a_{11} + (b_{12} + c_{12})a_{21} & (b_{11} + c_{11})a_{12} + (b_{12} + c_{12})a_{22} \\ (b_{21} + c_{22})a_{11} + (b_{22} + c_{22})a_{21} & (b_{21} + c_{22})a_{12} + (b_{22} + c_{22})a_{22} \end{bmatrix}$.

The element in the first row, first column is given by

$b_{11}a_{11} + c_{11}a_{21} + b_{12}a_{21} + c_{12}a_{21} = (b_{11}a_{11} + b_{12}a_{21}) + (c_{11}a_{21} + c_{12}a_{21})$.

When the remaining elements are treated in a similar manner,

$(B + C)A = \begin{bmatrix} (b_{11}a_{11} + b_{12}a_{21}) + (c_{11}a_{11} + c_{12}a_{21}) & (b_{11}a_{12} + b_{12}a_{22}) + (c_{11}a_{12} + c_{12}a_{22}) \\ (b_{21}a_{11} + b_{22}a_{21}) + (c_{21}a_{11} + c_{22}a_{21}) & (b_{21}a_{12} + b_{22}a_{22}) + (c_{21}a_{12} + c_{22}a_{22}) \end{bmatrix}$

$= \begin{bmatrix} b_{11}a_{11} + b_{12}a_{21} & b_{11}a_{12} + b_{12}a_{22} \\ b_{21}a_{11} + b_{22}a_{21} & b_{21}a_{12} + b_{22}a_{22} \end{bmatrix} + \begin{bmatrix} c_{11}a_{11} + c_{12}a_{21} & c_{11}a_{12} + c_{12}a_{22} \\ c_{21}a_{11} + c_{22}a_{21} & c_{21}a_{12} + c_{22}a_{22} \end{bmatrix}$

$= \begin{bmatrix} b_{11} & b_{12} \\ b_{21} & b_{22} \end{bmatrix} \cdot \begin{bmatrix} a_{11} & a_{12} \\ a_{21} & a_{22} \end{bmatrix} + \begin{bmatrix} c_{11} & c_{12} \\ c_{21} & c_{22} \end{bmatrix} \cdot \begin{bmatrix} a_{11} & a_{12} \\ a_{21} & a_{22} \end{bmatrix} = BC + CA$.

31. $(A_{2\times 2} \cdot B_{2\times 2}) = \begin{bmatrix} a_{11}b_{11} + a_{12}b_{21} & a_{11}b_{12} + a_{12}b_{22} \\ a_{21}b_{11} + a_{22}b_{21} & a_{21}b_{12} + a_{22}b_{22} \end{bmatrix}$.

Hence, $(A_{2\times 2} \cdot B_{2\times 2})^t = \begin{bmatrix} a_{11}b_{11} + a_{12}b_{21} & a_{21}b_{11} + a_{22}b_{21} \\ a_{11}b_{12} + a_{12}b_{22} & a_{21}b_{12} + a_{22}b_{22} \end{bmatrix}$.

By the commutative property of multiplication for real numbers,

$(A_{2\times 2} \cdot B_{2\times 2})^t = \begin{bmatrix} b_{11}a_{11} + b_{21}a_{12} & b_{11}a_{21} + b_{21}a_{22} \\ b_{12}a_{11} + b_{22}a_{12} & b_{12}a_{21} + b_{22}a_{22} \end{bmatrix} = \begin{bmatrix} b_{11} & b_{21} \\ b_{12} & b_{22} \end{bmatrix} \cdot \begin{bmatrix} a_{11} & a_{21} \\ a_{12} & a_{22} \end{bmatrix} = B_{2\times 2}^t \cdot A_{2\times 2}^t$

33. Let $A = \begin{bmatrix} a_{11} & a_{12} \\ a_{21} & a_{22} \end{bmatrix}$ and c be a scalar; then by Def. 9.6 and the closure property of multiplication of real

numbers, $c\begin{bmatrix} a_{11} & a_{12} \\ a_{21} & a_{22} \end{bmatrix} = \begin{bmatrix} ca_{11} & ca_{12} \\ ca_{21} & ca_{22} \end{bmatrix}$, which is an $m \times n$ matrix.

35. By Def. 9.6 and the distributive property for real numbers, and by Def. 9.3,

$(c+d)\begin{bmatrix} a_{11} & a_{12} \\ a_{21} & a_{22} \end{bmatrix} = \begin{bmatrix} (c+d)a_{11} & (c+d)a_{12} \\ (c+d)a_{21} & (c+d)a_{22} \end{bmatrix} = \begin{bmatrix} ca_{11} + da_{11} & ca_{12} + da_{12} \\ ca_{21} + da_{21} & ca_{22} + da_{22} \end{bmatrix}$

$= \begin{bmatrix} ca_{11} & ca_{12} \\ ca_{21} & ca_{22} \end{bmatrix} + \begin{bmatrix} da_{11} & da_{12} \\ da_{21} & da_{22} \end{bmatrix} = c\begin{bmatrix} a_{11} & a_{12} \\ a_{21} & a_{22} \end{bmatrix} + d\begin{bmatrix} a_{11} & a_{12} \\ a_{21} & a_{22} \end{bmatrix} = cA + dA$.

37. By Defs. 9.6 and 9.5, $(-1)A = -1\begin{bmatrix} a_{11} & a_{12} \\ a_{21} & a_{22} \end{bmatrix} = \begin{bmatrix} -a_{11} & -a_{12} \\ -a_{21} & -a_{22} \end{bmatrix} = -A$.

39. $0_{2\times 2} = \begin{bmatrix} 0 & 0 \\ 0 & 0 \end{bmatrix}$; hence, by Def. 9.6, $c \cdot 0_{2\times 2} = c\begin{bmatrix} 0 & 0 \\ 0 & 0 \end{bmatrix} = \begin{bmatrix} c \cdot 0 & c \cdot 0 \\ c \cdot 0 & c \cdot 0 \end{bmatrix} = \begin{bmatrix} 0 & 0 \\ 0 & 0 \end{bmatrix} = 0_{2\times 2}$.

Exercise 9.3 (page 223)   **1.** $\{(2, -1)\}$ **3.** $\{(5, 1)\}$ **5.** $\{(8, 1)\}$ **7.** $\{(-2, 2, 0)\}$

9. $\left\{\left(\frac{5}{4}, \frac{5}{2}, -\frac{1}{2}\right)\right\}$ **11.** $\left\{\left(-\frac{77}{27}, -\frac{8}{27}, \frac{29}{27}\right)\right\}$

13. $\begin{bmatrix} k & 0 \\ 0 & 1 \end{bmatrix}\begin{bmatrix} a & b \\ c & d \end{bmatrix} = \begin{bmatrix} k \cdot a + 0 \cdot c & k \cdot b + 0 \cdot d \\ 0 \cdot a + 1 \cdot c & 0 \cdot b + 1 \cdot d \end{bmatrix} = \begin{bmatrix} ka & kb \\ c & d \end{bmatrix}$

15. $\begin{bmatrix} 0 & 1 \\ 1 & 0 \end{bmatrix}\begin{bmatrix} a & b \\ c & d \end{bmatrix} = \begin{bmatrix} 0 \cdot a + 1 \cdot c & 0 \cdot b + 1 \cdot d \\ 1 \cdot a + 0 \cdot c & 1 \cdot b + 0 \cdot d \end{bmatrix} = \begin{bmatrix} c & d \\ a & b \end{bmatrix}$

17. $\begin{bmatrix} 1 & 0 \\ k & 1 \end{bmatrix}\begin{bmatrix} a & b \\ c & d \end{bmatrix} = \begin{bmatrix} 1 \cdot a + 0 \cdot c & 1 \cdot b + 0 \cdot d \\ k \cdot a + 1 \cdot c & k \cdot b + 1 \cdot d \end{bmatrix} = \begin{bmatrix} a & b \\ ka + c & kb + d \end{bmatrix}$

Exercise 9.4 (page 227)   **1.** 0   **3.** -6 **5.** -2

7. $M_{11} = \begin{vmatrix} 0 & 3 & -1 \\ 1 & 2 & 2 \\ -1 & 3 & 1 \end{vmatrix}$, $A_{11} = \begin{vmatrix} 0 & 3 & -1 \\ 1 & 2 & 2 \\ -1 & 3 & 1 \end{vmatrix}$

9. $M_{23} = \begin{vmatrix} 2 & 1 & 0 \\ -2 & 1 & 2 \\ 1 & -1 & 1 \end{vmatrix}$, $A_{23} = -\begin{vmatrix} 2 & 1 & 0 \\ -2 & 1 & 2 \\ 1 & -1 & 1 \end{vmatrix}$

11. $M_{31} = \begin{vmatrix} 1 & -2 & 0 \\ 0 & 3 & -1 \\ -1 & 3 & 1 \end{vmatrix}$, $A_{31} = \begin{vmatrix} 1 & -2 & 0 \\ 0 & 3 & -1 \\ -1 & 3 & 1 \end{vmatrix}$

13. $M_{44} = \begin{vmatrix} 2 & 1 & -2 \\ 1 & 0 & 3 \\ -2 & 1 & 2 \end{vmatrix}$, $A_{44} = \begin{vmatrix} 2 & 1 & -2 \\ 1 & 0 & 3 \\ -2 & 1 & 2 \end{vmatrix}$

15. 1   **17.** 0   **19.** -30 **21.** $3x + 1$ **23.** 3   **25.** 0   **27.** -1 **29.** 0

31. x^3 **33.** $x = 5$ **35.** $x = 3$

37. Expanding by elements in the first row yields $\begin{vmatrix} 0 & 1 & 0 & 0 \\ 1 & 0 & 3 & 2 \\ 5 & -1 & 2 & 1 \\ 1 & 0 & 1 & 1 \end{vmatrix} = -\begin{vmatrix} 1 & 3 & 2 \\ 5 & 2 & 1 \\ 1 & 1 & 1 \end{vmatrix}$;

expanding by elements in the third row yields $-\begin{vmatrix} 3 & 2 \\ 2 & 1 \end{vmatrix} + \begin{vmatrix} 1 & 2 \\ 5 & 1 \end{vmatrix} - \begin{vmatrix} 1 & 3 \\ 5 & 2 \end{vmatrix} = 1 - 9 + 13 = 5$

39. 2, 6, 24

41. Let $A = \begin{bmatrix} a_{11} & a_{12} \\ a_{21} & a_{22} \end{bmatrix}$; by Def. 9.6 and the fact that $\delta(A) = a_{11}a_{22} - a_{12}a_{21}$, it follows that

$aA = \begin{bmatrix} aa_{11} & aa_{12} \\ aa_{21} & aa_{22} \end{bmatrix}$ and $\delta(aA) = a^2 a_{11}a_{22} - a^2 a_{12}a_{21} = a^2(a_{11}a_{22} - a_{12}a_{21}) = a^2\delta(A)$.

43. Let $A = \begin{bmatrix} a_{11} & a_{12} \\ a_{21} & a_{22} \end{bmatrix}$ and $B = \begin{bmatrix} b_{11} & b_{12} \\ b_{21} & b_{22} \end{bmatrix}$; then $\delta(A) = a_{11}a_{22} - a_{12}a_{21}$

and $\delta(B) = b_{11}b_{22} - b_{12}b_{21}$, $AB = \begin{bmatrix} a_{11}b_{11} + a_{12}b_{21} & a_{11}b_{12} + a_{12}b_{22} \\ a_{21}b_{11} + a_{22}b_{21} & a_{21}b_{12} + a_{22}b_{22} \end{bmatrix}$

and $\delta(AB) = (a_{11}b_{11} + a_{12}b_{21})(a_{21}b_{12} + a_{22}b_{22}) - (a_{11}b_{12} + a_{12}b_{22})(a_{21}b_{11} + a_{22}b_{21})$, which simplifies to
$a_{11}b_{11}a_{22}b_{22} - a_{11}b_{12}a_{22}b_{21} - a_{12}b_{22}a_{21}b_{11} + a_{12}b_{21}a_{21}b_{12}$
$= a_{11}a_{22}(b_{11}b_{22} - b_{12}b_{21}) - a_{12}a_{21}(b_{11}b_{22} - b_{12}b_{21}) = (a_{11}a_{22} - a_{12}a_{21})(b_{11}b_{22} - b_{12}b_{21}) = \delta(A) \cdot \delta(B)$.

Exercise 9.5 (page 234)   **1.** Theorem 9.7   **3.** Theorem 9.9   **5.** Theorem 9.10
  7. Theorem 9.10   **9.** Theorem 9.12   **11.** Theorem 9.12

13. $\begin{vmatrix} 1 & 3 \\ 0 & -4 \end{vmatrix}$ **15.** $\begin{vmatrix} 1 & -2 & 1 \\ 0 & 7 & 1 \\ 0 & 2 & 1 \end{vmatrix}$ **17.** $\begin{vmatrix} 0 & 1 & -3 & -2 \\ 0 & 2 & 1 & 2 \\ 1 & 1 & 2 & 3 \\ 0 & 1 & 1 & 1 \end{vmatrix}$

19. $-1\begin{vmatrix} 2 & 1 \\ -1 & 2 \end{vmatrix} = -5$ **21.** $\begin{vmatrix} -1 & -5 \\ 2 & -2 \end{vmatrix} = 12$ **23.** $\begin{vmatrix} 4 & 4 \\ 3 & 7 \end{vmatrix} = 16$

25. $\begin{vmatrix} -1 & 1 \\ 4 & -1 \end{vmatrix} = -3$ **27.** $\begin{vmatrix} 3 & 1 \\ 5 & 4 \end{vmatrix} = 7$ **29.** $\begin{vmatrix} 3 & 6 \\ 0 & 0 \end{vmatrix} = 0$

31. $\begin{vmatrix} 6 & 1 \\ 0 & 3 \end{vmatrix} = 18$ **33.** $-16\begin{vmatrix} 1 & 2 \\ 2 & 3 \end{vmatrix} = 16$ **35.** $\begin{vmatrix} 4 & -4 \\ 3 & -9 \end{vmatrix} = -24$

37. Let $A = \begin{vmatrix} x & y & 1 \\ x_1 & y_1 & 1 \\ x_2 & y_2 & 1 \end{vmatrix} = 0$.

Expanding about the first row gives $\delta(A) = x\begin{vmatrix} y_1 & 1 \\ y_2 & 1 \end{vmatrix} - y\begin{vmatrix} x_1 & 1 \\ x_2 & 1 \end{vmatrix} + 1\begin{vmatrix} x_1 & y_1 \\ x_2 & y_2 \end{vmatrix} = 0$;
hence $(y_1 - y_2)x + (x_2 - x_1)y + (x_1y_2 - y_1x_2) = 0$. Further, y_1, y_2, x_1, x_2 are real numbers; hence there exist real numbers, a, b, c such that $(y_1 - y_2) = a$, $(x_2 - x_1) = b$, and $(x_1y_2 - y_1x_2) = c$ with a and b not both 0 since $(x_1, y_1) \neq (x_2, y_2)$. Substituting yields $ax + by + c = 0$, which is the equation of a straight line. By theorem 9.9, the line passes through (x_1, y_1) and (x_2, y_2),
39. Multiply column 1 by $(-a)$ and add result to column 2; also, multiply column 1 by $(-a^2)$ and add result to column 3, to obtain $\begin{vmatrix} 1 & a & a^2 \\ 1 & b & b^2 \\ 1 & c & c^2 \end{vmatrix} = \begin{vmatrix} 1 & 0 & 0 \\ 1 & b-a & b^2-a^2 \\ 1 & c-a & c^2-a^2 \end{vmatrix}$. Expand about the first row to obtain

$1\begin{vmatrix} b-a & b^2-a^2 \\ c-a & c^2-a^2 \end{vmatrix} = (b-a)[c^2 - a^2] - (c-a)[b^2 - a^2]$
$\qquad = (b-a)[(c-a)(c+a)] - (c-a) \cdot [(b-a)(b+a)]$
$\qquad = -(a-b)[(c-a)(c+a)] + (c-a)[(a-b)(a+b)]$
$\qquad = (a-b)(c-a)[-(c+a) \cdot (a+b)]$
$\qquad = (a-b)(c-a)(b-c) = (b-c)(c-a)(a-b)$.

Exercise 9.6 (page 239)   **1.** $\begin{bmatrix} 3 & -2 \\ -1 & 1 \end{bmatrix}$ **3.** $\dfrac{1}{5}\begin{bmatrix} 1 & 3 \\ -1 & 2 \end{bmatrix}$ **5.** $|A| = 0$;   no inverse

7. $-1\begin{bmatrix} 4 & -7 \\ -3 & 5 \end{bmatrix}$ **9.** $\dfrac{1}{2}\begin{bmatrix} -2 & -4 \\ 4 & 7 \end{bmatrix}$ **11.** $|A| = 0$;   no inverse

13. $\dfrac{1}{6}\begin{bmatrix} 2 & 2 & -5 \\ -4 & 2 & 1 \\ 0 & 0 & 3 \end{bmatrix}$ **15.** $\dfrac{1}{3}\begin{bmatrix} -2 & 3 & -1 \\ -1 & 0 & 1 \\ 6 & -6 & 3 \end{bmatrix}$ **17.** $|A| = 0$;   no inverse

19. $|A| = 0$; no inverse   **21.** $\begin{bmatrix} 2 & -3 & 11 \\ -2 & 4 & -13 \\ 1 & -2 & 7 \end{bmatrix}$ **23.** $-1\begin{bmatrix} 0 & 0 & -1 \\ 0 & -1 & 0 \\ -1 & 0 & 0 \end{bmatrix}$

25. Let $A \cdot B = \begin{bmatrix} 2 & 3 \\ 1 & -1 \end{bmatrix} \cdot \begin{bmatrix} 0 & 1 \\ 3 & 1 \end{bmatrix}$; $A \cdot B = \begin{bmatrix} 9 & 5 \\ -3 & 0 \end{bmatrix}$ and $\delta(AB) = 15$,

so $(A \cdot B)^{-1} = \dfrac{1}{15}\begin{bmatrix} 0 & -5 \\ 3 & 9 \end{bmatrix}$; also, since $\delta(A) = -5$ and $\delta(B) = -3$,

$B^{-1} = -\dfrac{1}{3}\begin{bmatrix} 1 & -1 \\ -3 & 0 \end{bmatrix}$ and $A^{-1} = \dfrac{1}{5}\begin{bmatrix} -1 & -3 \\ -1 & 2 \end{bmatrix}$; $B^{-1} \cdot A^{-1} = \dfrac{1}{15}\begin{bmatrix} 0 & -5 \\ 3 & 9 \end{bmatrix}$;

hence $(A \cdot B)^{-1} = B^{-1} \cdot A^{-1}$.

27. Let $A = \begin{bmatrix} 3 & 0 & 1 \\ 2 & 1 & 0 \\ 0 & 1 & 2 \end{bmatrix}$; then $\delta(A) = 8$ and $A^{-1} = \dfrac{1}{8}\begin{bmatrix} 2 & 1 & -1 \\ -4 & 6 & 2 \\ 2 & -3 & 3 \end{bmatrix}$.

Let $B = \begin{bmatrix} 2 & 1 & 0 \\ 1 & 1 & 2 \\ 0 & 1 & 0 \end{bmatrix}$; then $\delta(B) = -\dfrac{1}{4}$ and $B^{-1} = -\dfrac{1}{4}\begin{bmatrix} -2 & 0 & 2 \\ 0 & 0 & -4 \\ 1 & -2 & 1 \end{bmatrix}$;

$A \cdot B = \begin{bmatrix} 6 & 4 & 0 \\ 5 & 3 & 2 \\ 1 & 3 & 2 \end{bmatrix}$, $\delta(A \cdot B) = -32$,   and $(A \cdot B)^{-1} = -\dfrac{1}{32}\begin{bmatrix} 0 & -8 & 8 \\ -8 & 12 & -12 \\ 12 & -14 & -2 \end{bmatrix}$;

$B^{-1} \cdot A^{-1} = -\dfrac{1}{4}\begin{bmatrix} -2 & 0 & 2 \\ 0 & 0 & -4 \\ 1 & -2 & 1 \end{bmatrix} \cdot \dfrac{1}{8}\begin{bmatrix} 2 & 1 & -1 \\ -4 & 6 & 2 \\ 2 & -3 & 3 \end{bmatrix} = -\dfrac{1}{32}\begin{bmatrix} 0 & -8 & 8 \\ -8 & 12 & -12 \\ 12 & -14 & -2 \end{bmatrix}$;

hence $(A \cdot B)^{-1} = B^{-1} \cdot A^{-1}$.

29. Let $A = \begin{bmatrix} a_{11} & a_{12} \\ a_{21} & a_{22} \end{bmatrix}$; then $\delta(A) = (a_{11}a_{22} - a_{12}a_{21}) \neq 0$, since A is nonsingular, and hence $\dfrac{1}{\delta(A)}$ is defined and A^{-1} exists.

$$A^{-1} = \frac{1}{\delta(A)}\begin{bmatrix} a_{22} & -a_{12} \\ -a_{21} & a_{11} \end{bmatrix} = \begin{bmatrix} \dfrac{a_{22}}{\delta(A)} & \dfrac{-a_{12}}{\delta(A)} \\ \dfrac{-a_{21}}{\delta(A)} & \dfrac{a_{11}}{\delta(A)} \end{bmatrix}$$

and $\delta(A^{-1}) = \dfrac{a_{22}a_{11}}{[\delta(A)]^2} - \dfrac{a_{12}a_{21}}{[\delta(A)]^2} = \dfrac{a_{22}a_{11} - a_{12}a_{21}}{[\delta(A)]^2} = \dfrac{\delta(A)}{[\delta(A)]^2} = \dfrac{1}{\delta(A)}$.

31. From the results of Problem 43, Ex. 9.4, and Problem 29 above,

$$\delta[B^{-1}AB] = \delta[B^{-1}(AB)] = \delta(B^{-1}) \cdot \delta(AB) = \delta(B^{-1}) \cdot \delta(A) \cdot \delta(B) = \frac{1}{\delta(B)} \cdot \delta(A) \cdot \delta(B) = \delta(A).$$

Exercise 9.7 (page 242)   **1.** $\{(1, 1)\}$ **3.** $\{(2, 2)\}$ **5.** $\{(6, 4)\}$ **7.** $\{(1, 1, 1)\}$ **9.** $\{(1, 1, 0)\}$
11. $\{(1, -2, 3)\}$ **13.** $\{(3, -1, -2)\}$

Exercise 9.8 (page 246)   **1.** $\left\{\left(\dfrac{13}{5}, \dfrac{3}{5}\right)\right\}$ **3.** $\left\{\left(\dfrac{22}{7}, \dfrac{20}{7}\right)\right\}$ **5.** $\{(6, 4)\}$ **7.** Inconsistent

9. $\{(4, 1)\}$ **11.** $\left\{\left(\dfrac{1}{a+b}, \dfrac{1}{a+b}\right)\right\}$ $(a \neq -b)$ **13.** $\{(1, 1, 0)\}$ **15.** $\{(1, -2, 3)\}$

17. $\{(3, -1, -2)\}$ **19.** $\left\{\left(-\dfrac{1}{3}, -\dfrac{25}{24}, -\dfrac{5}{8}\right)\right\}$ **21.** $\left\{\left(1, -\dfrac{1}{3}, \dfrac{1}{2}\right)\right\}$

23. $\{(w, x, y, z)\} = \{(2, -1, 1, 0)\}$ **25.** $A = \begin{vmatrix} a_1 & b_1 \\ a_2 & b_2 \end{vmatrix}$ and $\delta(A) = a_1 b_2 - a_2 b_1$;

$A_y = \begin{vmatrix} a_1 & c_1 \\ a_2 & c_2 \end{vmatrix}$ and $\delta(A_y) = a_1 c_2 - a_2 c_1 = 0$, so $a_1 c_2 = a_2 c_1$;

$A_x = \begin{vmatrix} c_1 & b_1 \\ c_2 & b_2 \end{vmatrix}$ and $\delta(A_x) = b_2 c_1 - b_1 c_2 = 0$, so $b_1 c_2 = b_2 c_1$;

hence, $\dfrac{a_1 c_2}{b_1 c_2} = \dfrac{a_2 c_1}{b_2 c_1}; \dfrac{a_1}{b_1} = \dfrac{a_2}{b_2}$ and $a_1 b_2 = a_2 b_1$, so $a_1 b_2 - a_2 b_1 = 0$; therefore $\delta(A) = 0$.

Chapter 9 Review (page 247)   **1.** $\begin{bmatrix} 1 & -1 \\ 1 & 1 \end{bmatrix}$ **2.** $\begin{bmatrix} 2 & 5 & -2 \\ 14 & -1 & 12 \end{bmatrix}$ **3.** $\begin{bmatrix} -2 & -3 & 1 \\ 3 & -7 & 6 \\ -5 & 2 & 5 \end{bmatrix}$

4. $\begin{bmatrix} -8 & -3 & -6 \\ 6 & -3 & 0 \\ -9 & 5 & 5 \end{bmatrix}$ **5.** $\begin{bmatrix} -21 & 7 \\ -14 & 0 \\ -7 & -7 \end{bmatrix}$ **6.** $[13]$ **7.** $\begin{bmatrix} -13 & 3 \\ -19 & 27 \end{bmatrix}$ **8.** $\begin{bmatrix} 18 & 7 & 25 \\ 8 & -1 & 11 \\ 3 & 0 & 3 \end{bmatrix}$

9. $\{(2, -1)\}$ **10.** $\{(2, 1, 1)\}$ **11.** -3 **12.** 7 **13.** -3 **14.** 14 **15.** 15

16. -1578 **17.** $\dfrac{1}{34}\begin{bmatrix} -3 & 2 \\ 11 & 4 \end{bmatrix}$ **18.** $\dfrac{1}{6}\begin{bmatrix} 1 & 3 & -2 \\ -3 & -3 & 6 \\ 1 & -3 & 4 \end{bmatrix}$ **19.** $\{(-2, 1)\}$ **20.** $\{(3, -1, 1)\}$

21. $\left\{\left(-\dfrac{16}{7}, -\dfrac{13}{7}\right)\right\}$ **22.** $\{(2, -1, 0)\}$

Exercise 10.1 (page 252)   **1.** $(5, 7)$ **3.** $(-6, -1)$ **5.** $(3, 7)$ **7.** $(3, 4)$ **9.** $(0, 2)$
11. $(5, 14)$ **13.** $(-2, 14)$ **15.** $(1, -5)$ **17.** $(-1, 7)$ **19.** $(a_1 + a_2, 0); (a_1 a_2, 0)$
21.

	If $a, b, c \in R$	If $(a, b), (c, d), (e, f) \in C$	
F-1	$a + b \in R$	$(a, b) + (c, d) \in C$	Closure law for addition
F-2	$(a + b) + c = a + (b + c)$	$[(a, b) + (c, d)] + (e, f) = (a, b) + [(c, d) + (e, f)]$	Associative law for addition
F-3	There exists $0 \in R$ such that $a + 0 = 0$ and $0 + a = a$	There exists $(0, 0) \in C$ such that $(a, b) + (0, 0) = (a, b)$ and $(0, 0) + (a, b) = (a, b)$	Additive-identity law
F-4	For each $a \in R$, there exists $-a \in R$ such that $a + (-a) = 0$ and $(-a) + a = 0$	For each $(a, b) \in C$, there exists $-(a, b) \in C$ such that $(a, b) + [-(a, b)] = (0, 0)$ and $[-(a, b)] + (a, b) = (0, 0)$	Additive-inverse law

	If $a, b, c \in R$	If $(a, b), (c, d), (e, f) \in C$	
F-5	$a + b = b + a$	$(a, b) + (c, d) = (c, d) + (a, b)$	Commutative law for addition
F-6	$ab \in R$	$(a, b) \cdot (c, d) \in C$	Closure law for multiplication
F-7	$(ab)c = a(bc)$	$[(a, b) \cdot (c, d)] \cdot (e, f)$ $= (a, b) \cdot [(c, d) \cdot (e, f)]$	Associative law for multiplication
F-8	There exists $1 \in R$ such that $a \cdot 1 = a$ and $1 \cdot a = a$	There exists $(1, 0) \in C$ such that $(a, b) \cdot (1, 0) = (a, b)$ and $(1, 0) \cdot (a, b) = (a, b)$	Multiplicative-identity law
F-9	For each $a \in R$, $a \neq 0$, there exists $a^{-1} \in R$, such that $aa^{-1} = 1$ and $a^{-1}a = 1$	For each $(a, b) \in C$, $(a, b) \neq (0, 0)$, there exists $(a, b)^{-1} \in C$, such that $(a, b) \cdot (a, b)^{-1} = (1, 0)$ and $(a, b)^{-1} \cdot (a, b) = (1, 0)$	Multiplicative-inverse law
F-10	$ab = ba$	$(a, b) \cdot (c, d) = (c, d) \cdot (a, b)$	Commutative law for multiplication
F-11	$a(b + c) = ab + ac$ and $(b + c)a = ba + ca$	$(a, b) \cdot [(c, d) + (e, f)]$ $= (a, b) \cdot (c, d) + (a, b) \cdot (e, f)$ and $[(c, d) + (e, f)] \cdot (a, b)$ $= (c, d) \cdot (a, b) + (e, f) \cdot (a, b)$	Distributive law

23. Let $z_1 = (a_1, b_1)$ and $z_2 = (a_2, b_2)$; then by Def. 10.3-I and F-5 for real numbers, $z_1 + z_2 = (a_1, b_1) + (a_2, b_2) = (a_1 + a_2, b_1 + b_2) = (a_2 + a_1, b_2 + b_1) = z_2 + z_1$.

25. Let $z_1 = (a_1, b_1)$, $z_2 = (a_2, b_2)$, and $z_3 = (a_3, b_3)$; then by Def. 10.3-II and F-11, F-2, F-5 for real numbers, $(z_1 \cdot z_2)z_3 = [(a_1, b_1) \cdot (a_2, b_2)] \cdot (a_3, b_3) = [(a_1a_2 - b_1b_2, a_1b_2 + a_2b_1)] \cdot (a_3, b_3)$
$= [(a_1a_2 - b_1b_2)a_3 - (a_1b_2 + a_2b_1)b_3, (a_1a_2 - b_1b_2)b_3 + (a_1b_2 + a_2b_1)a_3]$
$= [a_1(a_2a_3 - b_2b_3) - b_1(a_2b_3 + b_2a_3), a_1(a_2b_3 + b_2a_3) + b_1(a_2a_3 - b_2b_3)]$
$= (a_1, b_1) \cdot [a_2a_3 - b_2b_3, a_2b_3 + b_2a_3] = (a_1, b_1) \cdot [(a_2, b_2) \cdot (a_3, b_3)]$.

27. $\left(\dfrac{a}{a^2 + b^2}, \dfrac{-b}{a^2 + b^2} \right)$

Exercise 10.2 (page 255) **1.** $(3, 1)$ **3.** $(-9, 1)$ **5.** $(8, -8)$ **7.** $\left(\dfrac{7}{4}, -\dfrac{1}{4} \right)$

9. $\left(\dfrac{1}{10}, -\dfrac{7}{10} \right)$ **11.** $(-2, 0)$ **13.** $\left(\dfrac{23}{37}, \dfrac{27}{37} \right)$ **15.** $\left(-\dfrac{2}{5}, -\dfrac{9}{5} \right)$ **17.** $\left(-\dfrac{5}{2}, -\dfrac{1}{2} \right)$

19. $(2, 0)$

21. Let $z = (a, b)$; then $\bar{z} = (a, -b)$. From this result and Ths. 10.3 and 10.4,
$\dfrac{(a, b)}{(a, b)} = \dfrac{(a, b) \cdot (a, -b)}{(a, b) \cdot (a, -b)} = \dfrac{(a^2 + b^2, 0)}{(a^2 + b^2, 0)} = \left(\dfrac{a^2 + b^2}{a^2 + b^2}, \dfrac{0}{a^2 + b^2} \right) = (1, 0)$.

23. Let $z_1 = (a_1, b_1)$, $z_2 = (a_2, b_2)$; then $\bar{z}_1 = (a_1, -b_1)$, $\bar{z}_2 = (a_2, -b_2)$,
$\bar{z}_1 + \bar{z}_2 = (a_1, -b_1) + (a_2, -b_2) = [(a_1 + a_2), -(b_1 + b_2)] = \overline{z_1 + z_2}$.

25. For F-1: $(a, 0) + (b, 0) = (a + b, 0) \in C$.
For F-2: $[(a, 0) + (b, 0)] + (c, 0) = (a + b, 0) + (c, 0) = [(a + b) + c, 0]$
$= [a + (b + c), 0] = (a, 0) + [b + c, 0] = (a, 0) + [(b, 0) + (c, 0)]$.

For F-3: $(0, 0)$ is the additive-identity law, since $(a, 0) + (0, 0) = (a + 0, 0) = (a, 0)$ and $(0, 0) + (a, 0) = (0 + a, 0) = (a, 0)$.

For F-4: $(-a, 0)$ is the additive-inverse law, since
$(a, 0) + (-a, 0) = (a - a, 0) = (0, 0)$ and $(-a, 0) + (a, 0) = (-a + a, 0) = (0, 0)$.

For F-5: $(a, 0) + (b, 0) = (a + b, 0) = (b + a, 0) = (b, 0) + (a, 0)$.

For F-6: $(a, 0) \cdot (b, 0) = (ab - 0, 0 + 0) = (ab, 0) \in C$.

For F-7: $[(a, 0) \cdot (b, 0)] \cdot (c, 0) = (ab, 0) \cdot (c, 0) = [(ab)c, 0] = [a(bc), 0] = (a, 0) \cdot (bc, 0)$
$= (a, 0) \cdot [(b, 0) \cdot (c, 0)]$.

For F-8: $(1, 0)$ is the multiplicative-identity law since $(a, 0) \cdot (1, 0) = (a, 0)$ and $(1, 0) \cdot (a, 0) = (a, 0)$.

For F-9: use $\left(\dfrac{1}{a}, 0\right)$ for multiplicative-inverse law, since if

$a \neq 0$, $(a, 0) \cdot \left(\dfrac{1}{a}, 0\right) = (1, 0)$ and $\left(\dfrac{1}{a}, 0\right) \cdot (a, 0) = (1, 0)$.

For F-10: $(a, 0) \cdot (b, 0) = (ab, 0) = (ba, 0) = (b, 0) \cdot (a, 0)$.

For F-11: $(a, 0) \cdot [(b, 0) + (c, 0)] = (a, 0) \cdot [b + c, 0] = (ab + ac, 0) = (ab, 0) + (ac, 0)$
$= (a, 0) \cdot (b, 0) + (a, 0) \cdot (c, 0)$; $[(b, 0) + (c, 0)] \cdot (a, 0) = (b + c, 0) \cdot (a, 0)$.
$= (ba + ca, 0) = (ba, 0) + (ca, 0) = (b, 0) \cdot (a, 0) + (c, 0) \cdot (a, 0)$.

Exercise 10.3 (page 259)   **1.** $2 + 6i$ **3.** $5 - 2i$ **5.** $-7 - 3i$ **7.** $4 + 0i$ **9.** $(2, 3)$
11. $(-3, 1)$ **13.** $(0, 4)$ **15.** $(7, 0)$ **17.** $2 - 3i$ **19.** $-3 - i$ **21.** $0 - 4i$ **23.** $7 - 0i$

25. $x = \dfrac{3}{2}, y = -2$ **27.** $x = 2, y = -2$; $x = -2, y = 2$ **29.** $x = 3, y = 9$; $x = -3, y = -9$

31. $5 + 5i$ **33.** $5 - 3i$ **35.** $-1 - 2i$ **37.** $1 + i$ **39.** $-\dfrac{1}{10} + \dfrac{7}{10}i$ **41.** $-8 - 6i$

43. $2 - 2i$ **45.** $4 - i\sqrt{7}$ **47.** $5 - 2i$ **49.** $-\sqrt{7}$ **51.** $1 - i$ **53.** $\dfrac{4}{13} + \dfrac{7}{13}i$

55. 4 **57.** 2 **59.** $\sqrt{5}$ **61.** $\sqrt{5}$
63.   **65.**

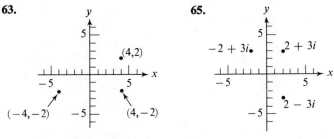

67. Let $z = (a, b)$; then $\bar{z} = (a, -b)$ and $|\bar{z}| = |a - bi| = \sqrt{a^2 + (-b)^2} = \sqrt{a^2 + b^2} = |z|$.
69. Let $z_1 = (a, b)$ and $z_2 = (a, -b)$; then $z_1 + z_2 = (a + a, b - b) = (2a, 0) = 2a$ and
$z_1 \cdot z_2 = (a^2 + b^2, -ab + ab) = (a^2 + b^2, 0) = a^2 + b^2$.

Exercise 10.4 (page 261)   **1.** $\left\{\dfrac{-1 - 2i}{3}\right\}$ **3.** $\left\{\dfrac{6 - 12i}{5}\right\}$ **5.** $\left\{\dfrac{5 - 15i}{2}\right\}$ **7.** $\{2i, -2i\}$

9. $\{i\sqrt{3}, -i\sqrt{3}\}$ **11.** $\{-2 + i, -2 - i\}$ **13.** $\{-5 + 4i, -5 - 4i\}$ **15.** $\{-2 + i\sqrt{3}, -2 - i\sqrt{3}\}$

17. $\{1 + i\sqrt{2}, 1 - i\sqrt{2}\}$ **19.** $\left\{\dfrac{-1 + i\sqrt{83}}{6}, \dfrac{-1 - i\sqrt{83}}{6}\right\}$

21. $\left\{\dfrac{-10 + i + \sqrt{99 - 16i}}{2}, \dfrac{-10 + i - \sqrt{99 - 16i}}{2}\right\}$ **23.** $\left\{\dfrac{4 + i + \sqrt{23 + 8i}}{4}, \dfrac{4 + i - \sqrt{23 + 8i}}{4}\right\}$

25. $\left\{ \dfrac{-3i + \sqrt{32i - 17}}{4}, \dfrac{-3i - \sqrt{32i - 17}}{4} \right\}$ **27.** $x^2 - 4x + 5 = 0$ **29.** $x^2 + 2 = 0$

31. $x^2 + 3x - xi - 3i = 0$ **33.** $x^2 - 3x + xi - 3i + 4 = 0$ **35.** $x^3 - 3x^2 + x - 3$

37. $x^3 - (5 + i)x^2 + (6 + 5i)x - 6i.$

Exercise 10.5 (page 265)

1. $(8, 5);\ \sqrt{89}$

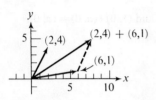

3. $(-1, 3);\ \sqrt{10}$

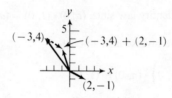

5. $(5, 9);\ \sqrt{106}$

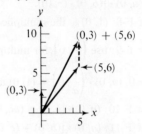

7. $(8, 6);\ 10$

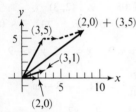

9. $(8, 1);\ \sqrt{65}$

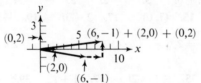

11. $(3, 1);\ \sqrt{10}$

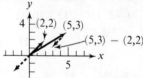

13. $(-4, 5);\ \sqrt{41}$

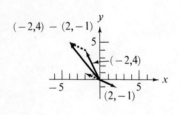

15. $(0, 8);\ 8$

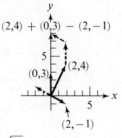

17. $(-1, 1);\ \sqrt{2}$

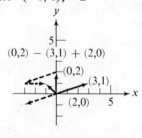

19. $(11, 11);\ 11\sqrt{2}$

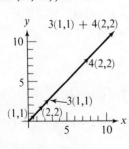

21. $(-6, 1);\ \sqrt{37}$

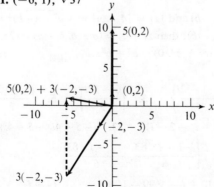

23. $(3, 10);\ \sqrt{109}$

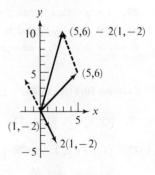

25. $(-8, 12); 4\sqrt{13}$

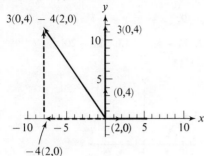

27. $(10, 4); 2\sqrt{29}$

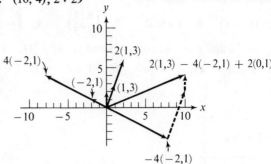

29. $(15, -5); 5\sqrt{10}$

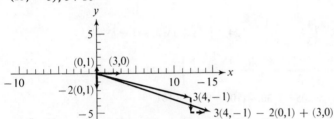

31. $\mathbf{v}_1 + \mathbf{v}_2 = (a_1, b_1) + (a_2, b_2)$ Hyp.
        $= (a_1 + a_2, b_1 + b_2)$ Def. 10.16-I

By the closure law for R, $a_1 + a_2 \in R$ and $b_1 + b_2 \in R$, so that by Def. 10.13 $(a_1 + a_2, b_1 + b_2) \in V$.

33. $(\mathbf{v}_1 + \mathbf{v}_2) + \mathbf{v}_3$
        $= [(a_1, b_1) + (a_2, b_2)] + (a_3, b_3)$ Hyp.
        $= (a_1 + a_2, b_1 + b_2) + (a_3, b_3)$ Def. 10.16-I
        $= [(a_1 + a_2) + a_3, (b_1 + b_2) + b_3]$ Def. 10.16-I
        $= [a_1 + (a_2 + a_3), b_1 + (b_2 + b_3)]$ Assoc. law of add. for R
        $= (a_1, b_1) + (a_2 + a_3, b_2 + b_3)$ Def. 10.16-I
        $= (a_1, b_1) + [(a_2, b_2) + (a_3, b_3)]$ Def. 10.16-I
        $= \mathbf{v}_1 + (\mathbf{v}_2 + \mathbf{v}_3)$ Hyp.

35. $\mathbf{v}_1 + (-\mathbf{v}_1)$
        $= (a_1, b_1) + (-a_1, -b_1)$ Hyp., Def. 10.18
        $= (a_1 + (-a_1), b_1 + (-b_1))$ Def. 10.16-I
        $= (a_1 - a_1, b_1 - b_1)$ Def. 1.12
        $= (0, 0)$ Add. inverse law for R
        $= \mathbf{0}$ Def. 10.17

37. $c(d\mathbf{v}_1)$
        $= (cd(a_1, b_1))$ Hyp.
        $= c(da_1, db_1)$ Def. 10.16-II
        $= (cda_1, cdb_1)$ Def. 10.16-II
        $= cd(a_1, b_1)$ Closure mult. law for R, Def. 10.16-II
        $= (cd)\mathbf{v}_1$

39. $c(\mathbf{v}_1 + \mathbf{v}_2)$
        $= c[(a_1, b_1) + (a_2, b_2)]$ Hyp.
        $= c(a_1 + a_2, b_1 + b_2)$ Def. 10.16-I
        $= [c(a_1 + a_2), c(b_1 + b_2)]$ Def. 10.16-II
        $= (ca_1 + ca_2, cb_1 + cb_2)$ Distrib. law for R
        $= (ca_1, cb_1) + (ca_2, cb_2)$ Def. 10.16-I
        $= c\mathbf{v}_1 + c\mathbf{v}_2$ Def. 10.16-II

Chapter 10 Review (page 266)   **1.** $(-1, 11)$ **2.** $(-4, 4)$ **3.** $(7, -1)$ **4.** $(-2, -14)$

5. $(8, -1)$ **6.** $(-5, 3)$ **7.** $\left(\dfrac{8}{5}, \dfrac{1}{5}\right)$ **8.** $\left(-\dfrac{1}{2}, -\dfrac{1}{2}\right)$ **9.** $x = \dfrac{2}{3}, y = -3$

10. $x = 3$ and $y = -6$; $x = -3$ and $y = 6$ **11.** $5 + i$ **12.** $1 - 5i$ **13.** $21 - 20i$

14. $1 + 0i$ **15.** $-\sqrt{14}$ **16.** $1 + i$ **17.** $2\sqrt{5}$ **18.** $\sqrt{13}$

19.

20.

21. $\left\{\dfrac{1 + 2i}{4}\right\}$ **22.** $\left\{\dfrac{3 - 4i}{2}\right\}$ **23.** $\left\{\dfrac{4}{3}i, -\dfrac{4}{3}i\right\}$ **24.** $\{-1 + i\sqrt{2}, -1 - i\sqrt{2}\}$

25. $\{-i + \sqrt{2}, -i - \sqrt{2}\}$ **26.** $\left\{\dfrac{3i + i\sqrt{5 + 8i}}{2}, \dfrac{3i - i\sqrt{5 + 8i}}{2}\right\}$ **27.** $(-1, 1)$; $\sqrt{2}$

28. $(5, -7)$; $\sqrt{74}$ **29.** $(3, 14)$; $\sqrt{205}$ **30.** $(10, -9)$; $\sqrt{181}$

Exercise 11.1 (page 270)   **1.** 40 **3.** $-41 + 33i$ **5.** 36 **7.** $-11 + 40i$ **9.** $-2 - 7i$

11. $Q = 2x^2 + 5x + 18$; $R = 58$ **13.** $Q = x^2 + (1 + 2i)x - 7 + 2i$; $R = -1 - 14i$

15. $Q = x^3 - ix^2 - 4x + 1 + 4i$; $R = 3 - i$ **17.** $Q = x^2 + (i - 1)x - 1$; $R = -i$

19. Divide $x^3 + 2x^2 - 5x - 6$ by $x - 2$. Remainder is 0, hence $x - 2$ is a factor.

21. $(-2i)^3 - (-2i)^2 + 4(-2i) - 4 = 0$. Hence $-2i$ is a root.   **23.** $Q(x) = x^2 + (-3 + i)x - 3i$

25. 80 **27.** -22

Exercise 11.2 (page 272)   **1.** $-2i$ **3.** $-i$ **5.** $i, -2i$ **7.** $3i, \dfrac{-3 + \sqrt{29}}{2}, \dfrac{-3 - \sqrt{29}}{2}$

9. $1 - i$; $x^3 - 2x + 4 = 0$ **11.** $P(x) = (2x - 3)(x - 2 + 2i)(x - 2 - 2i)$ **13.** $4 + i, 1 + i, 1 - i$

15. Let $P(x) = 0$ be a polynomial equation with real coefficients of degree n, where n is odd; then, $P(x) = 0$ must have at least n zeros and, since $P(x)$ has only real coefficients, any complex zeros must occur as conjugate pairs; since there must always be at least one real zero, then $P(x) = 0$ must have at least one real root.

17. $-1, \dfrac{1 + \sqrt{3}\,i}{2}, \dfrac{1 - \sqrt{3}\,i}{2}$

Exercise 11.3 (page 277)   **1.** 2 positive real, 2 negative real; or 0 positive real, 2 negative real, 2 complex; or 0 positive real, 0 negative real, 4 complex

3. 0 positive real, 1 negative real, 4 complex   **5.** 0 positive real, 0 negative real, 4 complex

7. Upper bound 3, lower bound -4 **9.** Upper bound 4, lower bound -3

11. Upper bound 2, lower bound -3 **13.** Upper bound 1, lower bound -2

15. $f(0) = 1$ and $f(1) = -1$ **17.** $g(-3) = 10$ and $g(-2) = -33$

19. $P(-3) = 2$ and $P(-2) = -4$; $P(0) = -4$ and $P(1) = 2$

Exercise 11.4 (page 280)   **1.** $\{4\}$ **3.** $\{1, -2\}$ **5.** None   **7.** $\left\{-\dfrac{3}{2}\right\}$ **9.** $\left\{2, -\dfrac{7}{4}\right\}$ **11.** $\left\{\dfrac{3}{2}\right\}$

13. $\left\{-\dfrac{1}{3}, 1 + \sqrt{5}, 1 - \sqrt{5}\right\}$ **15.** $\left\{-1, \dfrac{1}{2}, \dfrac{-1 + i\sqrt{3}}{2}, \dfrac{-1 - i\sqrt{3}}{2}\right\}$ **17.** $\left\{\dfrac{3}{4}, -\dfrac{4}{3}, i, -i\right\}$

19. $(2x + 3)(x - 1)(x + 1)$

21. Consider $x^2 - 3 = 0$. If the equation has real zeros, then they are either rational or irrational. If rational, by Th. 11.11, the only possibilities are ± 1 and ± 3; however, direct substitution shows none of these satisfies the equation. Now, $\sqrt{3}$ is a real number, and since $(\sqrt{3})^2 - 3 = 3 - 3 = 0$, $\sqrt{3}$ is a real zero of the equation, and since it is not among the rational zeros, it must be irrational.

Exercise 11.5 (page 283)   **1.**   1.5   **3.** 2.1   **5.** 1.7   **7.** 0.7   **9.** 1.7

Chapter 11 Review (page 283)   **1.** -132 **2.** $-13+18i$

3. $x^2 + (-2+i)x + (2-2i)$;   remainder: $-3+2i$

4.   Upon division by $x-i$, the remainder is 0.   **5.** $\{2, 2i, -2i\}$ **6.** $\left\{\frac{3}{2}, 2+2i, 2-2i\right\}$

7.   There are two sign changes and therefore the equation has two or no positive real solutions. The equation also has two or no negative real solutions.

8.   Integral least upper bound: 3; integral greatest lower bound: -4

9. $P(1) = -3$ and $P(2) = 10$; hence there is a zero between 1 and 2.   **10.** $\{4\}$ **11.** $\left\{-\frac{1}{2}, 3\right\}$

12. $\{-2, i\sqrt{2}, -i\sqrt{2}\}$ **13.** $\{1.2\}$ **14.** $\{0.6\}$

Exercise 12.1 (page 287)   **1.**   **a.**   For $n=1: \dfrac{n}{2} = \dfrac{1}{2}$; $\dfrac{n(n+1)}{4} = \dfrac{1(1+1)}{4} = \dfrac{1}{2}$.

   b.   For $n=k: \dfrac{1}{2} + \dfrac{2}{2} + \dfrac{3}{2} + \cdots + \dfrac{k}{2} = \dfrac{k(k+1)}{4}$ and $(k+1)$th term $= \dfrac{k+1}{2}$;

hence $\dfrac{1}{2} + \dfrac{2}{2} + \dfrac{3}{2} + \cdots + \dfrac{k}{2} + \dfrac{k+1}{2} = \dfrac{k(k+1)}{4} + \dfrac{k+1}{2} = \dfrac{k^2+k+2k+2}{4} = \dfrac{k^2+3k+2}{4} = \dfrac{(k+1)(k+2)}{4}$.

3.   **a.**   For $n=1: 2n = 2(1) = 2$; $n(n+1) = 1(1+1) = 2$.

   b.   For $n=k: 2+4+6+\cdots+2k = k(k+1)$ and $(k+1)$th term is $2(k+1)$;

hence $2+4+6+\cdots+2k+2(k+1) = k(k+1) + 2(k+1) = (k+1)(k+2)$.

5.   **a.**   For $n=1: n^2 = 1^2 = 1$; $\dfrac{n(n+1)(2n+1)}{6} = \dfrac{1(2)(3)}{6} = 1$.

   b.   For $n=k: 1^2 + 2^2 + 3^2 + \cdots + k^2 = \dfrac{k(k+1)(2k+1)}{6}$ and $(k+1)$th term is $(k+1)^2$;

hence $1^2 + 2^2 + 3^2 + \cdots + k^2 + (k+1)^2 = \dfrac{k(k+1)(2k+1)}{6} + (k+1)^2 = \dfrac{k(k+1)(2k+1) + 6(k+1)^2}{6}$

$= \dfrac{(k+1)[k(2k+1) + 6(k+1)]}{6} = \dfrac{(k+1)(2k^2 + 7k + 6)}{6} = \dfrac{(k+1)(k+2)(2k+3)}{6}$

$= \dfrac{(k+1)[(k+1)+1][2(k+1)+1]}{6}$.

7.   **a.**   For $n=1: (2n-1)^3 = (2-1)^3 = 1^3 = 1$; $n^2(2n^2-1) = 1(2-1) = 1(1) = 1$.

   b.   for $n=k: 1^3 + 3^3 + 5^3 + \cdots + (2k-1)^3 = k^2(2k^2-1)$ and the $(k+1)$th term is

$[2(k+1)-1]^3 = (2k+1)^3$; hence $1^3 + 3^3 + 5^3 + \cdots + (2k-1)^3 + (2k+1)^3 = k^2(2k^2-1) + (2k-1)^3 =$

$2k^4 + 8k^3 + 11k^2 + 6k + 1$; by use of the factor theorem and synthetic division,

$2k^4 + 8k^3 + 11k^2 + 6k + 1 = (k+1) \times (k+1)(2k^2 + 4k + 1)$;

also, $2k^2 + 4k + 1 = 2(k^2 + 2k + 1) - 2 + 1 = 2(k+1)^2 - 1$;

hence $2k^4 + 8k^3 + 11k^2 + 6k + 1 = (k+1)^2[2(k+1)^2 - 1]$.

9.   **a.**   For $n=1: n(n+1) = 1(2) = 2$; $\dfrac{n(n+1)(n+2)}{3} = \dfrac{1(2)(3)}{3} = 2$.

   b.   For $n=k: 1\cdot 2 + 2\cdot 3 + 3\cdot 4 + \cdots + k(k+1) = \dfrac{k(k+1)(k+2)}{3}$ and the $(k+1)$th term is

$(k+1)[(k+1)+1] = (k+1)(k+2)$; hence $1\cdot 2 + 2\cdot 3 + 3\cdot 4 + \cdots + k(k+1) + [(k+1)(k+2)]$

$= \dfrac{k(k+1)(k+2)}{3} + (k+1)(k+2) = \dfrac{[k(k+1)(k+2)] + [3(k+1)(k+2)]}{3}$

$= \dfrac{(k+1)(k+2)(k+3)}{3} = \dfrac{(k+1)[(k+1)+1][(k+1)+2]}{3}$.

11.   For $n=k: 2+4+6+\cdots+2k = k(k+1) + 2$ and $(k+1)$th term is $2(k+1)$;

hence $2+4+6+\cdots+2k+2(k+1) = k(k+1) + 2 + 2(k+1) = (k^2 + 3k + 2) + 2 = (k+1)(k+2) + 2$

$= (k+1)[(k+1)+1] + 2$.

However, for $n=1: 2n = 2(1) = 2$; $n(n+1) + 2 = 1(2) + 2 = 4$. Hence not true for every $n \in N$.

Exercise 12.2 (page 290)   **1.** $-4, -3, -2, -1$ **3.** $-\dfrac{1}{2}, 1, \dfrac{7}{2}, 7$ **5.** $2, \dfrac{3}{2}, \dfrac{4}{3}, \dfrac{5}{4}$ **7.** 0, 1, 3, 6

9. $-1, 1, -1, 1$ **11.** $1, 0, -\dfrac{1}{3}, \dfrac{1}{2}$ **13.** 11, 15, 19   **15.** $x+2, x+3, x+4$

17. $2x + 7, 2x + 10, 2x + 13$ **19.** 32, 128, 512, 2048   **21.** $\dfrac{8}{3}, \dfrac{16}{3}, \dfrac{32}{3}, \dfrac{64}{3}$

23. $\dfrac{x}{a}, -\dfrac{x^2}{a^2}, \dfrac{x^3}{a^3}, -\dfrac{x^4}{a^4}$ **25.** $4n + 3, 31$ **27.** $-5n + 8, -92$ **29.** $48(2)^{n-1}, 1536$

31. $-\dfrac{1}{3}(-3)^{n-1}, -243$ **33.** 2; 3; 41   **35.** 28th   **37.** 3

39. **a.** For $n = 1: s_n = s_1 = a;$ $ar^{n-1} = a(r)^0 = a(1) = a.$
   b. For $n = k: s_k = ar^{k-1};$ to obtain s_{k+1} multiply s_k by $r;$ hence, multiplying both sides by r, we obtain
$s_k(r) = s_{k+1} = (ar^{k-1})(r) = ar^{k-1+1} = ar^{(k+1)-1}.$

Exercise 12.3 (page 295)   **1.** $1 + 4 + 9 + 16$ **3.** $-\dfrac{1}{2} + \dfrac{1}{4} - \dfrac{1}{8}$ **5.** $1 + \dfrac{1}{2} + \dfrac{1}{4} + \cdots$

7. $\displaystyle\sum_{j=1}^{4} x^{2j-1}$ **9.** $\displaystyle\sum_{j=1}^{5} j^2$ **11.** $\displaystyle\sum_{j=1}^{\infty} j(j+1)$ **13.** $\displaystyle\sum_{j=1}^{\infty} \dfrac{j+1}{j}$ **15.** 63   **17.** 806   **19.** -6

21. 1092   **23.** $\dfrac{31}{32}$ **25.** $\dfrac{364}{729}$ **27.** 168   **29.** 196   **31.** $\dfrac{3}{4}, \dfrac{7}{8}, \dfrac{15}{16}, \dfrac{31}{32}; 1$

33. $p = 4, q = -3$
35. Since $s_n = a + (n - 1)d$ and $S_{n+1} = S_n + s_n$, it follows that
   a. For $n = 1: S_n = s_n = s_1 = a;$ $\dfrac{n}{2}(a + s_n) = \dfrac{1}{2}(2a) = a.$

   b. For $n = k: S_k = \dfrac{k}{2}(a + s_k), s_k = a + (k - 1)d;$ now $s_{k+1} = a + [(k + 1) - 1]d = a + kd;$

adding produces $S_k + s_{k+1} = S_{k+1} = \dfrac{k}{2}(a + s_k) + (a + kd) = \dfrac{k}{2}[a + a + (k - 1)d] + \dfrac{2}{2}(a + kd)$

$= \dfrac{2ka + k^2d - kd + 2a + 2kd}{2} = \dfrac{2ka + 2a + k^2d + kd}{2} = \dfrac{(k + 1)2a + k(k + 1)d}{2}$

$= \dfrac{(k + 1)}{2}[2a + kd] = \dfrac{k + 1}{2}[a + (a + kd)] = \dfrac{k + 1}{2}[a + s_{n+1}].$

37. Let $s_1, s_2, s_3, \ldots, s_n$ and $t_1, t_2, t_3, \ldots, t_n$ be two sequences with terms in arithmetic progression; then, by Problem 36, $(s_1 + t_1), (s_2 + t_2), \ldots, (s_n + t_n)$ is also a sequence with terms in arithmetic progression.
$S_1 = s_1 + s_2 + s_3 + \cdots + s_n = \displaystyle\sum_{j=1}^{n} s_j, S_2 = t_1 + t_2 + t_3 + \cdots + t_n = \displaystyle\sum_{j=1}^{n} t_j$, and $S_1 + S_2 = \displaystyle\sum_{j=1}^{n} s_j + \displaystyle\sum_{j=1}^{n} t_j;$
however, $(S_1 + S_2) = (s_1 + t_1) + (s_2 + t_2) + \cdots + (s_n + t_n)$, and $S_1 + S_2 = \displaystyle\sum_{j=1}^{n} ((s_j + t_j));$
hence $\displaystyle\sum_{j=1}^{n} s_j + \displaystyle\sum_{j=1}^{n} t_j = \displaystyle\sum_{j=1}^{n} (s_j + t_j).$

Exercise 12.4 (page 301)   **1.** $\displaystyle\lim_{n \to \infty} s_n = 0$ **3.** $\displaystyle\lim_{n \to \infty} s_n = 1$ **5.** $\displaystyle\lim_{n \to \infty} s_n$ is undefined   **7.** $\displaystyle\lim_{n \to \infty} s_n = 0$

9. Convergent, $\displaystyle\lim_{n \to \infty} \left| 0 - \dfrac{1}{2^n} \right| = 0$ **11.** Divergent, $\displaystyle\lim_{n \to \infty} n$ is undefined

13. Convergent, $\displaystyle\lim_{n \to \infty} \left| 0 - (-1)^{n+1} \dfrac{1}{2^{n-1}} \right| = 0$ **15.** 24   **17.** No sum   **19.** 2   **21.** $\dfrac{31}{99}$

23. $2\dfrac{410}{999}$ **25.** $\dfrac{29}{225}$ **27.** 20 cm

Exercise 12.5 (page 306) **1.** $8 \cdot 7 \cdot 6 \cdot 5 \cdot 4 \cdot 3 \cdot 2 \cdot 1$ **3.** $6 \cdot 5 \cdot 4 \cdot 3 \cdot 2 \cdot 1$

5. $5 \cdot 4 \cdot 3 \cdot 2 \cdot 1 = 120$ **7.** $\dfrac{9 \cdot 8 \cdot 7!}{7!} = 72$ **9.** $\dfrac{5 \cdot 4 \cdot 3 \cdot 2 \cdot 1 \cdot 7}{8 \cdot 7!} = 15$ **11.** $\dfrac{8 \cdot 7 \cdot 6!}{2 \cdot 1 \cdot 6!} = 28$

13. $3!$ **15.** $\dfrac{6!}{2!}$ **17.** $\dfrac{8!}{5!}$ **19.** $\dfrac{6!}{5! \, 1!} = 6$ **21.** $\dfrac{3!}{3! \, 0!} = 1$ **23.** $\dfrac{7!}{0! \, 7!} = 1$ **25.** $\dfrac{5!}{2! \, 3!} = 10$

27. $(n)(n-1)(n-2) \cdot \cdots \cdot 3 \cdot 2 \cdot 1$ **29.** $(3n)(3n-1)(3n-2) \cdot \cdots \cdot 3 \cdot 2 \cdot 1$

31. $(n-2)(n-3)(n-4) \cdot \cdots \cdot 3 \cdot 2 \cdot 1$ **33.** $(n+2)(n+1)$ **35.** $\dfrac{n+1}{n+3}$ **37.** $\dfrac{2n-1}{2n-2}$

39. $x^5 + 15x^4 + 90x^3 + 270x^2 + 405x + 243$ **41.** $x^4 - 12x^3 + 54x^2 - 108x + 81$

43. $8x^3 - 6x^2y + \dfrac{3}{2} xy^2 - \dfrac{1}{8} y^3$ **45.** $\dfrac{1}{64} x^6 + \dfrac{3}{8} x^5 + \dfrac{15}{4} x^4 + 20x^3 + 60x^2 + 96x + 64$

47. $x^{20} + 20x^{19}y + \dfrac{20 \cdot 19}{2!} x^{18}y^2 + \dfrac{20 \cdot 19 \cdot 18}{3!} x^{17}y^3,$ or

$$\binom{20}{0} x^{20} + \binom{20}{1} x^{19}y + \binom{20}{2} x^{18}y^2 + \binom{20}{3} x^{17}y^3$$

49. $a^{12} + 12a^{11}(-2b) + \dfrac{12 \cdot 11}{2!} a^{10}(-2b)^2 + \dfrac{12 \cdot 11 \cdot 10}{3!} a^9(-2b)^3,$ or

$$\binom{12}{0} a^{12} + \binom{12}{1} a^{11}(-2b) + \binom{12}{2} a^{10}(-2b)^2 + \binom{12}{3} a^9(-2b)^3$$

51. $x^{10} + 10x^9(-\sqrt{2}) + \dfrac{10 \cdot 9}{2!} x^8(-\sqrt{2})^2 + \dfrac{10 \cdot 9 \cdot 8}{3!} x^7(-\sqrt{2})^3,$ or

$$\binom{10}{0} x^{10} + \binom{10}{1} x^9(-\sqrt{2}) + \binom{10}{2} x^8(-\sqrt{2})^2 + \binom{10}{3} x^7(-\sqrt{2})^3$$ **53.** 1.22 **55.** \$1480

57. $-3003a^{10}b^5$ **59.** $3360x^6y^4$ **61. a.** $1 - x + x^2 - x^3 + \cdots$ **b.** $1 - x + x^2 - x^3 + \cdots$

Chapter 12 Review (page 310) **1.** Formula holds for 1. Assume formula holds for n; test for $n + 1$:

$$3 + 6 + 9 + \cdots + 3n + 3(n+1) = \frac{3n(n+1)}{2} + 3(n+1).$$

Right-hand member is equivalent to:

$$\frac{3n^2 + 3n}{2} + \frac{6(n+1)}{2} = \frac{3n^2 + 9n + 6}{2} = \frac{3(n^2 + 3n + 2)}{2} = \frac{3(n+1)(n+2)}{2} = \frac{3(n+1)((n+1)+1)}{2}.$$

2. Formula holds for 1. Assume formula holds for n; test for $n + 1$:

$$\frac{1}{2} + \frac{1}{4} + \frac{1}{8} + \cdots + \frac{1}{2^n} + \frac{1}{2^{n+1}} = 1 - \frac{1}{2^n} + \frac{1}{2^{n+1}}.$$

Right-hand member is equivalent to: $\dfrac{2^{n+1} - 2 + 1}{2^{n+1}} = \dfrac{2^{n+1} - 1}{2^{n+1}} = 1 - \dfrac{1}{2^{n+1}}.$

3. 13, 16, 19 **4.** $a - 4, a - 6, a - 8$ **5.** $-18, 54, -162$ **6.** $\dfrac{3}{2}, \dfrac{9}{4}, \dfrac{27}{8}$

7. $s_n = 5n - 8; \; s_7 = 27$ **8.** $s_n = (-2)\left(\dfrac{-1}{3}\right)^{n-1}; \; s_5 = \dfrac{-2}{81}$ **9.** 25 **10.** 6th term

11. $2 + 6 + 12 + 20$ **12.** $\displaystyle\sum_{k=1}^{\infty} x^{k+1}$ **13.** 119 **14.** $\dfrac{121}{243}$ **15.** 3 **16.** $\dfrac{8}{3}$ **17.** $\dfrac{1}{2}$

18. $\dfrac{4}{9}$ **19.** $5 \cdot (2 \cdot 1)$ **20.** 48 **21.** 21 **22.** $\dfrac{1}{n(n+1)!}$

23. $x^{10} - 20x^9y + 180x^8y^2 - 960x^7y^3$ **24.** $-15,360x^3y^7$

Exercise 13.1 (page 314)   **1.** 1, 5, 8   **3.** 2, 6, 16   **5.** 2, 2, 4   **7.** 4   **9.** 24   **11.** 16

13. 64   **15.** 24   **17.** 216   **19.** 375   **21.** 30   **23.** 10   **25.** 48   **27.** $\dfrac{5!}{2!}$, or 60

29. $\dfrac{8!}{3!}$, or 6720   **31.** $P_{5,3} = \dfrac{5!}{2!} = \dfrac{5 \cdot 4!}{2!} = 5\left(\dfrac{4!}{2!}\right) = 5(P_{4,2})$

33. $P_{n,3} = \dfrac{n!}{(n-3)!} = \dfrac{n(n-1)!}{(n-3)!} = n\left[\dfrac{(n-1)!}{(n-3)!}\right] = n(P_{n-1,1})$ **35.** 9   **37.** 24   **39.** 48

Exercise 13.2 (page 318)   **1.** 7   **3.** 15   **5.** $\dbinom{52}{5}$ **7.** $\dbinom{13}{5} \cdot \dbinom{13}{5} \cdot \dbinom{13}{3}$ **9.** $4 \cdot \dbinom{13}{5}$

11. 164   **13.** 10   **15.** 210   **17.** 12

Exercise 13.3 (page 322)   **1.** {1, 2, 3, 4, 5, 6}, {3, 4, 5, 6}, $\dfrac{2}{3}$

3. {(H, H), (H, T), (T, H), (T, T)}, {(H, H), (T, T)}, $\dfrac{1}{2}$ **5.** $\dfrac{1}{6}$ **7.** $\dfrac{1}{18}$ **9.** $\dfrac{5}{9}$ **11.** $\dfrac{1}{52}$

13. $\dfrac{3}{26}$ **15.** $\dfrac{1}{17}$ **17.** $\dfrac{11}{221}$ **19.** $\dfrac{1}{190}$ **21.** $\dfrac{3}{38}$

Exercise 13.4 (page 325)   **1.** $\dfrac{1}{6}$ **3.** $\dfrac{7}{18}$ **5.** $\dfrac{13}{18}$ **7.** $\dfrac{5}{33}$ **9.** $\dfrac{1}{11}$ **11.** $\dfrac{5}{22}$ **13.** $\dfrac{14}{33}$

15. $\dfrac{15}{22}$ **17.** $\dfrac{3}{13}$ **19.** $2.14; less   **21.** 5.3 cents   **23.** $3.29

Exercise 13.5 (page 328)   **1.** $\dfrac{20}{91}$; no   **3. a.** $\dfrac{2}{45}$ **b.** $\dfrac{28}{75}$ **c.** $\dfrac{4}{225}$ **d.** $\dfrac{1}{9}$

5. a. $\dfrac{11}{36}$ **b.** $\dfrac{5}{36}$ **c.** $\dfrac{2}{11}$ **d.** No   **7. a.** $\dfrac{1}{2}$ **b.** $\dfrac{1}{2}$ **c.** $\dfrac{1}{4}$ **d.** $\dfrac{1}{4}$ **e.** $\dfrac{1}{8}$

9. a. $\dfrac{71}{72}$ **b.** $\dfrac{5}{9}$ **c.** $\dfrac{5}{36}$ **d.** $\dfrac{61}{72}$ **11. a.** $\dfrac{1}{210}$ **b.** $\dfrac{29}{210}$ **c.** $\dfrac{29}{70}$ **d.** $\dfrac{29}{30}$; yes

13. $\dfrac{1}{4}$ **15. a.** $\dfrac{15}{77}$ **b.** $\dfrac{16}{77}$ **c.** $\dfrac{46}{77}$

Chapter 13 Review (page 330)   **1.** 16   **2.** 64   **3.** 128   **4.** 360   **5.** 792

6. 1,033,885,600   **7.** 200   **8.** 84   **9.** $\dfrac{1}{17}$ **10.** $\dfrac{1}{221}$ **11.** $\dfrac{10}{17}$ **12.** $\dfrac{25}{102}$ **13.** $\dfrac{26}{51}$

14. $\dfrac{80}{221}$ **15.** $\dfrac{41}{663}$ **16.** $\dfrac{8}{7}$ **17.** $\dfrac{25}{1326}$ **18.** $\dfrac{20}{221}$ **19.** $\dfrac{25}{221}$ **20.** $\dfrac{95}{663}$

Index

Supplemental Index
for
Topics of Analytic Geometry

[5.4]	f^{-1}	the inverse function of f
[5.5]	$[x]$	the greatest integer not greater than x
[7.2]	$\log_b x$	the logarithm of x to the base b
[7.3]	$\text{antilog}_b x$	the antilogarithm of x to the base b
[7.6]	e	an irrational number, approximately equal to 2.7182818
[8.2]	(x, y, z)	the ordered triple of numbers whose first component is x, second component is y, and third component is z
[8.5]	$\mathscr{R}, \mathscr{S}$, etc.	a set of points
[9.1]	$\begin{bmatrix} a_1 & b_1 & c_1 \\ a_2 & b_2 & c_2 \end{bmatrix}$, etc.	matrix
	$A_{m \times n}$	m by n matrix
	$a_{i,j}$	the element in the ith row and jth column of the matrix A
	A^t	the transpose of the matrix A
	$0_{m \times n}$	the m by n zero matrix
	$-A_{m \times n}$	the negative of $A_{m \times n}$
[9.2]	$I_{n \times n}$	the identity matrix for all n by n matrices
[9.3]	$A \sim B$	A is equivalent to B (for matrices)
[9.4]	$\begin{vmatrix} a_1 & b_1 & c_1 \\ a_2 & b_2 & c_2 \\ a_3 & b_3 & c_3 \end{vmatrix}$, etc.	determinant
	M_{ij}	the minor of the element a_{ij}
	A_{ij}	the cofactor of a_{ij}
	$\delta(A)$	the determinant of A
[9.6]	A^{-1}	the inverse of A
[10.1]	z	complex number
	$-z$	the additive inverse, or negative, of z
	z^{-1}	the multiplicative inverse, or reciprocal, of z